VOLUME FIVE HUNDRED AND FORTY THREE

METHODS IN ENZYMOLOGY

Cell-wide Metabolic Alterations Associated with Malignancy

METHODS IN ENZYMOLOGY

VOLUME FIVE HUNDRED AND FORTY THREE

METHODS IN ENZYMOLOGY

Cell-wide Metabolic Alterations Associated with Malignancy

Edited by

LORENZO GALLUZZI
Université Paris Descartes,
Institut Gustave Roussy, France

GUIDO KROEMER
INSERM UMRS 872,
Cordeliers Research Center,
Hôpital Européen Georges Pompidou,
Université Paris Descartes, France

AMSTERDAM • BOSTON • HEIDELBERG • LONDON
NEW YORK • OXFORD • PARIS • SAN DIEGO
SAN FRANCISCO • SINGAPORE • SYDNEY • TOKYO
Academic Press is an imprint of Elsevier

Academic Press is an imprint of Elsevier
525 B Street, Suite 1800, San Diego, CA 92101-4495, USA
225 Wyman Street, Waltham, MA 02451, USA
Radarweg 29, PO Box 211, 1000 AE Amsterdam, The Netherlands
The Boulevard, Langford Lane, Kidlington, Oxford, OX5 1GB, UK
32 Jamestown Road, London NW1 7BY, UK

First edition 2014

For information on all Academic Press publications
visit our website at store.elsevier.com

ISBN: 978-0-12-801329-8
ISSN: 0076-6879

14 15 16 17 11 10 9 8 7 6 5 4 3 2 1

CONTENTS

CONTRIBUTORS

Claire Aukim-Hastie
Faculty of Health & Medical Sciences, University of Surrey, Guildford, and Faculty of Medicine, Cancer Sciences and CES Units, Institute for Life Sciences, University of Southampton, Southampton, United Kingdom

Chantal Bauvy
INSERM U1151-CNRS UMR 8253, Institut Necker Enfants-Malades, University Paris Descartes, Paris cedex 14, France

Marisa Brini
Department of Biology, University of Padova, Padova, Italy

Tito Calì
Department of Biology, University of Padova, Padova, Italy

Jin-Lian Chen
Department of Gastroenterology, Shanghai East Hospital, Tongji University, School of Medicine, Shanghai, China

Alexis Chery
Metabolomics and Cell Biology Platform, Gustave Roussy, and INSERM, U848, Villejuif, France

Elisa Ciraolo
Molecular Biotechnology Center, Department of Molecular Biotechnology and Health Sciences, University of Torino, Torino, Italy

Patrice Codogno
INSERM U1151-CNRS UMR 8253, Institut Necker Enfants-Malades, University Paris Descartes, Paris cedex 14, France

Guido R.Y. De Meyer
Laboratory of Physiopharmacology, University of Antwerp, Antwerp, Belgium

Diego De Stefani
Department of Biomedical Sciences, CNR Neuroscience Institute, University of Padua, Padua, Italy

Nicolas Dupont
INSERM U1151-CNRS UMR 8253, Institut Necker Enfants-Malades, University Paris Descartes, Paris cedex 14, France

Sylvere Durand
Metabolomics and Cell Biology Platform, Gustave Roussy, and INSERM, U848, Villejuif, France

Lorenzo Galluzzi
Université Paris Descartes/Paris V, Sorbonne Paris Cité; Equipe 11 labellisée par la Ligue Nationale contre le Cancer, Centre de Recherche des Cordeliers, Paris, and Gustave Roussy, Villejuif, France

Sheila Ganti
Department of Comparative Medicine, University of Washington, Seattle, Washington, USA

Spiros D. Garbis
Faculty of Medicine, Cancer Sciences and CES Units, Institute for Life Sciences, University of Southampton, Southampton, United Kingdom

Federico Gulluni
Molecular Biotechnology Center, Department of Molecular Biotechnology and Health Sciences, University of Torino, Torino, Italy

Yousang Gwack
Department of Physiology, David Geffen School of Medicine at UCLA, Los Angeles, California, USA

Marcia C. Haigis
Department of Cell Biology, Harvard Medical School, Boston, Massachusetts, USA

Julia M. Hill
Department of Cell and Developmental Biology, Consortium for Mitochondrial Research, University College London, London, United Kingdom

Emilio Hirsch
Molecular Biotechnology Center, Department of Molecular Biotechnology and Health Sciences, University of Torino, Torino, Italy

Alfredo J. Ibáñez
Department of Chemistry and Applied Biosciences, Eidgenössiche Technische Hochschule Zurich, Zurich, Switzerland

Li-Juan Ji
Department of Rehabilitation, The Second People's Hospital of Huai'an, Huai'an, China

Aleck W.E. Jones
Department of Cell and Developmental Biology, Consortium for Mitochondrial Research, University College London, London, United Kingdom

Stefan Kempa
Berlin Institute for Medical Systems Biology at the MDC Berlin-Buch, Berlin, Germany

Hector Keun
Department of Surgery and Cancer, Imperial College London, South Kensington, London, United Kingdom

Kyun-Do Kim
Department of Physiology, David Geffen School of Medicine at UCLA, Los Angeles, California, USA

Imhoi Koo
Department of Chemistry, University of Louisville, Louisville, Kentucky, USA

Guido Kroemer
Metabolomics and Cell Biology Platform, Gustave Roussy; INSERM, U848; Villejuif, France; Université Paris Descartes/Paris V, Sorbonne Paris Cité; Equipe 11 labellisée par la

Ligue Nationale contre le Cancer, Centre de Recherche des Cordeliers, and Pôle de Biologie, Hôpital Européen Georges Pompidou, AP-HP, Paris, France

Xiaojie Lu
Department of Gastroenterology, Shanghai East Hospital, Tongji University, School of Medicine, Shanghai, China

Wim Martinet
Laboratory of Physiopharmacology, University of Antwerp, Antwerp, Belgium

Idil Orhon
INSERM U1151-CNRS UMR 8253, Institut Necker Enfants-Malades, University Paris Descartes, Paris cedex 14, France

Denis Ottolini
Department of Biology, University of Padova, Padova, Italy

Matthias Pietzke
Berlin Institute for Medical Systems Biology at the MDC Berlin-Buch, Berlin, Germany

Juan Ramon Hernandez-Fernaud
Vascular Proteomics Group, Cancer Research UK Beatson Institute, Glasgow, United Kingdom

Steven Reid
Vascular Proteomics Group, Cancer Research UK Beatson Institute, Glasgow, United Kingdom

Rosario Rizzuto
Department of Biomedical Sciences, CNR Neuroscience Institute, University of Padua, Padua, Italy

Asier Ruiz
Department of Neurosciences, University of the Basque Country (UPV/EHU), Achúcarro Basque Center for Neuroscience-UPV/EHU, Leioa, Spain; Instituto de Salud Carlos III, Centro de Investigación Biomédica en Red de Enfermedades Neurodegenerativas, (CIBERNED), Madrid, Spain

F. Kyle Satterstrom
Harvard School of Engineering and Applied Sciences, Cambridge, and Department of Cell Biology, Harvard Medical School, Boston, Massachusetts, USA

Marie Scoazec
Metabolomics and Cell Biology Platform, Gustave Roussy, and INSERM, U848, Villejuif, France

Sonal Srikanth
Department of Physiology, David Geffen School of Medicine at UCLA, Los Angeles, California, USA

Aleš Svatoš
Mass Spectrometry/Proteomic Research Group, Max Planck institute, Jena, Germany

Gyorgy Szabadkai
Department of Cell and Developmental Biology, Consortium for Mitochondrial Research, University College London, London, United Kingdom, and Department of Biomedical Sciences, CNR Neuroscience Institute, University of Padua, Padua, Italy

Jean-Pierre Timmermans
Laboratory of Cell Biology and Histology, University of Antwerp, Antwerp, Belgium

Xiaoli Wei
Department of Chemistry, University of Louisville, Louisville, Kentucky, USA

Robert H. Weiss
Division of Nephrology, Department of Internal Medicine, University of California, Davis, and Medical Service, Sacramento VA Medical Center, Sacramento, California, USA

Hiromi I. Wettersten
Division of Nephrology, Department of Internal Medicine, University of California, Davis, California, USA

Sara Zanivan
Vascular Proteomics Group, Cancer Research UK Beatson Institute, Glasgow, United Kingdom

Xiang Zhang
Department of Chemistry, University of Louisville, Louisville, Kentucky, USA

PREFACE: ONCOMETABOLISM: A NEW FIELD OF RESEARCH WITH PROFOUND THERAPEUTIC IMPLICATIONS

In 1924, the German physiologist Otto Heinrich Warburg was the first to report the propensity of neoplastic cells to metabolize glucose via aerobic glycolysis rather than via the citric acid cycle (also known as Krebs cycle) as a fuel for mitochondrial respiration (Koppenol, Bounds, & Dang, 2011; Warburg, 1924). For a long time since then, however, much greater attention has been attracted by the discovery of the genetic and epigenetic alterations that characterize cancer cells than by their metabolic profile. Such a tendency was so pronounced that, at the end of the twentieth century, several chemotherapeutic agents specifically targeting oncogene addiction, i.e., the process whereby cancer cells rely for their survival and growth on the constitutive activation of oncogenic signaling pathways and/or on the permanent inactivation of oncosuppressive mechanisms, were licensed for use in cancer patients (Luo, Solimini, & Elledge, 2009). Conversely, no chemotherapeutic agent specifically devised to target the metabolism of cancer cells was available then, nor it is now, although (1) several widely employed and effective chemotherapeutics including methotrexate, 5-fluorouracil, gemcitabine, and many others (which are cumulatively known as antimetabolites), *de facto* exert antineoplastic effects by operating as metabolic inhibitors (but were discovered and developed based on empirical, as opposed to mechanistic, grounds); and (2) the safety and therapeutic profile of many of these agents are being evaluated in a growing number of clinical trials (Chabner & Roberts, 2005; Galluzzi, Kepp, Vander Heiden, & Kroemer, 2013). Indeed, it is only over the past decade that the complexity and prominent therapeutic implications of oncometabolism, which can be defined as the ensemble of metabolic rearrangements that accompany oncogenesis and tumor progression, have been fully recognized as a central aspect of malignant transformation (Hanahan & Weinberg, 2011). As a result of such a refocus in the interest of researchers and clinicians, the metabolic rewiring of cancer cells is now viewed as a rich source of targets for the development of novel chemotherapeutic agents, and an intense wave of investigation currently explores this possibility (Galluzzi et al., 2013; Vander Heiden, 2011).

At odds with long-standing beliefs, it is now clear that the so-called Warburg effect represents only the tip of the iceberg of metabolic alterations associated with oncogenesis, which also encompass an increased flux through the pentose phosphate pathway, elevated rates of lipid biosynthesis, intense glutamine consumption, an improved control of redox homeostasis, and (at least in the initial steps of malignant transformation) decreased levels of macroautophagy (Schulze & Harris, 2012; Vander Heiden, Cantley, & Thompson, 2009; White, 2012). A few other common misconceptions about oncometabolism are in the process of being reconsidered based on preclinical and clinical findings from several laboratories worldwide.

First, the metabolic rewiring of neoplastic cells should not be considered as a self-standing hallmark of malignancy, but rather as a phenomenon that intimately accompanies, allows for and cannot be mechanistically separated from many, if not all, aspects of oncogenesis (Galluzzi et al., 2013; Locasale & Cantley, 2011; Wellen & Thompson, 2012). Accumulating evidence indicates indeed that (1) several metabolic intermediates such as ATP, acetyl-CoA, α-ketoglutarate, and reactive oxygen species play a major role in cell-intrinsic as well as cell-extrinsic signaling pathways (Galluzzi, Kepp, & Kroemer, 2012; Locasale & Cantley, 2011; Wellen & Thompson, 2012); (2) multiple proteins with prominent metabolic functions such as cytochrome *c* (which operates as an electron shuttle in the mitochondrial respiratory chain) and the M2 isoform of pyruvate kinase (PKM2, which catalyzes the last step glycolysis) participate in signal transduction (Galluzzi, Kepp, & Kroemer, 2012; Galluzzi, Kepp, Trojel-Hansen, & Kroemer, 2012; Gao, Wang, Yang, Liu, & Liu, 2012; Luo et al., 2011; Yang et al., 2011); and (3) several proteins initially viewed as "pure" signal transducers including (but not limited to) the antiapoptotic Bcl-2 family members BCL-X_L and MCL1 also impact on metabolic functions such as the handling of Ca^{2+} ions at the endoplasmic reticulum and the enzymatic activity of the F_1F_0 ATP synthase (Alavian et al., 2011; Perciavalle et al., 2012; Rong & Distelhorst, 2008).

Second, the metabolic changes linked to malignant transformation should not be considered as a general property shared by all types of cancer. It has indeed been clearly demonstrated that several variables including tissue type and oncogenic driver (and presumably many others) determine the metabolic profile of developing tumors (Yuneva et al., 2012). This has obvious implications for the use of metabolic inhibitors in cancer therapy.

Third, it should be kept in mind that the metabolic profile of neoplastic cells is far less specific than previously thought, but (with some exceptions) resemble that of highly proliferating nontransformed cells (Altman & Dang, 2012; Michalek & Rathmell, 2010). This notion is corroborated by the fact that the most severe side effects of antimetabolites involve highly proliferating normal tissues, such as the intestinal epithelium and bone marrow (Chabner & Roberts, 2005). Nonetheless, the clinical success of these widely employed chemotherapeutic agents points to the existence of a therapeutic window for the use of metabolic inhibitors in cancer patients (Galluzzi et al., 2013; Vander Heiden, 2011). As it stands, multiple facets of oncometabolism can be considered as forms of "nononcogene addiction," a term referring to the fact that the survival of malignant cells relies not only on the constitutive activation of oncogenes and/or the permanent inactivation of oncosuppressive mechanisms but also on a wide array of genes and functions that are not inherently tumorigenic (Luo et al., 2009).

Finally, tumors (in particular solid neoplasms but also hematological malignancies) should no longer be considered as homogenous entities predominantly composed of malignant cells. Indeed, it is now clear that neoplastic lesions contain a large amount of nontransformed cells, including endothelial, stromal, and immune cells, and that oncogenesis takes place in the context of a complex network of physical and functional interactions among the malignant and nonmalignant components of the tumor microenvironment (Nagaraj & Gabrilovich, 2010; Pietras & Ostman, 2010). These interactions, part of which have direct metabolic implications (Nieman et al., 2011; Sonveaux et al., 2008; Whitaker-Menezes et al., 2011), are also attracting attention as targets for the development of novel antineoplastic agents (Galluzzi, Senovilla, Zitvogel, & Kroemer, 2012; Zitvogel, Galluzzi, Smyth, & Kroemer, 2013).

In *OncoMetabolism*, a thematic collection covering two volumes of the successful *Methods in Enzymology* series, leading researchers summarize the current state of the field from both a conceptual and methodological standpoint. The first volume, entitled "*Conceptual background and bioenergetic/mitochondrial aspects of oncometabolism*," provides a robust theoretical background on cancer-associated metabolic alterations, discussing how these relate to other aspects of oncogenesis such as the relentless proliferation and resistance to death exhibited by neoplastic cells. Thereafter, this volume offers a collection of techniques that can be employed to study the major bioenergetic and mitochondrial aspects of oncometabolism, including (but not limited to) alterations in glycolysis and oxidative phosphorylation. The second volume, entitled "*Cell-wide metabolic alterations associated with malignancy*," proposes a series of methods for the investigation of global facets of oncometabolism, including (but not limited to) deregulations in Ca^{2+} fluxes and autophagy, as well as (malignant) cell- or tissue-wide metabolomic alterations.

OncoMetabolism is expected to provide a comprehensive and reliable methodological guide to beginners and experts in this exciting and rapidly expanding area of cancer research.

ACKNOWLEDGMENTS

Lorenzo Galluzzi and Guido Kroemer are supported by the Ligue contre le Cancer (équipe labellisée), Agence National de la Recherche (ANR), AXA Chair for Longevity Research, ARC, Cancéropôle Ile-de-France, Institut National du Cancer (INCa), Fondation Bettencourt-Schueller, Fondation de France, Fondation pour la Recherche Médicale (FRM), the European Commission (ArtForce), the European Research Council (ERC), the LabEx Immuno-Oncology, the SIRIC Stratified Oncology Cell DNA Repair and Tumor Immune Elimination (SOCRATE), the SIRIC Cancer Research and Personalized Medicine (CARPEM), and the Paris Alliance of Cancer Research Institutes (PACRI).

REFERENCES

Alavian, K. N., Li, H., Collis, L., Bonanni, L., Zeng, L., Sacchetti, S., et al. (2011). Bcl-xL regulates metabolic efficiency of neurons through interaction with the mitochondrial F1FO ATP synthase. *Nature Cell Biology*, *13*, 1224–1233.

Altman, B. J., & Dang, C. V. (2012). Normal and cancer cell metabolism: Lymphocytes and lymphoma. *The FEBS Journal*, *279*, 2598–2609.

Chabner, B. A., & Roberts, T. G., Jr. (2005). Timeline: Chemotherapy and the war on cancer. *Nature Reviews. Cancer*, *5*, 65–72.

Galluzzi, L., Kepp, O., & Kroemer, G. (2012). Mitochondria: Master regulators of danger signalling. *Nature Reviews. Molecular Cell Biology*, *13*, 780–788.

Galluzzi, L., Kepp, O., Trojel-Hansen, C., & Kroemer, G. (2012). Non-apoptotic functions of apoptosis-regulatory proteins. *EMBO Reports*, *13*, 322–330.

Galluzzi, L., Kepp, O., Vander Heiden, M. G., & Kroemer, G. (2013). Metabolic targets for cancer therapy. *Nature Reviews. Drug Discovery*, *12*, 829–846.

Galluzzi, L., Senovilla, L., Zitvogel, L., & Kroemer, G. (2012). The secret ally: Immunostimulation by anticancer drugs. *Nature Reviews. Drug Discovery*, *11*, 215–233.

Gao, X., Wang, H., Yang, J. J., Liu, X., & Liu, Z. R. (2012). Pyruvate kinase M2 regulates gene transcription by acting as a protein kinase. *Molecular Cell*, *45*, 598–609.

Hanahan, D., & Weinberg, R. A. (2011). Hallmarks of cancer: The next generation. *Cell*, *144*, 646–674.

Koppenol, W. H., Bounds, P. L., & Dang, C. V. (2011). Otto Warburg's contributions to current concepts of cancer metabolism. *Nature Reviews. Cancer*, *11*, 325–337.

Locasale, J. W., & Cantley, L. C. (2011). Metabolic flux and the regulation of mammalian cell growth. *Cell Metabolism*, *14*, 443–451.

Luo, J., Solimini, N. L., & Elledge, S. J. (2009). Principles of cancer therapy: Oncogene and non-oncogene addiction. *Cell*, *136*, 823–837.

Luo, W., Hu, H., Chang, R., Zhong, J., Knabel, M., O'Meally, R., et al. (2011). Pyruvate kinase M2 is a PHD3-stimulated coactivator for hypoxia-inducible factor 1. *Cell*, *145*, 732–744.

Michalek, R. D., & Rathmell, J. C. (2010). The metabolic life and times of a T-cell. *Immunological Reviews*, *236*, 190–202.

Nagaraj, S., & Gabrilovich, D. I. (2010). Myeloid-derived suppressor cells in human cancer. *Cancer Journal*, *16*, 348–353.

Nieman, K. M., Kenny, H. A., Penicka, C. V., Ladanyi, A., Buell-Gutbrod, R., Zillhardt, M. R., et al. (2011). Adipocytes promote ovarian cancer metastasis and provide energy for rapid tumor growth. *Nature Medicine*, *17*, 1498–1503.

Perciavalle, R. M., Stewart, D. P., Koss, B., Lynch, J., Milasta, S., Bathina, M., et al. (2012). Anti-apoptotic MCL-1 localizes to the mitochondrial matrix and couples mitochondrial fusion to respiration. *Nature Cell Biology*, *14*, 575–583.

Pietras, K., & Ostman, A. (2010). Hallmarks of cancer: Interactions with the tumor stroma. *Experimental Cell Research*, *316*, 1324–1331.

Rong, Y., & Distelhorst, C. W. (2008). Bcl-2 protein family members: Versatile regulators of calcium signaling in cell survival and apoptosis. *Annual Review of Physiology*, *70*, 73–91.

Schulze, A., & Harris, A. L. (2012). How cancer metabolism is tuned for proliferation and vulnerable to disruption. *Nature*, *491*, 364–373.

Sonveaux, P., Vegran, F., Schroeder, T., Wergin, M. C., Verrax, J., Rabbani, Z. N., et al. (2008). Targeting lactate-fueled respiration selectively kills hypoxic tumor cells in mice. *The Journal of Clinical Investigation*, *118*, 3930–3942.

Vander Heiden, M. G. (2011). Targeting cancer metabolism: A therapeutic window opens. *Nature Reviews. Drug Discovery*, *10*, 671–684.

Vander Heiden, M. G., Cantley, L. C., & Thompson, C. B. (2009). Understanding the Warburg effect: The metabolic requirements of cell proliferation. *Science*, *324*, 1029–1033.

Warburg, O. (1924). Über den Stoffwechsel der Carcinomzelle. *Biochemische Zeitschrift*, *152*, 309–344.

Wellen, K. E., & Thompson, C. B. (2012). A two-way street: Reciprocal regulation of metabolism and signalling. *Nature Reviews. Molecular Cell Biology*, *13*, 270–276.

Whitaker-Menezes, D., Martinez-Outschoorn, U. E., Lin, Z., Ertel, A., Flomenberg, N., Witkiewicz, A. K., et al. (2011). Evidence for a stromal-epithelial "lactate shuttle" in human tumors: MCT4 is a marker of oxidative stress in cancer-associated fibroblasts. *Cell Cycle*, *10*, 1772–1783.

White, E. (2012). Deconvoluting the context-dependent role for autophagy in cancer. *Nature Reviews. Cancer*, *12*, 401–410.

Yang, W., Xia, Y., Ji, H., Zheng, Y., Liang, J., Huang, W., et al. (2011). Nuclear PKM2 regulates beta-catenin transactivation upon EGFR activation. *Nature*, *480*, 118–122.

Yuneva, M. O., Fan, T. W., Allen, T. D., Higashi, R. M., Ferraris, D. V., Tsukamoto, T., et al. (2012). The metabolic profile of tumors depends on both the responsible genetic lesion and tissue type. *Cell Metabolism*, *15*, 157–170.

Zitvogel, L., Galluzzi, L., Smyth, M. J., & Kroemer, G. (2013). Mechanism of action of conventional and targeted anticancer therapies: Reinstating immunosurveillance. *Immunity*, *39*, 74–88.

Lorenzo Galluzzi

Guido Kroemer

CHAPTER ONE

Methods to Measure Cytoplasmic and Mitochondrial Ca^{2+} Concentration Using Ca^{2+}-Sensitive Dyes

Sonal Srikanth[1], Kyun-Do Kim, Yousang Gwack

Department of Physiology, David Geffen School of Medicine at UCLA, Los Angeles, California, USA
[1]Corresponding author: e-mail address: ssrikanth@mednet.ucla.edu

Contents

Abstract

Ca^{2+} is a ubiquitous second messenger that is involved in regulation of various signaling pathways. Cytoplasmic Ca^{2+} is maintained at low concentrations (~100 n*M*) by many active mechanisms. Increases in intracellular Ca^{2+} concentration ($[Ca^{2+}]_i$) indeed can initiate multiple signaling pathways, depending both on their pattern and subcellular

Methods in Enzymology, Volume 543
ISSN 0076-6879
http://dx.doi.org/10.1016/B978-0-12-801329-8.00001-5

localization. In T cells, the stimulation of T-cell receptor leads to an increase in $[Ca^{2+}]_i$ upon the opening of Ca^{2+} release-activated calcium (CRAC) channels. T cells can actually sustain high $[Ca^{2+}]_i$ for several hours, resulting in the activation of transcriptional programs orchestrated by members of the nuclear factor of activated T-cell (NFAT) protein family. Here, we describe an imaging method widely employed to measure cytoplasmic $[Ca^{2+}]$ in naive and effector T cells based on the ratiometric dye Fura-2. Furthermore, we discuss a pharmacological method relying on an inhibitor of CRAC channels, 2-aminoethyldiphenyl borate, to validate the role of CRAC channels in cytoplasmic Ca^{2+} elevation. Finally, we describe an approach to measure mitochondrial $[Ca^{2+}]$ based on another fluorescent dye, Rhod-2. With appropriate variations, our methodological approach can be employed to assess the effect and regulation of cytosolic and mitochondrial Ca^{2+} waves in multiple experimental settings, including cultured cancer cells.

1. INTRODUCTION

Ca^{2+} signaling plays a pivotal role in activation, proliferation, and cytokine production in T cells (Cahalan & Chandy, 2009; Lewis, 2011). Stimulation of T-cell receptor (TCR) by ligation with peptide-major histocompatibility antigen complex presented by antigen-presenting cells, activates multiple enzymes including protein kinases and phospholipase C gamma 1 (PLCγ1) (Srikanth & Gwack, 2013b). PLCγ1 hydrolyzes its substrate phosphatidylinositol 4,5-bisphosphate (PIP_2) to generate inositol 1,4,5 trisphosphate ($InsP_3$) and diacyl glycerol (DAG). $InsP_3$ binds to its cognate receptor, $InsP_3R$ on the endoplasmic reticulum (ER) membrane and causes depletion of Ca^{2+} from the ER stores (Fig. 1.1). This depletion of the ER Ca^{2+} stores activates Ca^{2+} entry via Ca^{2+} release-activated calcium (CRAC) channels in a process termed as store-operated Ca^{2+} entry (SOCE) (Putney, 1986). For productive activation of T cells, this elevation of $[Ca^{2+}]_i$ is sustained for a long time of several hours to influence transcriptional programs.

Recent breakthroughs in reverse-genomics techniques have allowed for identification of STIM1 and Orai1 proteins as molecular components of CRAC channels (Feske et al., 2006; Gwack et al., 2007; Liou et al., 2005; Roos et al., 2005; Vig et al., 2006; Zhang et al., 2006). Currently, it is commonly accepted that Orai1 residing on the plasma membrane (PM) forms the pore subunit of the CRAC channels. STIM1, residing on the ER membrane acts as a signaling molecule to activate Orai1 after store

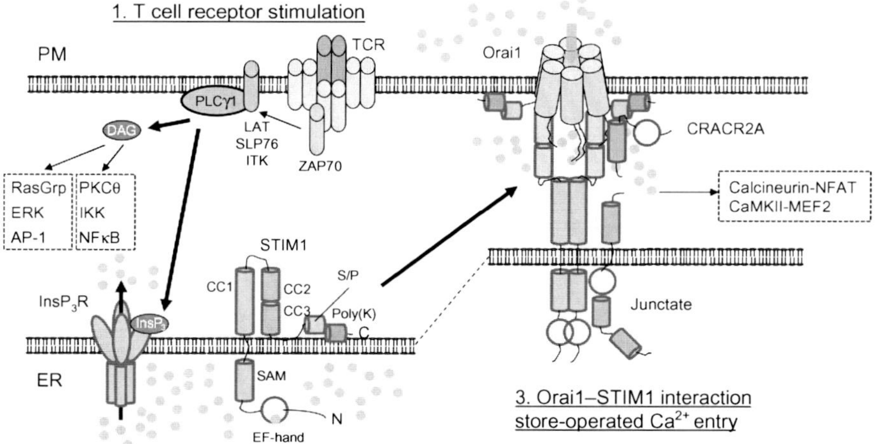

Figure 1.1 *Activation mechanism of CRAC channel-mediated Ca^{2+} signaling.* Schematic showing current understanding of CRAC channel activation. Under resting conditions, Orai1 and STIM1 are distributed at the PM and the ER membranes. The subunit stoichiometry of Orai1 under resting and stimulated conditions is currently unclear. For convenience, a tetrameric assembly of Orai1 is depicted here. T-cell receptor stimulation induces assembly and phosphorylation of key signaling molecules including ZAP70, LAT, SLP76, and ITK resulting in activation of PLCγ1. PLCγ1 hydrolyzes PIP_2 into DAG and $InsP_3$. Upon store depletion triggered by $InsP_3$ production, STIM1 oligomerizes by sensing ER Ca^{2+} depletion with its ER-luminal EF-hand domain, clusters, and translocates to the ER–PM junctions. STIM1 contains an ER-luminal region comprising the EF-hand and SAM domains, a single transmembrane segment, and a cytoplasmic region. The cytoplasmic region has three coiled-coil domains (CC1, CC2, and CC3), serine/proline-rich domain containing the residues involved in posttranslational modifications, and a polybasic region (poly-K) at the C-terminus that interacts with phosphoinositides. By physical interaction with the cytoplasmic, N- and C- terminus of tetramembrane spanning PM Orai1 through the CAD/SOAR domain (coiled-coil domains 2 and 3) clustered STIM1 recruits and activates Orai1 in the ER–PM junctions. *Figure modified from Srikanth and Gwack (2013b).*

depletion and can also affect the ion selectivity of CRAC channels. STIM proteins are single pass membrane-spanning EF-hand domain-containing Ca^{2+}-binding proteins predominantly localized to the ER membranes. The ER-luminal N-terminus of STIM proteins contains two EF-hand domains and a sterile-alpha motif (SAM) domain, whereas its cytoplasmic region includes coiled-coil domains, a serine/proline-rich domain, and a lysine-rich C-terminal tail (Fig. 1.1). STIM family contains two members,

STIM1 and STIM2 both of which have different affinities for Ca^{2+} binding via their EF-hands and correspondingly play an important role in activation of Ca^{2+} entry and Ca^{2+} homeostasis under resting conditions, respectively (Brandman, Liou, Park, & Meyer, 2007; Cahalan, 2009; Soboloff, Rothberg, Madesh, & Gill, 2012). Under resting conditions, STIM1 binds Ca^{2+} and remains in a predominantly mono- or dimeric form. Upon ER Ca^{2+} depletion, STIM1 loses Ca^{2+} binding, multimerizes via its ER-luminal SAM domain and cytoplasmic coiled-coil domain 1 (CC1) and translocates to ER–PM junctions. In these junctions, STIM1 physically interacts with Orai1 via its coiled-coil domains CC2 and CC3 to activate Ca^{2+} entry (Soboloff et al., 2012).

Orai family comprises of Orai1, Orai2, and Orai3 proteins all of which are four transmembrane-spanning PM proteins with their N- and C- termini facing the cytoplasm (Feske et al., 2006; Gwack et al., 2007). In T cells, CRAC channels are predominantly composed of multimers of Orai1 gated by STIM1. T cells lacking expression of Orai1 or STIM1 are severely impaired in CRAC channel-mediated Ca^{2+} entry (Gwack et al., 2008; Oh-Hora et al., 2008). All the three Orai proteins physically interact with and can be gated by STIM1 in a store-depletion-dependent manner (Gwack et al., 2007; Lewis, 2011). When coexpressed with STIM1, all three Orai proteins give rise to CRAC currents with broadly similar biophysical characteristics of high Ca^{2+} selectivity and low conductance of large cations including Cs^{+}. CRAC channel-mediated Ca^{2+} entry is dually regulated by 2-aminoethyldiphenyl borate (2-APB), a widely used nonspecific small molecule blocker. Low micromolar concentrations of 2-APB can enhance SOCE via the CRAC channels, however, higher concentrations of 2-APB (>30 μM) block Ca^{2+} influx via the CRAC channels (Prakriya & Lewis, 2001). This biphasic response to 2-APB is often used to isolate Orai–STIM protein-mediated SOCE from that mediated by other channels including the TrpC family of store-operated channels in various cell types. High concentration of 2-APB is suggested to block CRAC channel-mediated SOCE via at least two independent mechanisms, first via physical occlusion of the ion conduction pore and second via dissociating Orai and STIM proteins (Dehaven, Smyth, Boyles, Bird, & Putney, 2008; Peinelt, Lis, Beck, Fleig, & Penner, 2008; Prakriya & Lewis, 2001). In addition to STIM1-mediated gating, Orai proteins, specifically Orai3 and to a lesser extent Orai1 can also be gated in a STIM1-independent manner by 2-APB (Dehaven et al., 2008; Peinelt et al., 2008; Schindl et al., 2008; Yamashita, Somasundaram, & Prakriya, 2011; Zhang et al., 2008). STIM1-independent 2-APB-mediated currents via Orai3 and Orai1 exhibit lower Ca^{2+} selectivity and increased conductance to Cs^{+} (Dehaven et al., 2008; Peinelt et al., 2008; Schindl et al., 2008; Yamashita et al., 2011; Zhang et al., 2008).

CRAC channels are primarily responsible for elevation of $[Ca^{2+}]_i$ and thereby activation of T cells, as exemplified by occurrence of severe combined immune deficiency in patients lacking functional Orai1 or STIM1 proteins (Hogan, Lewis, & Rao, 2010). However, molecular identification of Orai and STIM proteins have helped in identifying important functions of these proteins in a variety of other tissues including smooth muscle cells, skeletal muscle cells, and cardiac myocytes (Collins, Zhu-Mauldin, Marchase, & Chatham, 2013; Stiber & Rosenberg, 2011; Trebak, 2012). In addition, SOCE via Orai1 and STIM1 has been shown to play an important role in tumor migration and metastasis of various cancers including breast and cervical cancers (Prevarskaya, Skryma, & Shuba, 2011). One of the first reports identified important function for both Orai1 and STIM1 in breast tumor cell migration *in vitro* and in tumor xenograft model *in vivo* (Yang, Zhang, & Huang, 2009). In an entirely different mechanism, STIM1-independent, secretary pathway Ca^{2+}-ATPase, SPCA2-mediated gating of Orai1 was proposed to play an important role in human breast cancer tumorigenesis (Feng et al., 2010). In another study, STIM1 overexpression was observed in 71% of the cases of early stage cervical cancer (Chen et al., 2011). STIM1 silencing significantly inhibited cervical cancer cell proliferation (Chen et al., 2011). Tumor metastasis is the primary reason for mortality of various cancers. The requirement of SOCE during cancer cell migration identifies Orai1 and STIM1 as potential therapeutic targets for cancer.

Mitochondria are known to accumulate high concentrations of Ca^{2+} within their lumen via the mitochondrial Ca^{2+} uniporter (MCU) located on the inner mitochondrial membrane (Baughman et al., 2011; De Stefani et al., 2011). It was shown that in addition to Orai1 and STIM1, mitochondria also translocate to the immunological synapse in T cells (Quintana et al., 2007, 2011). The authors showed that local $[Ca^{2+}]$ near the CRAC channels was kept low by mitochondrial Ca^{2+} buffering, thereby preventing slow Ca^{2+}-dependent inactivation and maintaining increased $[Ca^{2+}]_i$ for prolonged time periods required for activation of T cells (Quintana et al., 2011). In addition, mitochondrial Ca^{2+}-accumulation is known to play an important role in T-cell death. T-cell death induced by TCR stimulation is critical for homeostasis of peripheral T cells after antigen clearance and for negative selection of auto reactive T cells in the thymus (Budd, 2001; Krammer, Arnold, & Lavrik, 2007; Strasser, 2005). Abrogation of T-cell death leads to hypersensitive immune reaction and autoimmune disorders. Activation induced T-cell death occurs through the death receptor- (e.g., Fas) and mitochondria-mediated pathways. Death receptor-mediated apoptosis involves the Fas ligand/Fas signaling pathway, primarily regulated by NFAT (Hodge et al., 1996; Macian et al., 2002; Serfling et al., 2006). Mitochondria-mediated cell death occurs due to loss of mitochondrial membrane potential (Marsden & Strasser, 2003; Strasser, 2005). Mitochondria-mediated cell death pathway involving the Bcl-2 family members (e.g., Bcl-2 and Bcl-X_L) and the BH3-only proteins (e.g., Bad, Bik, Bim, and Noxa) play an important role in T-cell death and survival as seen in *in vitro* and *in vivo* models (Budd, 2001; Hildeman, Jorgensen, Kappler, & Marrack, 2007; Hildeman et al., 2002; Marrack & Kappler, 2004; Strasser, 2005; Strasser & Pellegrini, 2004). It was recently shown that Orai1-deficient T cells are strongly resistant to cell death due to reduction in death receptor- and mitochondria-mediated cell death mechanisms by decreasing expression of proapoptotic genes mediated by NFAT and mitochondrial Ca^{2+} uptake (Kim et al., 2011).

Study of CRAC channel-mediated SOCE entails measurement of changes in intracellular $[Ca^{2+}]$ in live cells. In previous studies, we identified a novel cytoplasmic EF-hand-containing protein, CRACR2A and a junctional protein junctate as important regulators of Orai1–STIM1 interaction (Srikanth et al., 2010, 2012). By using single-cell Ca^{2+} measurement technique, we showed that an EF-hand mutant of CRACR2A elevated cytoplasmic $[Ca^{2+}]$, enhanced STIM1 clustering, and induced cell death. Single-cell Ca^{2+} imaging technique can determine $[Ca^{2+}]_i$ in 20–100 cells simultaneously and allows to detect even small changes in $[Ca^{2+}]_i$ in unstimulated cells. We have recently described measurement of $[Ca^{2+}]_i$ in commonly used cell lines using ratiometric single-cell Ca^{2+} imaging technique (Srikanth & Gwack, 2013a). In this protocol, we focus on ratiometric Ca^{2+} measurement in primary naïve and effector T cells. Additionally, we describe a protocol to measure mitochondrial Ca^{2+} accumulation in primary T cells (Quintana et al., 2007). These techniques provide basic platform for accurate estimation of changes in cytoplasmic and mitochondrial $[Ca^{2+}]$ in different physiological settings.

2. EXPERIMENTAL COMPONENTS AND CONSIDERATIONS

2.1. Isolation and differentiation of $CD4^{+}CD25^{-}$ naïve T cells from wild-type and Orai1-deficient ($Orai1^{-/-}$) animals

1. Dissection instruments—forceps and scissors.
2. Dynabeads Mouse CD4 (Invitrogen, Carlsbad, CA).
3. DETACHaBEAD Mouse CD4 (Invitrogen, Carlsbad, CA).
4. $CD4^{+}CD25^{+}$ Regulatory T-cell isolation kit (Miltenyi Biotec, Auburn, CA).
5. Complete Medium—DMEM (Mediatech) + 10% fetal bovine serum (Hyclone) + 1% penecillin–streptomycin (Mediatech) + 1 m*M* HEPES (Mediatech) + 1 m*M* glutamine (Mediatech) + 1 × nonessential amino acids (Hyclone) + 1 × minimum essential vitamins (MEM vitamins, Mediatech) + 1 m*M* sodium pyruvate (Mediatech) + 50 μ*M* 2-mecaptoethanol (Sigma).
6. Phosphate buffered saline (PBS, Mediatech) + 1% fetal bovine serum.
7. Cell strainers (70 μm, BD biosciences).
8. 3 ml syringes (VWR international).
9. Petri dishes, 100 mm (VWR international).
10. 15 ml Falcon tubes (VWR international).
11. 6- and 12-well tissue culture dishes (VWR international).
12. Centrifuge to spin down cells in 15 or 50 ml tubes.
13. Hemocytometer slide (VWR International).
14. Anti-mouse CD3 and anti-mouse CD28 antibodies (2 mg/ml stock, BioXcell, West Lebanon, NH).
15. Goat anti-hamster antibody (0.2 mg/ml stock, MP Biomedicals, OH).
16. Human Recombinant Interleukin 2 (IL-2), 10,000 U/ml (National Cancer Institute repository).

2.2. Reagents for ratiometric Ca^{2+} imaging

1. Fura 2-acetoxymethyl ester (AM, Invitrogen, Carlsbad, CA) resuspended in dimethyl sulfoxide (DMSO) to generate a 1 m*M* stock solution.
2. Ca^{2+}-free Ringer's solution (in m*M*): 145 NaCl, 4.5 KCl, 3 $MgCl_2$, 10 D-glucose, and 10 HEPES (pH 7.35).

3. 2 mM Ca^{2+}-containing Ringer's solution (in mM): 145 NaCl, 4.5 KCl, 1 $MgCl_2$, 2 $CaCl_2$, 10 D-glucose, and 10 HEPES (pH 7.35).
4. Thapsigargin (from EMD biosciences), final concentration of 1 μM in Ca^{2+}-free Ringer's solution.
5. 2-APB (from Sigma), final concentration of 100 mM in DMSO.
6. Fura-2 Ca^{2+} Imaging Calibration kit (Invitrogen).

2.3. Software for data acquisition and analysis

1. Slidebook (Intelligent Imaging Innovations, Inc.).
2. Origin Pro (Origin, Northampton, MA).

2.4. Equipment used for imaging

1. Olympus 1X-51 inverted microscope (Olympus Inc.) connected to a Lambda LB-LS/30 (Sutter instruments) light source housing a 300 W xenon-arc lamp (Perkin-Elmer) and an excitation filter-wheel, regulated by an external controller (figure depicted in Srikanth & Gwack, 2013a). Image acquisition is done using Fura-2 (z340/15 and z380/15 excitations, T400LP dichroic and D510/40 emission) filter cube (Chroma Technology Corp.).
2. Microscope is attached to a deep-cooled CCD camera (ORCA-ER-AG; Hamamatsu). Filter switching and image acquisition are controlled by scripts written in Slidebook (Intelligent Imaging Innovations, Inc.). All fluorescence image acquisition is done with 2×2 pixel binning at 5-s frame intervals and all the experiments are performed at room temperature (22–25 °C; Srikanth & Gwack, 2013a).
3. 15 mm round cover slips (Warner Instruments).
4. Poly-L-lysine (Sigma)
5. SA-OLY/2 (Warner Instruments), stage adapter for series 20 chamber platform.
6. Series 20 chamber platform P-5 (Warner instruments).
7. Cell culture/Imaging Chamber RC-20H (Warner Instruments).
8. Perfusion system (6-channel multi-valve perfusion system, Warner Instruments, Hamden, CT).
9. A 3-ml syringe filled with petroleum jelly and a 26-gauge needle attached to the tip of the syringe.
10. Philips #1 screwdriver.

3. ISOLATION AND DIFFERENTIATION OF NAÏVE CD4$^+$ CD25$^-$ T CELLS FROM SPLEEN AND LYMPH NODES OF WILD-TYPE AND ORAI1$^{-/-}$ MICE

3.1. Isolation of naïve CD4$^+$CD25$^-$ T cells from spleen and lymph nodes

1. Use sterile plastic wares, reagents, and medium. Perform all the techniques under aseptic conditions in a biosafety cabinet.
2. Genotype and identify wild-type and knock out animals of your interest, Orai1$^{-/-}$ mice in this case, that are about 6–8 weeks of age (Gwack et al., 2008).
3. Add 10 ml of PBS + 1% FBS in two 100 mm Petri dishes and label one as WT and other one as KO.
4. Euthanize the animals following the protocol approved by the animal research committee of your Institute or University.
5. Make a superficial cut in the abdomen and open up the skin leaving the peritoneal cavity intact. Quickly dissect the inguinal, axillary, brachial, and cervical lymph nodes (LNs) and put them in Petri dish containing PBS + 1% FBS. Readers are referred to a video for visual depiction of LN isolation (Matheu, Parker, & Cahalan, 2007). Open the peritoneal cavity and dissect the spleen and add it in the same dish containing the LNs. You can then collect the lumbar LNs. Repeat the same procedure with the knockout animal.
6. Transfer the dissected LNs onto a cell strainer. Using the plunger of a 3-ml syringe, grind the LNs completely to make a single-cell suspension. Then add the spleen onto the same strainer and grind the spleen sample as well. One can mix the spleen and LN tissues or keep them separate for isolation of spleen- or LN-specific T cells. Here, we describe isolating T cells from spleen and LNs together. Collect the single-cell suspension and transfer it to 15-ml conical tube. Repeat the same procedure for knockout animal tissues.
7. Spin the cells at 300 g for 5 min. Aspirate the supernatant and resuspend the pellet in 2 ml PBS + 1% FBS.
8. Isolate CD4$^+$ T cells using mouse CD4 dynabeads and DETACHaBEAD using manufacturer's instructions. One can easily obtain 10–14 × 10^6 CD4$^+$ T cells from one mouse.
9. Using these purified T cells, isolate CD4$^+$CD25$^-$ T cells using the regulatory T-cell isolation kit following manufacturer's instructions with

just one difference. In the last step, collect the flow through from the column, which contains $CD4^+CD25^-$ naïve T cells. The cells attached onto the column are $CD4^+CD25^+$ regulatory T cells.

10. Spin the cells at 300 g for 5 min, aspirate the supernatant, and resuspend the cells in 5 ml complete T-cell medium.

11. Count the naïve T cells using hemocytometer slide. One can easily isolate 9–12 × 10^6 naïve T cells from one mouse.

12. For imaging experiments, resuspend the purified T cells at 1 × 10^6/ml in complete medium and add 6 ml of cells in a 6-well plate (3 ml per well). So there will be two wells each for WT and $Orai1^{-/-}$ naïve T cells. Since these T cells were isolated using positive selection, which is known to activate the cells, an overnight rest is recommended before Ca^{2+} imaging. Alternatively, one can isolate untouched $CD4^+$ T cells using negative selection kits (sold by various companies including Invitrogen and Miltenyi Biotec) and proceed directly to Fura-2 imaging (Section 3.2). Negative isolation routinely yields 75–85% purity of cells, hence it is recommended to sort $CD4^+CD25^-CD62L^{high}$ T cells by flow cytometry for >99% purity of cells. Positive selection as described above yields $CD4^+$ cells with >95% purity routinely and hence is a preferred choice for higher purity of cells if easy access to flow sorting is not available.

13. To differentiate naïve T cells into effector T cells, first coat 12-well plate with goat anti-hamster antibody. Add 1 ml of goat anti-hamster antibody/well (two wells each for WT and KO cells) and incubate in 37 °C incubator for 1 h. Wash off the antibody with PBS 3 × and add 1–1.5 × 10^6 cells/well in T-cell medium containing 1 μg/ml each of anti-CD3 and anti-CD28 antibodies in a total volume of 2 ml. Incubate the cells in 37 °C incubator for 48 h. After 24 h cell size increases and by 48 h actively dividing cells are visible.

14. After 48 h, remove the cells off the plate by gentle pipetting and transfer into 6-well plates. Add fresh medium with recombinant hIL-2 at a final concentration of 10 U/ml. Grow the cells for a further 3–4 days by replenishing more medium every alternate day and keeping the cells at a density of 1 × 10^6 cells/ml. On days 5 or 6 cell proliferation reduces and cells start shrinking and become quiescent. This is the appropriate time for measurement of SOCE in effector T cells as described in Section 3.2.

3.2. Fura-2 imaging

1. Coat coverslips with 0.01% of poly-L-Lysine (stock diluted 1:10 in sterile water) by immersing them in solution containing poly-L-Lysine for 1 h at room temperature with occasional shaking. Remove the poly-L-Lysine solution and wash the coverslips with sterile water for 3 × with incubation of 10 min in between the washes. Air dry the coverslips in a biosafety cabinet and store them at 4 °C. They can be used for up to 6 months.

2. Naïve or effector T cells are loaded at 1×10^6 cells/ml with 1 μ*M* Fura 2-AM for 30 min in medium at 25 °C (in dark). The cell pellet is resuspended in 2 m*M* Ca^{2+}-containing Ringer's solution (50–100 μl volume) and attached to poly-L-Lysine-coated 15 mm coverslips for 10–15 min in dark.
3. The P-5 platform is inserted onto the stage adapter (SA-OLY/2) and a thin ring of petroleum jelly is applied around the edges for placement of coverslip on top. Once the coverslip containing the attached cells is placed on the platform, the imaging chamber (RC-20H) pretreated with petroleum jelly (as mentioned for the platform above) on both the sides is placed on top of the coverslip. Then a blank cover slip (without any cells) is attached on top of the imaging chamber. The imaging chamber is held onto the platform by four tiny screws. Once assembled, it forms the imaging area (Fig. 1.2A). Application of petroleum jelly ensures a leak free imaging chamber.
4. Slidebook 5.0 software is used for image acquisition, using settings for time-lapse ratiometric Fura-2 imaging, allowing for visualization of real-time pseudocolored images of intracellular Ca^{2+} during the acquisition. After chamber assembly, cells are visualized using a 20 × (NA 0.5) objective and a field containing a good number of cells and some blank area for background selection is selected (Fig. 1.2B, top panels). The cells are first perfused with Ringer solution containing 2 m*M* $CaCl_2$ to get a baseline reading for ~1 min (in the absence of any stimulus). Time course acquisition experiment is set up so that images are acquired once every 5 s. SOCE responses in T cells after depletion of intracellular stores are robust and long lasting, hence data are acquired every 5 s here; however, the time interval can be changed depending on the speed of response expected.
5. After acquiring the baseline, the chamber is perfused with Ca^{2+}-free Ringer's solution and measurements are continued for a further ~1 min.
6. For depletion of ER Ca^{2+}-stores, the chamber is perfused with 1 μ*M* thapsigargin-containing Ca^{2+}-free Ringer's solution for 5 min to observe an increase in $[Ca^{2+}]_i$ derived from depletion of ER Ca^{2+} store. Complete depletion of the store is ensured when $[Ca^{2+}]_i$ fluorescence levels return to the baseline.

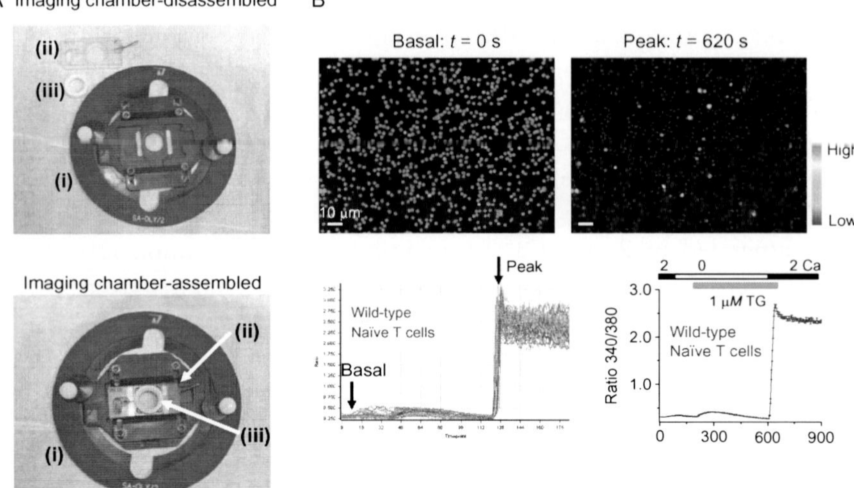

Figure 1.2 *Assembly of imaging chamber and measurement of store-operated Ca^{2+} entry from naïve T cells.* (A) Perfusion chamber setup. Top image depicts the disassembled chamber showing a series 20 chamber platform attached to Olympus stage adaptor (i) and the imaging chamber RC-20H (ii) with the white ring (iii). Bottom image shows the RC20H chamber assembled on the platform with the white ring on top. (B) Images of naïve T cells at different stages of SOCE measurement: resting and peak of SOCE after exchanging the Ca^{2+}-free extracellular solution with that containing 2 m*M* Ca^{2+} after store depletion with 1 μ*M* thapsigargin (TG). Scale bar 10 μm. Bottom images show raw traces of $[Ca^{2+}]_i$ in individual cells (left) or averaged ratio values of individual cells ± s.e.m. (right) at different time points.

7. SOCE is measured by subsequently perfusing the chamber with 2 m*M* Ca^{2+}-containing Ringer solution. A sharp increase in $[Ca^{2+}]_i$ is observed upon addition of 2 m*M* Ca^{2+}-containing Ringer's solution which is sustained (Fig. 1.2B, top right panel). Region of interests (ROIs) can be drawn around individual cells to observe changes in $[Ca^{2+}]_i$ in individual cells. This data is then processed using OriginPro 8.5 software to generate average ± s.e.m. from at least 50–60 individual T cells (Fig. 1.2B, bottom panel). As seen in Fig. 1.2B, naïve T cells are very small, and correspondingly their ER Ca^{2+} stores are very small.
8. Once naïve T cells get stimulated and differentiate into effector T cells, their size increases by more than fivefold. SOCE measurement in effector T cells is done at days 5–7 after stimulation using essentially the same method(s) as described for naïve T cells. We describe results from SOCE measurement in WT and $Orai1^{-/-}$ effector T cells here. Figure 1.3A

and B shows images of control and Orai1$^{-/-}$ effector cells under resting and SOCE induced conditions. After monitoring peak Ca^{2+}, cells were perfused with 100 μ*M* of 2-APB to block SOCE mediated by CRAC channels. Orai1 is the predominant mediator of SOCE in T cells; accordingly, we observed a profound reduction in SOCE in Orai1$^{-/-}$ T cells (Fig. 1.3B and C). As discussed above, while SOCE mediated by Orai1 and Orai2 can be readily blocked by 2-APB, that mediated by Orai3 is not blocked by 2-APB (Dehaven et al., 2008; Schindl et al., 2008; Zhang et al., 2008). We wanted to examine whether the residual

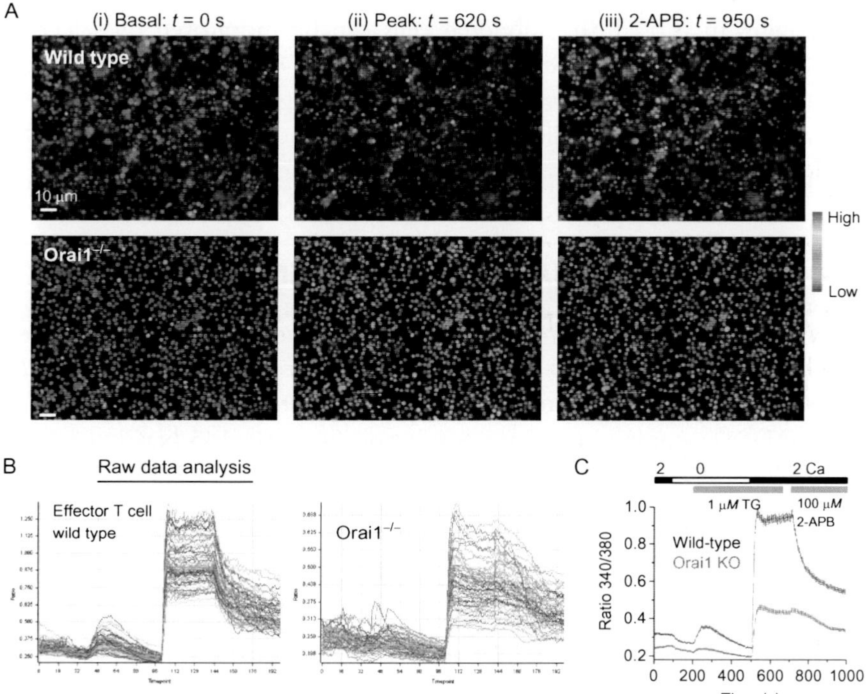

Figure 1.3 *Measurement of the intracellular* Ca^{2+} *concentrations using Fura-2 in control and Orai1*$^{-/-}$ *effector T cells.* (A) Images of wild-type (top panels) and Orai1$^{-/-}$ (bottom panels) effector T cells at different stages of SOCE measurement: (i) resting, (ii) peak of SOCE after exchanging the Ca^{2+}-free extracellular solution with that containing 2 m*M* Ca^{2+}, and (iii) residual SOCE after treatment with 100 μ*M* 2-APB. Scale bar 10 μm. (B) Traces of $[Ca^{2+}]_i$ in individual cells at different time points obtained by selecting individual region of interest (ROI) from cells depicted in panel A. (C) Graphs of average ± s.e.m. from several cells. 50–60 ROIs from WT and Orai1$^{-/-}$ cells were selected and their 340/380 ratio was averaged. Averaged ratio values of individual cells ± s.e.m. are plotted versus time using graph options of OriginPro.

SOCE in Orai1$^{-/-}$ cells is mediated by Orai2 or Orai3. Our results as seen in Fig. 1.3B and C show that residual SOCE in Orai1$^{-/-}$ cells can be slowly blocked by 2-APB, suggesting that it is most likely mediated by homomultimers of Orai2 or heteromultimers of Orai2 and Orai3, which may retain sensitivity to block by 2-APB. This data is supported by our previous observation, where we observed that expression of the dominant mutant of Orai1, Orai1^{E106Q} was able to further decrease residual SOCE in Orai1$^{-/-}$ T cells (Kim et al., 2011). Since all three Orai proteins can heteromultimerize, these results suggest that the residual SOCE in Orai1$^{-/-}$ T cells is most likely derived from activities of Orai2 or heteromultimers of Orai2 and Orai3 (Gwack et al., 2007).

9. Changes in $[Ca^{2+}]_i$ in individual cells can be monitored by selecting ROI around individual cells. By comparing SOCE from WT and Orai1$^{-/-}$ effector T cells, we can observe a strong reduction in SOCE in Orai1$^{-/-}$ effector T cells and the residual SOCE can be decreased by high concentration of 2-APB (Fig. 1.3C).

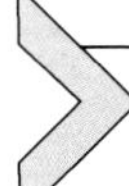

4. MITOCHONDRIAL Ca^{2+} MEASUREMENT

4.1. Reagents for mitochondrial Ca^{2+} imaging

1. Rhod-2-acetoxymethyl ester (AM, Invitrogen, Carlsbad, CA) resuspended in DMSO to generate a 10 m*M* stock solution.
2. 2 m*M* Ca^{2+}-containing Ringer's solution (in m*M*): 145 NaCl, 4.5 KCl, 1 $MgCl_2$, 2 $CaCl_2$, 10 D-glucose, and 10 HEPES (pH 7.35).
3. Thapsigargin (from EMD biosciences), final concentration of 1 μ*M* in Ca^{2+}-free Ringer's solution.
4. ECM-830 Electro Square Porator (Harvard apparatus)
5. Gene pulser 0.4 cm electroporation cuvettes (Biorad)
6. OptiMEM (Invitrogen) or plain DMEM (without additives or serum)
7. While all microscopy equipments are same as mentioned above for Fura-2 imaging, Rhod-2 images were acquired using triple-band ET—ECFP/EYFP/mCherry filter cube (Chroma Technology Corp. # 89006).

4.2. Mitochondrial Ca^{2+} imaging from effector T cells

1. Effector T cells ($2–3 \times 10^6$) are loaded with 10 μ*M* rhod-2/AM in culture medium at room temperature for 30 min in dark. Cells are washed twice with OptiMEM (or DMEM) and resuspended in OptiMEM (or

DMEM) at 5–6 × 10^6 cells/ml. A schematic of mitochondrial Ca^{2+} measurement is shown in Fig. 1.4A.

2. The cells (400 μl) are transferred into a 0.4 cm electroporation cuvette and electroporated at 200 V for 40 ms to remove residual cytoplasmic Rhod-2 dye (Fig. 1.4B). The exact condition for electroporation needs to be determined by control experiments targeting enough exclusion of Rhod-2 from the cytoplasm and minimum damage to the cells. This can also be achieved by cell-permeabilizing reagents in addition to electroporation. After electroporation, cells are washed twice with complete T-cell medium, incubated at room temperature for 10 min on poly-L-Lysine-coated cover slips and assembled onto the imaging chamber as discussed above for Fura-2 imaging. Cells are imaged using a 60 × objective to visualize mitochondria.

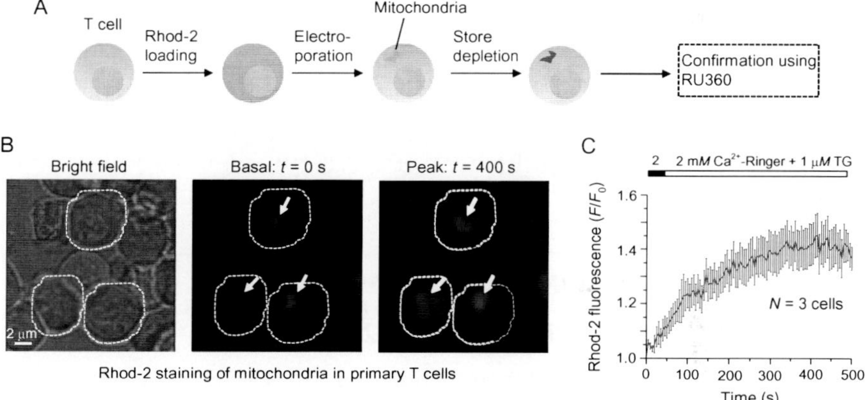

Figure 1.4 *Measurement of mitochondrial Ca^{2+} accumulation using Rhod-2.* (A) Schematic of mitochondrial Ca^{2+} measurement using Rhod-2. Rhod-2 is loaded into cells and electroporation permeabilizes the PM to exclude cytoplasmic Rhod-2 for specific mitochondrial loading. After reading the basal levels, cells are stimulated with TCR agonists or thapsigargin to induce Ca^{2+} entry into the mitochondria. Cells can be treated with mitochondrial uniporter inhibitor RU360 to confirm that the increase in Rhod-2 fluorescence was derived from mitochondrial Ca^{2+} accumulation. (B) Left most panel shows bright field image of Rhod-2 loaded primary effector T cells. Middle panel shows fluorescence image of resting cells while that on right shows image of the same cells after activation of SOCE using 1 μ*M* thapsigargin (TG). Scale bar 2 μm. Dotted lines show cells used for analysis and generation of the graph (panel C). (C) Background subtracted and intensity normalized graphs of average ± s.e.m. from three different cells. Three ROIs (as depicted in A) were selected and background was first subtracted followed by normalization to Rhod-2 intensity at time 0 s. Averaged normalized Rhod-2 fluorescence values ± s.e.m. are plotted versus time using graph options of Origin Pro.

3. The cells are first perfused with Ringer solution containing 2 mM $CaCl_2$ to get a baseline reading for ~1 min (in the absence of any stimulus, Fig. 1.4B, middle panel). Time course acquisition experiment is set up so that images are acquired once every 5 s.
4. After acquiring the baseline, the chamber is perfused with 2 mM Ca^{2+}-containing Ringer's solution with 1 μM thapsigargin to deplete the ER Ca^{2+} stores and measurements are continued for a further 7–8 min.
5. During data analysis, ROIs can be drawn around cells of interest and fluorescence intensity values monitored over time (Fig. 1.4B and C). Rhod-2 fluorescence is then background subtracted and normalized to the initial fluorescence values.
6. To confirm that the increase in fluorescence observed with rhod-2 is indeed from the mitochondria, it is recommended to use a pharmacological compound such as RU360, which inhibits the MCU. Addition of RU360 should block increase in rhod-2 fluorescence. Additionally, there are reports that measurement of mitochondrial Ca^{2+} using rhod-2 may affect mitochondrial morphology and response, hence data obtained with rhod-2 should be validated using genetically encoded Ca^{2+} indicators like ratiometric pericam or aequorin that are specifically targeted to mitochondria (Fonteriz et al., 2010).

5. DATA HANDLING AND PROCESSING

5.1. Ratio calibration for Fura-2

Usually changes in $[Ca^{2+}]_i$—for most purposes of SOCE measurement are just presented as ratio of 340/380 values for Fura-2. For sensitive analysis or for basal $[Ca^{2+}]$ measurements, ratio values can also be converted to absolute Ca^{2+} concentrations using two different methods as mentioned below.

1. In the first method, at the end of the experiment, cells are first perfused with a Ca^{2+}-free solution containing 2 mM ethylene glycol tetraacetic acid (EGTA, to obtain minimum ratio [R_{min}] value) and then with 1 μM Ionomycin in 2 mM Ca^{2+}-containing Ringer solution (to obtain maximum ratio [R_{max}] value). These values are then incorporated into the Grynkiewicz equation $[Ca^{2+}]_i = K \times (R - R_{min})(R_{max} - R)^{-1}$ (Grynkiewicz, Poenie, & Tsien, 1985). K represents the Ca^{2+} affinity of Fura-2 which is calculated as ~224 nM within cells (and 145 nM in extra cellular solutions). R_{min} and R_{max} are calculated at the end of the experiment as mentioned above.

2. In the second method, a calibration kit containing solutions with known amounts of free Ca^{2+} (Fura-2 Ca^{2+} Calibration Kit, Invitrogen) is used and fluorescence values at different Ca^{2+} concentrations are obtained as described in the manufacturer's protocol. Both the methods have advantages and disadvantages and the method of choice would depend on the sensitivity required for the experiments. While the first method is relatively easier and can be performed at the end of each experiment, it may not be the ideal method for measuring small changes in $[Ca^{2+}]_i$. The second method offers better accuracy in measurement of small changes in $[Ca^{2+}]_i$, however, it is more tedious.

5.2. Single-cell analysis

1. For each experiment, 50–60 T cells are selected by drawing ROIs around the cells using the analysis options of Slidebook. Individual cell traces are observed (Figs. 1.2B and 1.3B) and the results are converted into a text file displaying time course of ratio values of individual cells and imported into OriginPro (Originlab).
2. Ratio values of individual cells are averaged to obtain average ratios $\pm$ s.e.m. and plotted versus time using graph options of Origin Pro (Figs. 1.2B, 1.3C, and 1.4C).

6. NOTES

The intracellular Ca^{2+}-binding affinity for Fura-2 is ~224 nM. This means measurement of $[Ca^{2+}]_i$ higher than ~1 μM should be performed using low-affinity Ca^{2+} dyes such as Fura-4 F ($K_d[Ca^{2+}] = 770$ nM, information about alternative indicators is available at the Invitrogen Website). All the Ca^{2+} imaging experiments described in this study have been performed at room temperature since SOCE measurements in T cells are very robust and long lasting after store depletion using thapsigargin (Figs. 1.2 and 1.3). However, if one wants to measure SOCE in T cells using antigen-presenting cells loaded with appropriate antigen, the experiments should be performed at the physiological temperature of 37 °C.

7. CONCLUSION

In this protocol, we have demonstrated a widely used method for ratiometric measurement of SOCE in individual naïve and effector T cells using Fura-2 dye. Our data with SOCE measurement in WT and $Orai1^{-/-}$

effector T cells shows that Orai1$^{-/-}$ cells have a pronounced reduction in SOCE. Furthermore, treatment of 2-APB blocks SOCE in both WT and Orai1$^{-/-}$ T cells, indicating that the residual SOCE in Orai1$^{-/-}$ T cells is most likely due to activity of Orai2 or 2-APB sensitive heteromultimers of Orai2 and Orai3 proteins. We also described a protocol for measurement of changes in mitochondrial Ca^{2+} concentration in primary effector T cells using Rhod-2 dye. These methods will provide a basic tool to measure changes in $[Ca^{2+}]_i$ in a variety of cell types using diverse stimuli or agonists.

ACKNOWLEDGMENTS

This work was supported by National Institute of Health grants AI-083432 and AI-101569, a grant from the Lupus Research Institute (Y. G.), and Scientist Development Grant from the American Heart Association, 12SDG12040188 (S. S.).

REFERENCES

Baughman, J. M., Perocchi, F., Girgis, H. S., Plovanich, M., Belcher-Timme, C. A., Sancak, Y., et al. (2011). Integrative genomics identifies MCU as an essential component of the mitochondrial calcium uniporter. *Nature*, *476*(7360), 341–345.

Brandman, O., Liou, J., Park, W. S., & Meyer, T. (2007). STIM2 is a feedback regulator that stabilizes basal cytosolic and endoplasmic reticulum Ca2+ levels. *Cell*, *131*(7), 1327–1339.

Budd, R. C. (2001). Activation-induced cell death. *Current Opinion in Immunology*, *13*(3), 356–362.

Cahalan, M. D. (2009). STIMulating store-operated Ca(2+) entry. *Nature Cell Biology*, *11*(6), 669–677.

Cahalan, M. D., & Chandy, K. G. (2009). The functional network of ion channels in T lymphocytes. *Immunological Reviews*, *231*(1), 59–87.

Chen, Y. F., Chiu, W. T., Chen, Y. T., Lin, P. Y., Huang, H. J., Chou, C. Y., et al. (2011). Calcium store sensor stromal-interaction molecule 1-dependent signaling plays an important role in cervical cancer growth, migration, and angiogenesis. *Proceedings of the National Academy of Sciences of the United States of America*, *108*, 15225–15230.

Collins, H. E., Zhu-Mauldin, X., Marchase, R. B., & Chatham, J. C. (2013). STIM1/Orai1-mediated SOCE: Current perspectives and potential roles in cardiac function and pathology. *American Journal of Physiology. Heart and Circulatory Physiology*, *305*(4), H446–H458.

Dehaven, W. I., Smyth, J. T., Boyles, R. R., Bird, G. S., & Putney, J. W., Jr. (2008). Complex actions of 2-aminoethyldiphenyl borate on store-operated calcium entry. *Journal of Biological Chemistry*, *283*(28), 19265–19273.

De Stefani, D., Raffaello, A., Teardo, E., Szabo, I., & Rizzuto, R. (2011). A forty-kilodalton protein of the inner membrane is the mitochondrial calcium uniporter. *Nature*, *476*(7360), 336–340.

Feng, M., Grice, D. M., Faddy, H. M., Nguyen, N., Leitch, S., Wang, Y., et al. (2010). Store-independent activation of Orai1 by SPCA2 in mammary tumors. *Cell*, *143*(1), 84–98.

Feske, S., Gwack, Y., Prakriya, M., Srikanth, S., Puppel, S. H., Tanasa, B., et al. (2006). A mutation in Orai1 causes immune deficiency by abrogating CRAC channel function. *Nature*, *441*(7090), 179–185.

Fonteriz, R. I., de la Fuente, S., Moreno, A., Lobaton, C. D., Montero, M., & Alvarez, J. (2010). Monitoring mitochondrial [Ca(2+)] dynamics with rhod-2, ratiometric pericam and aequorin. *Cell Calcium*, *48*(1), 61–69.

Grynkiewicz, G., Poenie, M., & Tsien, R. Y. (1985). A new generation of Ca2+ indicators with greatly improved fluorescence properties. *Journal of Biological Chemistry*, *260*(6), 3440–3450.

Gwack, Y., Srikanth, S., Feske, S., Cruz-Guilloty, F., Oh-hora, M., Neems, D. S., et al. (2007). Biochemical and functional characterization of Orai proteins. *Journal of Biological Chemistry*, *282*(22), 16232–16243.

Gwack, Y., Srikanth, S., Oh-Hora, M., Hogan, P. G., Lamperti, E. D., Yamashita, M., et al. (2008). Hair loss and defective T and B cell function in mice lacking ORAI1. *Molecular and Cellular Biology*, *28*(17), 5209–5222.

Hildeman, D., Jorgensen, T., Kappler, J., & Marrack, P. (2007). Apoptosis and the homeostatic control of immune responses. *Current Opinion in Immunology*, *19*(5), 516–521.

Hildeman, D. A., Zhu, Y., Mitchell, T. C., Bouillet, P., Strasser, A., Kappler, J., et al. (2002). Activated T cell death in vivo mediated by proapoptotic bcl-2 family member bim. *Immunity*, *16*(6), 759–767.

Hodge, M. R., Ranger, A. M., Charles de la Brousse, F., Hoey, T., Grusby, M. J., & Glimcher, L. H. (1996). Hyperproliferation and dysregulation of IL-4 expression in NF-ATp-deficient mice. *Immunity*, *4*(4), 397–405.

Hogan, P. G., Lewis, R. S., & Rao, A. (2010). Molecular basis of calcium signaling in lymphocytes: STIM and ORAI. *Annual Review of Immunology*, *28*, 491–533.

Kim, K. D., Srikanth, S., Yee, M. K., Mock, D. C., Lawson, G. W., & Gwack, Y. (2011). ORAI1 deficiency impairs activated T cell death and enhances T cell survival. *Journal of Immunology*, *187*(7), 3620–3630.

Krammer, P. H., Arnold, R., & Lavrik, I. N. (2007). Life and death in peripheral T cells. *Nature Reviews. Immunology*, 7(7), 532–542.

Lewis, R. S. (2011). Store-operated calcium channels: New perspectives on mechanism and function. *Cold Spring Harbor Perspectives in Biology*, *3*(12), a003970.

Liou, J., Kim, M. L., Heo, W. D., Jones, J. T., Myers, J. W., Ferrell, J. E., Jr., et al. (2005). STIM is a Ca2+ sensor essential for Ca2+-store-depletion-triggered Ca2+ influx. *Current Biology*, *15*(13), 1235–1241.

Macian, F., Garcia-Cozar, F., Im, S. H., Horton, H. F., Byrne, M. C., & Rao, A. (2002). Transcriptional mechanisms underlying lymphocyte tolerance. *Cell*, *109*(6), 719–731.

Marrack, P., & Kappler, J. (2004). Control of T cell viability. *Annual Review of Immunology*, *22*, 765–787.

Marsden, V. S., & Strasser, A. (2003). Control of apoptosis in the immune system: Bcl-2, BH3-only proteins and more. *Annual Review of Immunology*, *21*, 71–105.

Matheu, M. P., Parker, I., & Cahalan, M. D. (2007). Dissection and 2-photon imaging of peripheral lymph nodes in mice. *Journal of Visualized Experiments*, (7), 265.

Oh-Hora, M., Yamashita, M., Hogan, P. G., Sharma, S., Lamperti, E., Chung, W., et al. (2008). Dual functions for the endoplasmic reticulum calcium sensors STIM1 and STIM2 in T cell activation and tolerance. *Nature Immunology*, *9*(4), 432–443.

Peinelt, C., Lis, A., Beck, A., Fleig, A., & Penner, R. (2008). 2-Aminoethoxydiphenyl borate directly facilitates and indirectly inhibits STIM1-dependent gating of CRAC channels. *Journal of Physiology*, *586*(13), 3061–3073.

Prakriya, M., & Lewis, R. S. (2001). Potentiation and inhibition of Ca(2+) release-activated Ca(2+) channels by 2-aminoethyldiphenyl borate (2-APB) occurs independently of IP(3) receptors. *Journal of Physiology*, *536*(Pt 1), 3–19.

Prevarskaya, N., Skryma, R., & Shuba, Y. (2011). Calcium in tumour metastasis: New roles for known actors. *Nature Reviews. Cancer*, *11*(8), 609–618.

Putney, J. W., Jr. (1986). A model for receptor-regulated calcium entry. *Cell Calcium*, 7(1), 1–12.

Quintana, A., Pasche, M., Junker, C., Al-Ansary, D., Rieger, H., Kummerow, C., et al. (2011). Calcium microdomains at the immunological synapse: How ORAI channels, mitochondria and calcium pumps generate local calcium signals for efficient T-cell activation. *EMBO Journal*, *30*(19), 3895–3912.

Quintana, A., Schwindling, C., Wenning, A. S., Becherer, U., Rettig, J., Schwarz, E. C., et al. (2007). T cell activation requires mitochondrial translocation to the immunological synapse. *Proceedings of the National Academy of Sciences of the United States of America*, *104*(36), 14418–14423.

Roos, J., DiGregorio, P. J., Yeromin, A. V., Ohlsen, K., Lioudyno, M., Zhang, S., et al. (2005). STIM1, an essential and conserved component of store-operated Ca2+ channel function. *Journal of Cell Biology*, *169*(3), 435–445.

Schindl, R., Bergsmann, J., Frischauf, I., Derler, I., Fahrner, M., Muik, M., et al. (2008). 2-Aminoethoxydiphenyl borate alters selectivity of orai3 channels by increasing their pore size. *Journal of Biological Chemistry*, *283*(29), 20261–20267.

Serfling, E., Klein-Hessling, S., Palmetshofer, A., Bopp, T., Stassen, M., & Schmitt, E. (2006). NFAT transcription factors in control of peripheral T cell tolerance. *European Journal of Immunology*, *36*(11), 2837–2843.

Soboloff, J., Rothberg, B. S., Madesh, M., & Gill, D. L. (2012). STIM proteins: Dynamic calcium signal transducers. *Nature Reviews. Molecular Cell Biology*, *13*(9), 549–565.

Srikanth, S., & Gwack, Y. (2013a). Measurement of intracellular Ca2+ concentration in single cells using ratiometric calcium dyes. *Methods in Molecular Biology*, *963*, 3–14.

Srikanth, S., & Gwack, Y. (2013b). Orai1-NFAT signalling pathway triggered by T cell receptor stimulation. *Molecules and Cells*, *35*(3), 182–194.

Srikanth, S., Jew, M., Kim, K. D., Yee, M. K., Abramson, J., & Gwack, Y. (2012). Junctate is a Ca2+-sensing structural component of Orai1 and stromal interaction molecule 1 (STIM1). *Proceedings of the National Academy of Sciences of the United States of America*, *109*(22), 8682–8687.

Srikanth, S., Jung, H. J., Kim, K. D., Souda, P., Whitelegge, J., & Gwack, Y. (2010). A novel EF-hand protein, CRACR2A, is a cytosolic Ca2+ sensor that stabilizes CRAC channels in T cells. *Nature Cell Biology*, *12*(5), 436–446.

Stiber, J. A., & Rosenberg, P. B. (2011). The role of store-operated calcium influx in skeletal muscle signaling. *Cell Calcium*, *49*(5), 341–349.

Strasser, A. (2005). The role of BH3-only proteins in the immune system. *Nature Reviews. Immunology*, *5*(3), 189–200.

Strasser, A., & Pellegrini, M. (2004). T-lymphocyte death during shutdown of an immune response. *Trends in Immunology*, *25*(11), 610–615.

Trebak, M. (2012). STIM/Orai signalling complexes in vascular smooth muscle. *Journal of Physiology*, *590*(Pt 17), 4201–4208.

Vig, M., Peinelt, C., Beck, A., Koomoa, D. L., Rabah, D., Koblan-Huberson, M., et al. (2006). CRACM1 is a plasma membrane protein essential for store-operated Ca2+ entry. *Science*, *312*(5777), 1220–1223.

Yamashita, M., Somasundaram, A., & Prakriya, M. (2011). Competitive modulation of Ca2+ release-activated Ca2+ channel gating by STIM1 and 2-aminoethyldiphenyl borate. *Journal of Biological Chemistry*, *286*(11), 9429–9442.

Yang, S., Zhang, J. J., & Huang, X. Y. (2009). Orai1 and STIM1 are critical for breast tumor cell migration and metastasis. *Cancer Cell*, *15*(2), 124–134.

Zhang, S. L., Kozak, J. A., Jiang, W., Yeromin, A. V., Chen, J., Yu, Y., et al. (2008). Store-dependent and -independent modes regulating Ca2+ release-activated Ca2+ channel activity of human Orai1 and Orai3. *Journal of Biological Chemistry*, *283*(25), 17662–17671.

Zhang, S. L., Yeromin, A. V., Zhang, X. H., Yu, Y., Safrina, O., Penna, A., et al. (2006). Genome-wide RNAi screen of Ca(2+) influx identifies genes that regulate Ca(2+) release-activated Ca(2+) channel activity. *Proceedings of the National Academy of Sciences of the United States of America*, *103*(24), 9357–9362.

CHAPTER TWO

Methods to Measure Intracellular Ca^{2+} Fluxes with Organelle-Targeted Aequorin-Based Probes

Denis Ottolini, Tito Calì, Marisa Brini[1]
Department of Biology, University of Padova, Padova, Italy
[1]Corresponding author: e-mail address: marisa.brini@unipd.it

Contents

Abstract

The photoprotein aequorin generates blue light upon binding of Ca^{2+} ions. Together with its very low Ca^{2+}-buffering capacity and the possibility to add specific targeting sequences, this property has rendered aequorin particularly suitable to monitor Ca^{2+} concentrations in specific subcellular compartments. Recently, a new generation of genetically encoded Ca^{2+} probes has been developed by fusing Ca^{2+}-responsive elements with the green fluorescent protein (GFP). Aequorin has also been employed to this aim, resulting in an aequorin-GFP chimera with the Ca^{2+} sensitivity of aequorin and the fluorescent properties of GFP. This setup has actually solved the major limitation of aequorin, for example, its poor ability to emit light, which rendered it inappropriate for the monitoring of Ca^{2+} waves at the single-cell level by imaging. In spite of the

Methods in Enzymology, Volume 543
ISSN 0076-6879
http://dx.doi.org/10.1016/B978-0-12-801329-8.00002-7

numerous genetically encoded Ca^{2+} indicators that are currently available, aequorin-based probes remain the method of election when an accurate quantification of Ca^{2+} levels is required. Here, we describe currently available aequorin variants and their use for monitoring Ca^{2+} waves in specific subcellular compartments. Among various applications, this method is relevant for the study of the alterations of Ca^{2+} homeostasis that accompany oncogenesis, tumor progression, and response to therapy.

1. INTRODUCTION

Ca^{2+} is the main second messenger of eukaryotic cells. It controls processes as diverse as fertilization, synaptic transmission, muscle contraction, gene expression, metabolism, proliferation, and cell death. The generation of Ca^{2+} signals relies primarily on the activation of Ca^{2+} influx at the plasma membrane and/or the release of Ca^{2+} from the intracellular organelles. Numerous Ca^{2+} membrane transporters and Ca^{2+}-binding proteins (the so-called "Ca^{2+} toolkit") orchestrate Ca^{2+} fluxes in a coordinated manner to guarantee proper cell function. Abnormalities in Ca^{2+} signaling have severe pathological consequences and can result in disorders of different nature, such as neurodegeneration, heart disease, skeletal muscle defects, and also cancer (Brini, Cali, Ottolini, & Carafoli, 2013).

The ability to monitor changes in the free Ca^{2+} concentration in living cells is fundamental to study the Ca^{2+}-regulated processes. The development, more than 30 years ago, of the first synthetic fluorescent dyes (fura 2 and quin 2) based on the structure of the Ca^{2+} chelator EGTA pioneered the field (Tsien, Pozzan, & Rink, 1982). The enormous improvement of their characteristics and of the technologies to monitor their fluorescence allowed to measure Ca^{2+} signals with extreme spatio-temporal resolution and to define numerous molecular mechanisms at the basis of Ca^{2+}-regulated cellular processes. More recently, starting from the cloning of the cDNA encoding some proteins with spontaneous luminescent or fluorescent properties, a new generation of genetically encoded Ca^{2+} indicators (usually referred as GECIs) has been developed. The photoprotein aequorin was the precursor: the cloning of its cDNA in 1985 (Inouye et al., 1985) opened the possibility to express it in different cell types and to target it into specific cellular districts. Other genetically encoded Ca^{2+} indicators have been developed after the cloning of the cDNA of the GFP in 1995 (Chalfie, 1995) and its manipulation to generate different variants with different spectral emission. These indicators, namely, camgaroos, G-CaMps, pericam, and cameleons, were constructed by inserting a Ca^{2+}-responsive element between two fluorescent proteins so that, an alteration in the efficiency of fluorescence resonance energy transfer occurs upon Ca^{2+} binding (McCombs & Palmer, 2008; Miyawaki et al., 1997; Romoser, Hinkle, & Persechini, 1997). These indicators are currently used to monitor Ca^{2+} fluxes with high spatial resolution using standard fluorescence microscope or laser-scanning confocal microscope. They are discussed in detail in other chapters of this issue.

In this chapter, we will focus on the properties of the photoprotein aequorin which, even if not yet perfectly suitable to perform Ca^{2+} measurements at single-cell level, still offer several advantages with respect to the GFP-based indicators. Among them, the very low Ca^{2+}-buffering activity and the ability to emit light upon Ca^{2+} binding without requirement of potentially damaging light excitation, guarantee minimal perturbation of the endogenous Ca^{2+} fluxes.

1.1. Why is it important to measure Ca^{2+} fluxes in the intracellular organelles?

In the last decades, it became evident that Ca^{2+} handling in the organelles is essential to their specific function but also for general processes such as apoptosis, proliferation, and metabolism which govern the balance between cell death and survival. In particular, ER and mitochondria strictly cooperate in this control (Ferri & Kroemer, 2001). Thus, the possibility to directly investigate their Ca^{2+} handling is of extreme importance for the understanding of the molecular basis of cellular dysfunctions. Thanks to numerous studies in this direction, both mitochondria and ER have been proposed as candidate targets for rational drug design and new therapeutic strategies in several pathologies including neurodegenerative diseases and cancer. Several oncogenes and tumor suppressors control the apoptotic process by modifying the activity of proteins involved in Ca^{2+} homeostasis. The analysis of the mitochondrial and ER Ca^{2+} levels led to the comprehension of many of the stages of the apoptotic cascade (Giorgi et al., 2012). One of the major trigger of the apoptotic process is the excessive mitochondrial Ca^{2+} accumulation: enhanced cytosolic Ca^{2+} levels and/or exaggerated Ca^{2+} release from the intracellular stores can be responsible for mitochondrial Ca^{2+} overload that may activate apoptotic process or sensitize the cells to different apoptotic stimuli (Giorgi, Romagnoli, Pinton, & Rizzuto, 2008). Numerous information on the action of the Bcl-2 protein family in terms of Ca^{2+} handling have been obtained by using targeted aequorin. The first evidence showing that the antiapoptotic protein Bcl-2 could limit ER Ca^{2+} content, thus preventing an excessive Ca^{2+} load within the mitochondria, came from Rizzuto and coworkers. By directly monitoring free Ca^{2+} levels in the lumen of the ER, they have shown that Bcl-2 overexpression caused a reduction of the ER Ca^{2+} concentration protecting cells against apoptotic cell death (Pinton et al., 2000). Later on, the Bcl-2 family proapoptotic members, Bax and Bak, have been shown to exert the opposite effect, that is, a reduction in the ER Ca^{2+} content and release was observed in Bax/Bak double KO cells, that also displayed resistance to apoptosis. Accordingly, the mitochondrially targeted aequorin revealed a reduction in mitochondrial Ca^{2+} accumulation in these cells (Scorrano et al., 2003). The direct monitoring of mitochondrial Ca^{2+} fluxes has been fundamental to establish a link between Ca^{2+} homeostasis and mitochondrial biology in cancer. In particular, the role of a specialized domain, the so-called mitochondria-associated ER-membranes (MAMs), is emerging as key determinant in the pathogenesis of numerous diseases including cancer. The promyelocytic leukemia protein has been demonstrated to regulate apoptosis by modulating the ER-mitochondria Ca^{2+} signaling (Giorgi et al., 2010), and a number of proteins known to participate in the induction of the tumorigenic process, that is, PTEN (Bononi et al., 2013; Garcia-Cao et al., 2012), Akt (Marchi et al., 2008, 2012), and H-Ras (Rimessi, Marchi, Patergnani, & Pinton, 2013), were found to localize at the MAMs. The importance of the MAMs is also indirectly confirmed by our recent work where DJ-1, an antiapoptotic protein related to familial Parkinson disease, has been shown to overcome the p53-induced mitochondrial impairment by modulating the ER-mitochondria interplay (Ottolini, Calì, Negro, & Brini, 2013).

Finally, the recent molecular identification of the channel responsible for the transport of Ca^{2+} within the mitochondrial matrix, the mitochondrial Ca^{2+} uniporter (MCU, Baughman et al., 2011; De Stefani, Raffaello, Teardo, Szabo, & Rizzuto, 2011), together with its regulatory components MICU1 (Perocchi et al., 2010), 2, and 3 (Plovanich et al., 2013), opened the possibility to directly manipulate mitochondrial Ca^{2+} fluxes and to study the consequences. Very recently, the role of MCU in cancer cells survival and in cell resistance to apoptotic stimuli has been investigated by Pinton and coworkers. They have identified a family of cancer-related miRNA (miRNA-25) able to regulate the expression level of MCU by promoting its degradation, thus conferring apoptotic resistance. Interestingly, MCU is downregulated and miR-25 is aberrantly expressed in human colon cancer samples, highlighting the importance of mitochondrial Ca^{2+} regulation in cancer cell survival (Marchi et al., 2013).

1.2. Aequorin: From the jellyfish to the transgenic mice

Aequorin is a 21 kDa photoprotein produced by the jellyfish *Aequorea victoria* composed by an apoprotein and a prosthetic group, the coelenterazine. It has been widely used in the 1960s and 1970s as a probe to measure cytosolic Ca^{2+} in living cells (Cobbold & Lee, 1991). The ability of the protein to bind Ca^{2+} is due to the presence of four helix-loop-helix EF-hand sites arranged in pairs to form the globular molecule. Upon Ca^{2+} binding to three high-affinity sites (the second EF-hand is not operative), aequorin undergoes irreversible reaction in which 1 photon is emitted and the oxidized coenzyme is released. Due to the cooperativity between the three binding sites, light emission is proportional to the second to third power of Ca^{2+} concentration. Native (wt) aequorin was originally extracted from the jellyfish and employed to measure cytosolic Ca^{2+} concentrations after microinjection in single cells. In the 1990s, the common use of molecular biology techniques replaced the need of microinjection with the possibility of recombinantly expressing aequorin, thus opening new applications for this probe, the possibility to target it to different cellular locations being probably the most interesting (see below). Aequorin is not toxic *per se* and after overexpression it reaches concentrations ranging from 10^{-6} to 10^{-7} *M* inside cells, a value that is 2–3 orders of magnitude lower than that reached by the classical loaded dyes. This gives to aequorin a very low Ca^{2+} buffer capacity compared to other probes, a characteristic that is appreciated when an accurate quantification of the Ca^{2+} levels is required. The steep relationship between Ca^{2+} and photon emission (see Fig. 2.1) requires a special comment: under conditions in which the Ca^{2+} concentration is dishomogeneous, as some time occurs in living cells, it may generally represent a disadvantage since the average signal is biased toward the highest values. However, it must be admitted that, in some particular circumstances, this characteristic, coupled with the specific targeting of the aequorin probe, has turned out to be useful to detect the presence of high Ca^{2+} concentration microdomains, even if their real Ca^{2+} concentration values are probably underestimated. This was the case of the aequorin probe targeted to the mitochondrial matrix that has revealed the tight Ca^{2+} coupling between ER and mitochondria (see below). The main disadvantage of aequorin is its very low quantum yield that prevents dynamic measurements of Ca^{2+} responses in single cells. However, recently, this problem has been overcome by the generation of GFP-aequorin fusion proteins that use a bioluminescence resonance energy transfer (BRET) reaction similar to the one that occurs naturally in the jellyfish. When Ca^{2+} binds to GFP-aequorin, the energy acquired by the excited coelenterazine is transferred to GFP via nonradioactive energy transfer

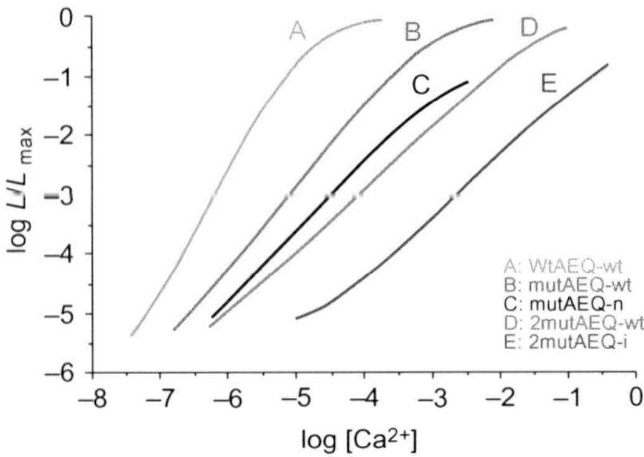

Figure 2.1 *Relationship between photon emission and Ca^{2+} concentration of different aequorin variants reconstituted with wt,* n *or* i *coelenterazine.* The curves were obtained by calculating the ratio between the luminescence obtained at a given [Ca^{2+}] (L) and the total remaining luminescent at that moment (L_{max}). L/L_{max} values are represented against the [Ca^{2+}]. (See the color plate.)

and without blue light being emitted. As the excited GFP returns to the ground state, green light ($\lambda_{max} = \sim 510$ nm) is emitted. These chimeras are more stable and have higher quantum yield in the presence of Ca^{2+} than aequorin alone and have been reported to be well suitable to monitor Ca^{2+} flux in whole animals *in vivo* (Baubet et al., 2000; Rogers et al., 2007).

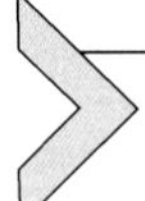

2. EXPERIMENTAL COMPONENTS AND CONSIDERATIONS

2.1. Aequorin variants with different Ca^{2+} affinity

Three different aequorin variants are available: native (wtAEQ), low-affinity (mutAEQ), and very low-affinity aequorin (2mutAEQ). The mutants originate from native aequorin that has been mutagenized in one or two of its Ca^{2+}-binding sites to reduce the Ca^{2+} affinity. The mutations generally have no effect on the luminescence yield of the protein thus preserving the quality of the signal.

Recombinant aequorin is expressed as an inactive apoprotein. The addition of the prosthetic group coelenterazine to the transfected cells is generally sufficient to reconstitute the active photoprotein (see Section 2.4). Synthetic coelenterazine is commercially available in several chemically modified variants, which differ in their Ca^{2+}-triggered reaction of the photoprotein. Wt coelenterazine is extensively used in cytosolic and mitochondrial Ca^{2+} measurements combined either with wt or mutant aequorin.

Coelenterazine *n* and *i* confer lower rate of luminescence emission than the wt form, therefore are used together with low-affinity aequorins for monitoring high Ca^{2+} level compartments, that is, the ER or the Golgi apparatus.

wtAEQ is suitable to measure Ca^{2+} concentrations ranging from 0.5 to 10 μM, which correspond to the Ca^{2+} values that are reached in the bulk cytosol of the large majority of the cells after stimulation. The introduction of mutations largely expanded the aequorin Ca^{2+} sensitivity range toward higher Ca^{2+} values and thus permitted to monitor values in the mM range. This wide dynamic range of aequorin sensitivity makes this probe particularly suitable to quantitatively measure large Ca^{2+} rises that occur in specific cell types or in compartments endowed with high Ca^{2+} concentration. The inaptness of aequorin to accurately monitor basal Ca^{2+} concentration in resting conditions is a major issue of this probe. It depends both on the low sensitivity of aequorin in the lower nM range (the relationship between light emission (L/L_{max}) and pCa is not linear for values below 10^{-7} M) and on the fact that, under condition of canonical transfection or transduction efficiency, the rate of photons emission at basal Ca^{2+} levels is very close to the background signal (on average about 10–20 counts per second) thus making a precise calibration almost impossible.

The first mutAEQ was generated by introducing an Asp119 → Ala substitution in the third EF-hand domain. The point mutation strongly impaired the ability of the photoprotein to bind Ca^{2+}, reducing it of approximately 50-fold (Montero et al., 1995). This chimera targeted to the lumen of the ER/SR by the addition of a proper signal sequence (see Section 2.2) together with the use of Sr^{2+} as a Ca^{2+} surrogate has permitted to quantitatively measure Ca^{2+} values of 200–500 μM in HeLa cells (Montero et al., 1995) or in primary skeletal muscles cultures (Brini et al., 1997). Then, the availability of coelenterazine *n* with lower relative luminescence intensity permitted Ca^{2+} measurements in living cells, avoiding the use of Sr^{2+}. MutAEQ reconstituted with coelenterazine *n* is suitable to measure Ca^{2+} concentrations ranging from 10 to 500 μM (Barrero, Montero, & Alvarez, 1997). Recently, a novel double mutated aequorin form has been generated by introducing a second point mutation (Asn28 → Leu) in the first Ca^{2+}-binding site. This additional mutation further reduced the Ca^{2+} affinity of the probe by a 1 order of magnitude when reconstituted with wild-type coelenterazine (de la Fuente, Fonteriz, de la Cruz, Montero, & Alvarez, 2012). The reconstitution with coelenterazine *n* instead produced an aequorin with an extremely low Ca^{2+} affinity but, unfortunately, with a poor luminescence yield. The use of a cofactor with intermediate sensitivity, namely, coelenterazine *i*, solved this problem and produced an aequorin with very good sensitivity in the m*M* range and good luminescence yield (de la Fuente, Fonteriz, Montero, & Alvarez, 2013).

The relationship between photon emission and Ca^{2+} concentration is graphically represented in Fig. 2.1. The curves were obtained by exposing lysates from cells expressing recombinant aequorins reconstituted with the different types of coelenterazine to increasing Ca^{2+} concentrations, under physiological conditions of pH, temperature, and ionic strength. The values are expressed as the ratio between the light intensity obtained at a given $[Ca^{2+}]$ (L) and the total remaining luminescence at that moment (L_{max}). As aequorin is continuously consumed, the value of L_{max} is not constant, but it is calculated point-by-point along the curve by subtracting the light output recorded at each point from the total light output recorded at the end of the experiment by exposing aequorin to saturating Ca^{2+} concentrations. These curves have been mathematically modeled to develop an algorithm that is currently employed in the calibration of light emission to Ca^{2+} concentration values (for details, see Alvarez & Montero, 2002; Barrero et al., 1997; Brini et al., 1995; de la Fuente et al., 2012, 2013). *In situ* calibration curve for mitochondrial targeted aequorin has been also determined by exposing digitonin permeabilized cells to increasing Ca^{2+} concentration and the calculations gave slightly lower Ca^{2+} values (Pitter, Maechler, Wollheim, & Spat, 2002).

2.2. Targeting aequorin to specific intracellular compartments

One of the major advantages of aequorin is that as a genetically encoded probe, it can be engineered by the addition of targeting sequences or by the fusion with resident proteins to reach a precise organelle or cellular compartment exposed to specific Ca^{2+} microdomains. The fusion with a protein of interest is extremely useful to monitor the concentration of Ca^{2+} to which it is exposed. Several aequorin chimeras have been developed.

The native recombinant protein is normally expressed in the cytosol of the transfected cells, and it does not need any modification to monitor Ca^{2+}. Cytosolic aequorin (*cytAEQ*) was only modified by adding the HA1 epitope tag at its N-terminus (Brini et al., 1995). All the aequorin variants are tagged with the same epitope to facilitate their localization by immunocytochemistry analysis.

Two different aequorins targeted to the nuclear compartment, specifically to the nucleoplasm, have been developed: a constitutive *nuAEQ* chimera, by fusing the DNA-binding domain and the nuclear localization signal (NLS) of the glucocorticoid receptor (GR) (Brini et al., 1993), and a nucleus/cytosol shuttling chimera (*nu/cytAEQ*) that allows to alternatively measure Ca^{2+} in both the compartments, thanks to the fusion of wt aequorin to the NLS and the hormone-binding domain of the GR. This chimera is normally cytosolic, but after cell treatment with glucocorticoids undergoes translocation to the nucleus (Brini, Marsault, Bastianutto, Pozzan, & Rizzuto, 1994).

Three mitochondrially targeted aequorins are currently available. All of them are targeted to the mitochondrial matrix by the fusion with the mitochondrial localization sequence of subunit VIII of human cytochrome *c* oxidase, but differ for their Ca^{2+} affinity. They permit to measure Ca^{2+} concentration in a broad range of values: the wt (*mtAEQ*) from 0.5 to 10 μ*M*, the low-affinity (*mtAEQ mut*) from 5 to 100 μ*M*, and the very low-affinity double mutant variant (*mtAEQ2mut*) from 100 μ*M* to m*M* values (de la Fuente et al., 2012; Rizzuto, Simpson, Brini, & Pozzan, 1992).

An aequorin localized in the mitochondrial intermembrane space (*mimsAEQ*) has also been developed by fusing, at the 5′ cDNA of aequorin, the 3′ cDNA of the glycerol phosphate dehydrogenase, an inner mitochondrial membrane with a large C-terminal tail protruding in the mitochondrial intermembrane space (Rizzuto et al., 1998). This chimera permits to selectively measure Ca^{2+} microdomains to which mitochondria are exposed.

An aequorin probe targeted to the intracellular side of the plasma membrane (*pmAEQ*) was created by fusing HA1-tagged aequorin with the SNAP-25 protein, a protein of the SNARE complex anchored to the cytosolic side of the plasma membrane after palmitoylation of specific cysteine residues. This probe was available both as wt and mutated aequorin (Marsault, Murgia, Pozzan, & Rizzuto, 1997).

The targeting of aequorin to the endoplasmic reticulum was less straightforward since the most common signal sequence KDEL is normally present at the C-terminal of the ER-resident proteins, but its addition to the C-terminus of aequorin impaired its luminescence properties (Nomura, Inouye, Ohmiya, & Tsuji, 1991; Watkins & Campbell, 1993). Despite this, one group produced an aequorin chimera with the canonical KDEL signal, but they reported very low Ca^{2+} level in the ER (Kendall, Dormer, & Campbell, 1992, Kendall, Sala-Newby, Ghalaut, Dormer, & Campbell, 1992). We have followed an alternative strategy. The N-terminal domain of the Igγ2b heavy chain, containing a leader sequence (L) and two domains called VDJ and CH1, respectively, responsible for the targeting and the retention of the nascent polypeptide chain in the ER lumen upon binding to the resident protein BiP, was fused in frame at the N-terminal of the

HA1-tagged mutated aequorin (Montero et al., 1995). The binding of CH1 domain to BiP is normally displaced only in the presence of the immunoglobulin light chain. Thus, in cells lacking it the chimera is retained in the ER lumen (*erAEQ*). Very recently, a second mutation was introduced in this chimera, generating a very low-affinity variant (*erAEQ2mut*) that allows prolonged measurements of ER Ca^{2+} at 37 °C without problems of aequorin consumption (de la Fuente et al., 2013).

A different variant of targeted aequorin has been constructed to specifically monitor Ca^{2+} concentration in the sarcoplasmic reticulum (SR). To this end, HA1-tagged mutated aequorin was added to the C-terminal of calsequestrin (*srAEQ*), a typical Ca^{2+} binding protein of the SR that prevalently distributes to the terminal cisternae (Brini et al., 1997). When transfected in primary skeletal muscle cells, this chimera displayed a different localization with respect to the erAEQ and it was reported to monitor different Ca^{2+} values, thus providing evidence for heterogeneities in Ca^{2+} concentration within the intracellular stores (Brini et al., 2005; Robert, De Giorgi, Massimino, Cantini, & Pozzan, 1998).

An aequorin probe targeted to the Golgi apparatus has been developed by adding a sequence of 17 amino acids corresponding to a portion of the transmembrane domain of sialyltransferase, a typical trans-Golgi resident protein (Pinton, Pozzan, & Rizzuto, 1998). Considering the high level of Ca^{2+} of this compartment, the low-affinity aequorin variant was used.

Recently, aequorin has been addressed to peroxisomes (peroxAEQ) by fusing the tripeptide SKL, a classical peroxisomal targeting signal, to its C-terminal without apparently altering the luminescent properties of the protein. Both wt and low-affinity aequorin chimeras were created, but considering the high level of Ca^{2+} inside these organelles following cell stimulation, only the mutated version can be successfully employed (Lasorsa et al., 2008).

Aequorin probes have also been developed to monitor Ca^{2+} in the secretory vesicles and the gap junctions. The strategy adopted in these cases was to create fusion proteins containing aequorin and a resident protein. The targeting to the secretory vesicles was achieved by the fusion with the VAMP2/synaptobrevin (Mitchell et al., 2001), that to the gap junction by the fusion with different connexins (i.e., 26, 32, and 43) (George, Kendall, Campbell, & Evans, 1998; Martin et al., 1998). This strategy can be ideally adopted to create any chimeric probe, however, one must be aware that the C-terminal of aequorin should not be modified and that the correct targeting of the fusion protein should be carefully controlled to avoid mislocalization of the probe that would affect the selectivity of the Ca^{2+} measurements.

2.3. Aequorin expression

2.3.1 Cells types

Aequorin-based probes can be used in almost every cell types: both stable mammalian cell lines (e.g., HeLa, PC12, CHO, SHSY-5Y, etc.) and primary cultures of different origin (e.g., neurons, fibroblasts, myocytes, etc.) can be successfully manipulated to introduce exogenous aequorin probes. Recently, the approach has been extended to *in vivo* application: transgenic animals (mice, zebrafish, and Drosophila) expressing aequorin alone or in fusion with GFP have been generated and employed to measure Ca^{2+} fluxes *in vivo* without toxicity or functional alterations (Ashworth & Brennan, 2005; Martin, Rogers, Chagneau, & Brulet, 2007; Rogers et al., 2007).

2.3.2 Transfection strategies

Different transfection protocols can be used to introduce the aequorin-encoding vectors in the cells. Calcium-phosphate (Ca-P) precipitation, electroporation, cationic lipids, gene-gun, viral infection, and injection are currently employed: the best choice is largely dependent on the cell type of interest (Brini, 2008; Ottolini, Cali, & Brini, 2013). The main achievement is not only to reach a relatively good expression level in order to reduce the background signal but also to avoid any possible toxic effect since Ca^{2+} signaling is extremely susceptible to cell conditions. A general suggestion is to perform the experiment at least 12–24 h after transfection, not only to optimize aequorin expression but also to guarantee full recovery of the transfected cells.

Both transient transfection and stable clones can be generated. To generate stable clones expressing aequorin, the selection with specific antibiotic is generally started 24 h after the transfection.

HeLa cells (or other stable cell lines such as HEK293, CHO, Cos7, etc.) are extensively used as cell model to measure intracellular Ca^{2+} fluxes with aequorin. They are currently transfected by Ca-P precipitation. Different protocols are available in literature (Ottolini, Cali, & Brini, 2013) and can be adjusted to reach optimal efficiency. We just highlight some general suggestions:

- Plate the cells to reach 50–60% of confluence at the day of transfection. Usually, we plate 30,000 to 40,000 cells for each 13 mm diameter glass coverslip.

- Make sure that cells are homogenously distributed on the coverslip, avoiding their accumulation at the center of the well.
- Change the medium just before the addition of Ca-P precipitates, this operation will improve the efficiency of the transfection. We have experienced that the presence of HEPES in the growth medium as well as its excessive alkalization can compromise the efficiency of the transfection.
- When aequorin expression plasmid is cotransfected with other plasmids, use a larger amount of cDNA of the protein of interest with respect to aequorin (a ratio of 2:1 is normally sufficient). This precaution avoids the possibility that some cells could be single transfected with aequorin probe. In this case, the effects of the protein of interest could be underestimated.
- After transfection, remove accurately the excess of Ca-P precipitates by washing twice with standard Phosphate Buffer Saline solution.

Neuroblastoma cell lines (e.g., SHSY-5Y), MEF, primary skeletal muscle, or neuronal cell cultures may be barely transfected by Ca-P coprecipitation. For these cells, cationic lipids are instead recommended. A good attachment of the cells to the coverslip is essential both for the transfection efficiency and the Ca^{2+} measurements that are generally performed in a perfusion system. To facilitate the attachment of the cells, the coverslips can be treated with gelatin, collagen, or poly-L-lysin. As for the cationic lipids, we routinely employ Lipofectamine 2000 (Invitrogen), Transfectin (Biorad), or Fugene (Roche). The protocol, based on the manufacturer's instruction, needs to be adjusted according to the cell type to achieve the optimal efficiency of transfection. Some suggestions:

- Plate the cells at 80–90% of confluence. Cationic lipids toxicity can cause a little percentage of cell death.
- Replace the growth medium with one without antibiotics before the transfection.
- Optimize the time of incubation with cationic lipid by reducing it as much as possible to avoid toxicity.
- Take care of the plastic tubes used for the preparation of transfection reactions: polypropylene tubes are recommended instead of polystyrene ones, considering the propensity of the latter to bind the cationic lipids.

2.4. Aequorin reconstitution

Aequorin reconstitution is achieved by adding the prosthetic group coelenterazine directly onto the transfected cells and incubating for 1–2 h at 37 °C in a 5% CO_2 atmosphere. Incubation at room temperature or

4 °C can also be performed when necessary (see below). Coelenterazine is highly lipophilic and it freely permeates through the membranes. 1–2 h of incubation are generally sufficient to get a good level of reconstituted aequorin.

Coelenterazine stock (100 ×) is prepared in methanol and stored at −80 °C protected from the light. It can be diluted to the working concentration of 5 μ*M* by directly adding it to the wells where the cells are seeded. The protocol of reconstitution differs according to the type of aequorin, the compartment that has to be measured, and the experimental protocol of Ca^{2+} measurements (Brini, 2008; Ottolini, Cali, & Brini, 2013). Reconstitution of cytosolic, mitochondrial (either wt or low-affinity variants), plasma membrane, and nuclear-targeted aequorin is usually performed in the medium where the cells are grown (e.g., DMEM, for HeLa cells), with the only carefulness to replace it with a fresh one and to reduce FBS concentration from 10% to 1%. Higher % of serum could sequestrate coelenterazine and thus reduce the efficiency of the reconstitution process. Reconstitution can be also achieved in Krebs–Ringer modified Buffer (KRB, 125 m*M* NaCl, 5 m*M* KCl, 1 m*M* Na_3PO_4, 1 m*M* $MgSO_4$, 5.5 m*M* glucose, 20 m*M* HEPES, pH 7.4, 37 °C). This protocol is recommended when it is necessary to control the extracellular Ca^{2+} concentration. 1 m*M* $CaCl_2$, or 0.1 m*M* EGTA can be alternatively added to the KRB depending on whether the Ca^{2+} measurement will be performed in the physiological or Ca^{2+}-free conditions.

Suggestions

- Add coelenterazine from the stock solution directly to the coverslip by dropping it in 300 μl of DMEM/1% FBS.
- Avoid to prepare a DMEM/1% FBS/coelenterazine solution in a tube and then distribute it on the wells.

The reconstitution of low-affinity aequorin (mutAEQ) targeted to the intracellular Ca^{2+} stores, that is, the ER/SR and the Golgi apparatus, requires previous depletion of the stores otherwise, because their high Ca^{2+} content, the rate of aequorin consumption would counteracts the rate of aequorin reconstitution. To avoid the discharge of the probe before the Ca^{2+} measurements, the reconstitution should be performed in the absence of Ca^{2+} and using low-affinity coelenterazine (i.e., coelenterazine *n* or *i*). The cells are thus incubated in KRB containing 600 μ*M* EGTA and the Ca^{2+} ionophore ionomycin (5 μ*M*). Inhibitors of the SERCA pump (i.e., 2,5 ditert-butyl benzohydroquinone (20 μ*M*) or cyclopiazonic acid (20 μ*M*)) and/or the application of extracellular agonists coupled to the generation of the second messenger inositol 1,4,5-trisphosphate ($InsP_3$) can be alternatively employed in particularly sensitive cells (e.g., SH-SY5Y) to discharge the intracellular Ca^{2+} stores. Cells are usually incubated for 1 h at 4 °C in the dark. Before the measurement, the cells are extensively washed with KRB supplemented with 1 m*M* EGTA and BSA 2% to completely remove ionomycin or SERCA pump blockers.

As mentioned above, a new double mutant very low-affinity aequorin variant (2mutAEQ) has been recently generated (de la Fuente et al., 2012, 2013). Thanks to its very low affinity, the reconstitution of this variant targeted to the ER lumen can be achieved with coelenterazine *i* in KRB containing 1 m*M* $CaCl_2$ at room temperature and avoiding experimental maneuvers to induce Ca^{2+} stores depletion. This represents a big improvement for the ER aequorin reconstitution process.

2.5. Aequorin measurements

2.5.1 Equipment

A typical aequorin measuring apparatus is built on the model described in (Cobbold & Lee, 1991). It is composed by a photomultiplier connected with a power supply and an amplifier/discriminator that generates the pulses and transfers them to a photon counting board installed in a compatible computer (Fig. 2.2A). During the measurements, the photomultiplier is placed in close proximity to the coverslip with the transfected cells (Fig. 2.2B and C).

2.5.2 Measurements and calibration

After aequorin reconstitution and immediately before performing the experiment, the coverslip with the transfected cells is transferred to the perfusion chamber of the luminometer. Cells are maintained at 37 °C and continuously perfused with KRB medium by a peristaltic pump connected with the thermostatted chamber. Luminescent data are stored every 50 ms, summed every second, and transformed in Ca^{2+} values through a custom-made software (Brini et al., 1995).

Figure 2.3 shows typical Ca^{2+} measurements obtained in HeLa cells transfected with cytAEQ, erAEQ, and mtAEQmut, respectively. In the monitoring of cytosolic and mitochondrial Ca^{2+}, transients cells were perfused with KRB medium containing 1 m*M* $CaCl_2$ for 20–30 s and then with the same saline solution containing histamine 100 μ*M*, an $InsP_3$-generating agonist. Cytosolic Ca^{2+} transients reach a peak value of about 2.5 μ*M* that gradually declines to basal levels, thanks to the action of Ca^{2+} pumps on the

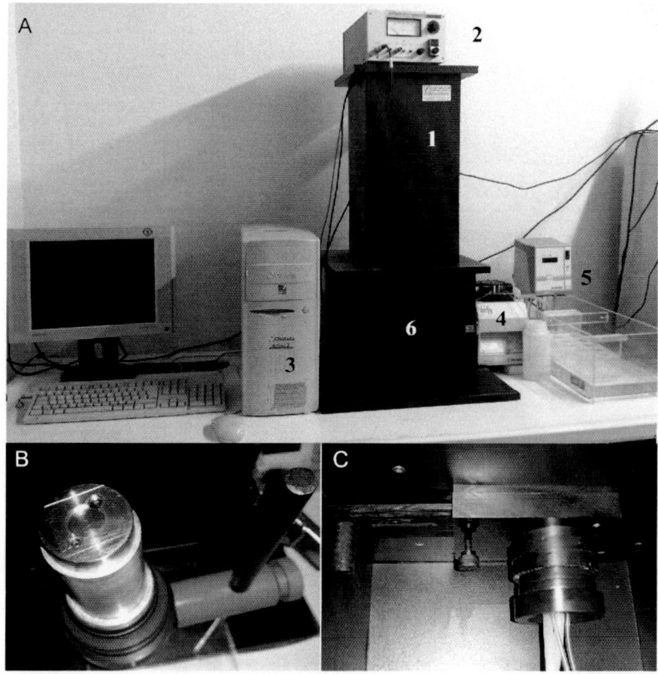

Figure 2.2 *The aequorin measuring system.* Panel A shows the photomultiplier connected with the amplifier/discriminator and kept in the dark by inserting it in a black box (1). It is connected to a power supply (2) and to a counting board inserted into a PC (3). A peristaltic perfusion pump (4) is connected with the chamber which accommodates the coverslip with the transfected cells (see it in details in panel B). The water jacket chamber is also connected with a thermostatic bath to keep it at the desired temperature (5). During the experiment, the perfusion chamber with the coverslip is placed in close proximity to the phototube by inserting it in a connected black box (6). Panel C shows in detail the position of the perfusion chamber during the Ca^{2+} measurements. (See the color plate.)

plasma membrane and on the intracellular Ca^{2+} stores. Mitochondrial Ca^{2+} transients reach higher values (about 100 μM at the peak) indicating that mitochondria rapidly accumulate large amounts of Ca^{2+} that are released by the opening of the $InsP_3$ channels of the ER. ER Ca^{2+} measurements are started in EGTA-containing medium to avoid aequorin discharge (see Section 2.4); when $CaCl_2$ 1 mM is added to the cells, the SERCA pump accumulates Ca^{2+} in the ER lumen until a plateau of 300–500 μM was reached in about 1 min from Ca^{2+} addition. The plateau indicates the maximum free Ca^{2+} content in the ER lumen. The subsequent perfusion of histamine induces the release of Ca^{2+} through the opening of the $InsP_3$ channels (in other types of cells, the discharge can be almost complete).

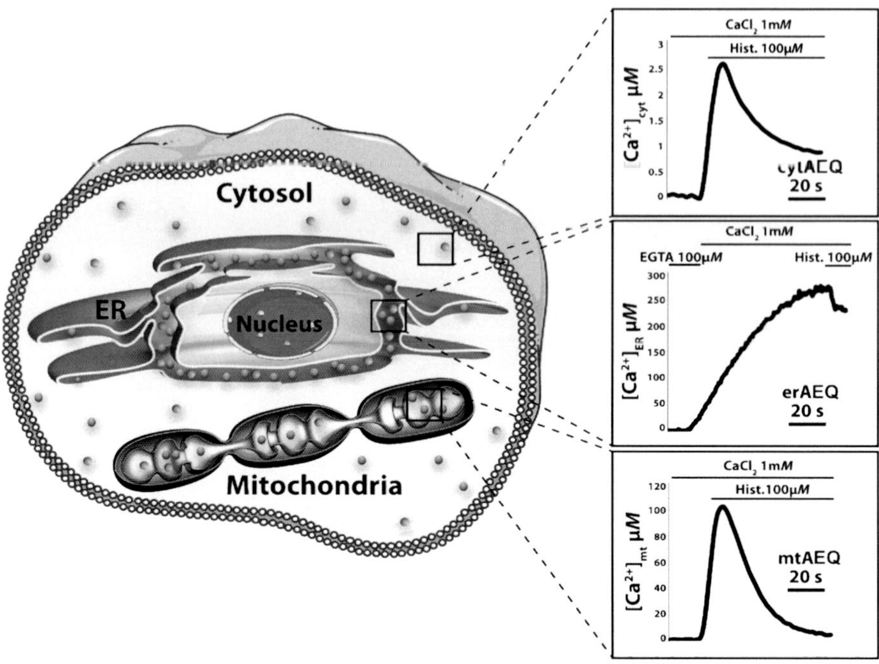

Figure 2.3 *Typical Ca^{2+} measurements in eukaryotic cells.* The traces refer to Ca^{2+} concentration transients monitored in HeLa cells expressing aequorin targeted to cytosol (cytAEQ), to endoplasmic reticulum (erAEQ) and to the mitochondrial matrix (mtAEQ). The cells were perfused with the indicated buffer and stimulated with histamine, an $InsP_3$-generating agonist. (See the color plate.)

At the end of each experiment, independently from probe used, the cells are perfused with a hypo-osmotic solution containing saturating Ca^{2+}-concentration (10 m*M* $CaCl_2$) and digitonin 100 μ*M* to discharge all the remaining active aequorin. This process is necessary to calibrate the luminescence signal in Ca^{2+} values (see Section 2.1).

3. APPLICATION EXAMPLES

3.1. Coexpression with a protein of interest to assess its role in Ca^{2+} signaling

A powerful approach to investigate the function of a protein consists in analyzing the effects of its overexpression or silencing on different key cellular parameters. Among them, Ca^{2+} signaling is one of the most investigated. This strategy requires the identification of the modified cells, which can be done either by cotransfection with suitable markers followed single-cell imaging or by selecting stable clones. However, both the procedures are high time consuming. The cotransfection of the cDNA or siRNA of the target protein with aequorin is a good alternative: according to the most common transfection protocols, the cotransfected plasmids enter in the same subset of cells. Because the signal comes only from aequorin-transfected cells, the Ca^{2+} measurements can be performed also in a dishomogeneous cell population.

3.2. Application to multiwell system

3.2.1 Equipment

In addition to the pioneering Ca^{2+} measurement with a purpose-built luminometer, aequorin-based measurements can be performed in multiwell plate systems by automatic acquisition of emitted luminescence. The improvement of the single-coverslip reading has been possible thanks to the development of high-sensitive multilabel plate readers. Specifically, we currently use the EnVision™ multilabel plate reader (Perkin Elmer). It is based on a PC-controlled platform that permits a quantitative detection of light-emitting or light-absorbing markers. It is equipped with a dispenser with two pumps and a temperature-controlled chamber that allows the housing of all types of microtitration plates (Fig. 2.4A). For aequorin measurements, the luminescence mode is applied, no excitation components are used. Briefly, the light coming from the sample is collected by the optics and directed through the aperture of the luminescence mirror module, which decreases the crosstalk between adjacent wells. The filter optics guides the light from the aperture to the photocathode of a photomultiplier tube (PMT, Fig. 2.4B) with spectral response from 230 to 850 nm. The presence of a high-sensitivity PMT as detector (e.g., with a spectral response from 300 up to 650 nm) offers the possibility to select an ultra-sensitive luminescence mode which is recommended because it guarantees extremely low background and high dynamic range. The detector has no optical components and the emission light is collected directly from the well. The position of the detector can be adjusted to be very close to the plate, thus reducing the crosstalk between wells. The single-photon counting system can be set to count the total number of photons emitted in a defined period of time (usually about 20–30 s).

3.2.2 Sample preparation and reading

At day zero, aequorin-transfected cells are washed, trypsinized, and seeded in a 96-well polystyrene white View Plate-96 (Product n. 6005181, Perkin

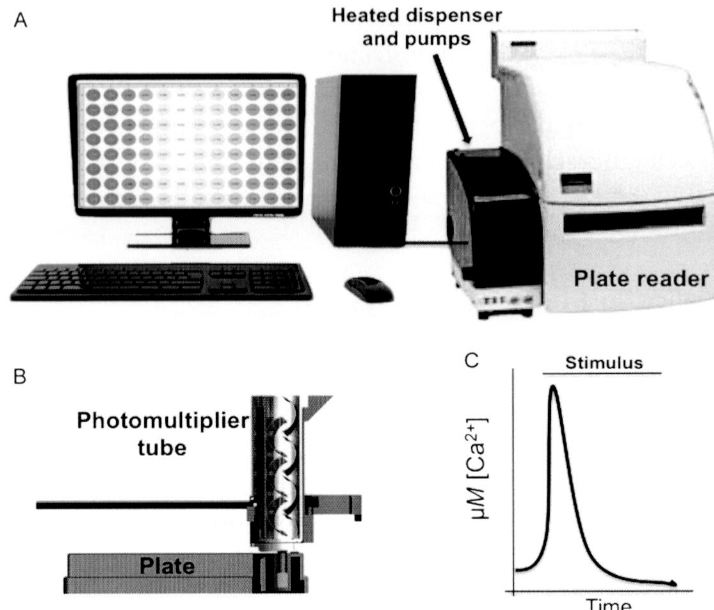

Figure 2.4 *Typical multiwell plate system for the automatic acquisition of luminescence.* The EnVision™ multilabel plate reader (A) (Perkin Elmer) is based on a PC-controlled platform that permits a quantitative detection of light-emitting or light-absorbing markers. The photomultiplier (B) is located in close proximity to the well and moves along the plate to collect the light from each single well. The output data can be visualized as luminescence intensity (A) or as calibrated Ca^{2+} transient trace (C). (See the color plate.)

Elmer) at about 80–90% of confluence ($1–2 \times 10^5$ cells/well) and maintained in their growth medium at 37 °C, 5% CO_2. After 24 h, the cells are washed two times with KRB saline solution and incubated in KRB with coelenterazine in a final volume of 40 μl/well to reconstitute active aequorin. After reconstitution, the cells are washed with KRB saline solution and kept in 70 μl/well of KRB saline solution to carry out aequorin measurement. The multiwell plate is loaded into the thermostatted chamber and the luminescent signal is read according to an aequorin standard protocol. The reader collects 35 readings/well (at 1 read/s rate) and then the first pump injects 30 μl of KRB saline solution supplemented with a threefold final concentration (3 ×) of an $InsP_3$-generating agonist (e.g., histamine, ATP, bradykinin, caffeine, etc.) to the first well. After 35 readings/well, a second pump injected 100 μl of the twofold concentrated discharging solution (digitonin 200 μM, $CaCl_2$ 20 mM in H_2O) to reach a final volume of

200 µl/well. All the solutions are kept at 37 °C in the built-in thermostatic bath. A second read of 35 s/well records the complete discharge of the remaining active aequorin pool. The total cps/well are collected in real time and the luminescence data are calibrated off-line in Ca^{2+} concentration values by using a custom-made macro-enabled Excel workbook (Fig. 2.4C). The operation is repeated for each well: using these settings, a 96-multiwell plate can be read in about 3 h.

3.3. High-throughput screening

The availability of automatized multi-plates readers for aequorin photon emission opens the possibility to perform aequorin-based high-throughput screening assays. The readers are extremely versatile since it is possible to read different format of plates (from 96 to 1566 wells). An example of this application has been recently reported: the translocation of the aequorin-containing chimera to the subplasmalemma rim was monitored with excellent signal-to-noise ratio and high responsiveness (Giorgi et al., 2011). The approach was based on the finding that subplasmalemma Ca^{2+} transients generated upon cell stimulation are higher than those generated in the bulk cytosol (at least 1 order of magnitude, Marsault et al., 1997). Aequorin is used as sensor to detect the movement of the tagged protein from a low Ca^{2+} concentration ambient (i.e., the cytosol) toward a Ca^{2+}-enriched environment (i.e., beneath the plasma membrane) within living cells. The automated multiwell detection could be easily employed to screen for thousands of compounds able to stimulate/inhibit protein translocation to the plasma membrane, for the pharmacological characterization of receptor agonists and the assessment of their functional properties (Fig. 2.5), thus providing important information about the time-course and the dose-dependence of their activation.

4. AEQUORIN IMPROVEMENT BY FUSION WITH GFP

One of the disadvantages of aequorin is its low amount of emitted light: only one photon per protein is released and only a small fraction of the aequorin pool ($<10^{-3}$) emits light during the experiment. These aspects do not represent a limit in Ca^{2+} measurements performed in cell population where the signal coming from the entire coverslip of transfected cells is integrated. Instead, at single-cell level, the light emitted is extremely low and does not permit to achieve sufficient spatial and time resolution. To overcome this restriction, a novel probe has been developed by coupling the

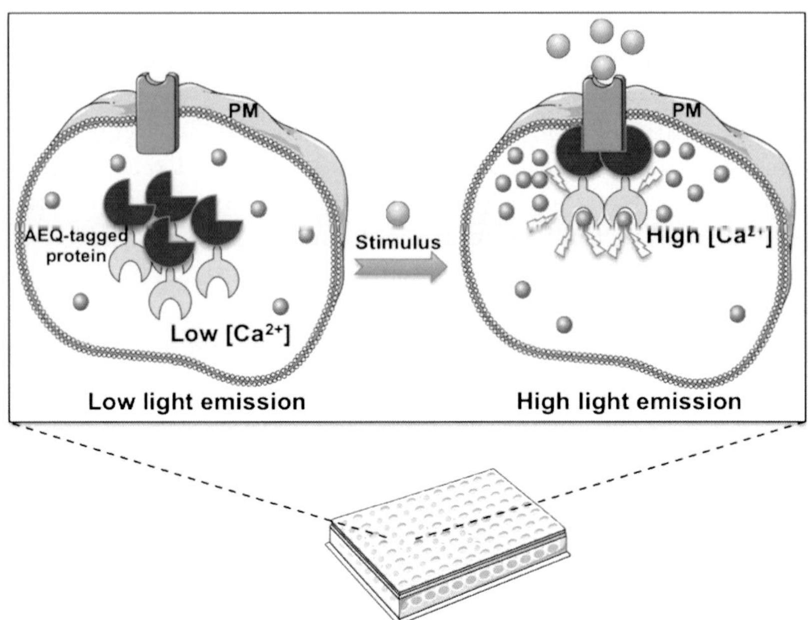

Figure 2.5 *Use of aequorin-tagged proteins to study the translocation of specific proteins in subcellular compartments characterized by different Ca^{2+} concentrations.* Upon cell stimulation a aequorin-tagged protein located in the cytosol translocates to the plasma membrane, where Ca^{2+} concentration is at least 1 order of magnitude higher than in the cytosol, giving rise to an easily detectable increase in light emission (yellow flashes). The changes in light emission reflect the translocation of the aequorin-tagged protein. (See the color plate.)

excellent fluorescent properties of GFP and the Ca^{2+} sensitivity of aequorin (Baubet et al., 2000). This novel sensor uses the BRET in a very similar way to what naturally occurs in the jellyfish: when Ca^{2+} binds to GFP-aequorin, the energy acquired by the excited coelenterazine is transferred to GFP that following excitation emits a fluorescence signal that can be detected and processed by conventional epifluorescence or confocal microscopes. This probe, thanks to its good stability (due to the GFP moiety) and its high signal-to-noise ratio (due to the aequorin moiety) is extremely reliable in single-cell experiments. Several GFP-AEQ chimeras targeted to different cellular compartments have been developed (Rogers et al., 2005). More recently, other fluorescent proteins with emission/excitation spectra different from GFP, like Venus (YFP), monomeric RFP or tandem-dimer Tomato (Bakayan, Vaquero, Picazo, & Llopis, 2011; Curie, Rogers,

Colasante, & Brulet, 2007), have been employed to improve the sensitivity of the probe. The use of "red-fluorescent" variants offers several advantages. Since their longer excitation wavelengths, they are less phototoxic and their signal penetrates deeper in the tissues of animals than that of the GFP or YFP-based probes. For these reasons, they should be preferred in BRET experiments in *in vivo* models (see Section 5, Martin, 2008; Webb & Miller, 2012). In addition, the monitoring of red fluorescence is particularly convenient with respect to the green signal because auto-fluorescence interferences are reduced and, finally, the development of targeted probes based on two different chromophores permits the simultaneous monitoring of Ca^{2+} signaling in two different domains (Manjarres et al., 2008).

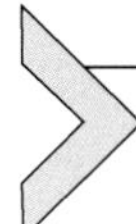

5. MONITORING CA^{2+} FLUXES IN LIVING ORGANISM BY BRET

The discovery that aequorin-GFP chimeras can be employed efficiently in BRET experiments open the possibility to use these probes for *in vivo* analysis (Webb & Miller, 2012). The development of transgenic mice (*Mus musculus*) expressing a mitochondrially targeted GFP-aequorin (mtGA) has permitted a number of studies in brain slices and in whole animal (Rogers et al., 2007). The mice have been used to monitor mitochondrial Ca^{2+} signaling during single-twitch and tetanic contractions of the muscle, in whole newborn mice and in the brains of freely moving or subjected to epileptic seizure adult animals. The results indicate that this technique could represent a noninvasive whole-animal imaging technology to provide key information regarding post-partum developmental events in the different organs during normal physiological conditions as well as during disease (Rogers et al., 2007). Interestingly, aequorin reconstitution can be successfully performed by injecting coelenterazine in the tail-vein in adults or intraperitoneal in neonates (Rogers et al., 2007; Roncali et al., 2008).

Other transgenic animal models that express aequorin have been generated. GFP-aequorin has been targeted to the mushroom bodies and antennal lobes of *Drosophila melanogaster* and has permitted to characterize Ca^{2+} response associated with cholinergic transmission in whole brain by imaging of specific neural structures (Martin et al., 2007). In zebrafish (*Danio rerio*), GFP-aequorin has been transiently and ubiquitously expressed in the embryos via the injection mRNA and has permitted to analyze Ca^{2+} transient during the development (Ashworth & Brennan, 2005).

ACKNOWLEDGMENTS

The original work by the authors has been supported over the years by grants from the Italian Ministry of University and Research (PRIN 2003, 2005, and 2008), the Telethon Foundation (Project GGP04169), the Italian National Research Council (Agenzia 2000, CNR), from the University of Padova (Progetto di Ateneo 2008 CPDA082825) to M. B. T. C. is supported by the University of Padova (Progetto Giovani GRIC128SP0, Bando 2012). The Figures 2.3 and 2.5 contain illustrations made by Servier Medical Art http://www.servier.fr/servier-medical-art.

REFERENCES

Alvarez, J., & Montero, M. (2002). Measuring [Ca^{2+}] in the endoplasmic reticulum with aequorin. *Cell Calcium*, *32*(5–6), 251–260.

Ashworth, R., & Brennan, C. (2005). Use of transgenic zebrafish reporter lines to study calcium signalling in development. *Briefings in Functional Genomics & Proteomics*, *4*(2), 186–193.

Bakayan, A., Vaquero, C. F., Picazo, F., & Llopis, J. (2011). Red fluorescent protein-aequorin fusions as improved bioluminescent Ca^{2+} reporters in single cells and mice. *PLoS One*, *6*(5), e19520.

Barrero, M. J., Montero, M., & Alvarez, J. (1997). Dynamics of [Ca^{2+}] in the endoplasmic reticulum and cytoplasm of intact HeLa cells. A comparative study. *Journal of Biological Chemistry*, *272*(44), 27694–27699.

Baubet, V., Le Mouellic, H., Campbell, A. K., Lucas-Meunier, E., Fossier, P., & Brulet, P. (2000). Chimeric green fluorescent protein-aequorin as bioluminescent Ca^{2+} reporters at the single-cell level. *Proceedings of the National Academy of Sciences of the United States of America*, *97*(13), 7260–7265.

Baughman, J. M., Perocchi, F., Girgis, H. S., Plovanich, M., Belcher-Timme, C. A., Sancak, Y., et al. (2011). Integrative genomics identifies MCU as an essential component of the mitochondrial calcium uniporter. *Nature*, *476*(7360), 341–345.

Bononi, A., Bonora, M., Marchi, S., Missiroli, S., Poletti, F., Giorgi, C., et al. (2013). Identification of PTEN at the ER and MAMs and its regulation of Ca signaling and apoptosis in a protein phosphatase-dependent manner. *Cell Death and Differentiation*, *20*(12), 1631–1643.

Brini, M. (2008). Calcium-sensitive photoproteins. *Methods*, *46*(3), 160–166.

Brini, M., Cali, T., Ottolini, D., & Carafoli, E. (2013). Intracellular calcium homeostasis and signaling. *Metal Ions in Life Sciences*, *12*, 119–168.

Brini, M., De Giorgi, F., Murgia, M., Marsault, R., Massimino, M. L., Cantini, M., et al. (1997). Subcellular analysis of Ca^{2+} homeostasis in primary cultures of skeletal muscle myotubes. *Molecular Biology of the Cell*, *8*(1), 129–143.

Brini, M., Manni, S., Pierobon, N., Du, G. G., Sharma, P., MacLennan, D. H., et al. (2005). Ca^{2+} signaling in HEK-293 and skeletal muscle cells expressing recombinant ryanodine receptors harboring malignant hyperthermia and central core disease mutations. *Journal of Biological Chemistry*, *280*(15), 15380–15389.

Brini, M., Marsault, R., Bastianutto, C., Alvarez, J., Pozzan, T., & Rizzuto, R. (1995). Transfected aequorin in the measurement of cytosolic Ca^{2+} concentration ([Ca^{2+}]$_c$). A critical evaluation. *Journal of Biological Chemistry*, *270*(17), 9896–9903.

Brini, M., Marsault, R., Bastianutto, C., Pozzan, T., & Rizzuto, R. (1994). Nuclear targeting of aequorin. A new approach for measuring nuclear Ca^{2+} concentration in intact cells. *Cell Calcium*, *16*(4), 259–268.

Brini, M., Murgia, M., Pasti, L., Picard, D., Pozzan, T., & Rizzuto, R. (1993). Nuclear Ca^{2+} concentration measured with specifically targeted recombinant aequorin. *EMBO Journal, 12*(12), 4813–4819.

Chalfie, M. (1995). Green fluorescent protein. *Photochemistry and Photobiology, 62*(4), 651–656.

Cobbold, P. H., & Lee, J. A. C. (1991). Aequorin measurements of cytoplasmic free calcium. In J. G. McCormack & P. H. Cobbold (Eds.), *Cellular calcium: A practical approach* (pp. 55–81). Oxford, New York, Tokyo: Oxford University Press.

Curie, T., Rogers, K. L., Colasante, C., & Brulet, P. (2007). Red-shifted aequorin-based bioluminescent reporters for in vivo imaging of Ca2 signaling. *Molecular Imaging, 6*(1), 30–42.

de la Fuente, S., Fonteriz, R. I., de la Cruz, P. J., Montero, M., & Alvarez, J. (2012). Mitochondrial free [Ca($^{2+}$)] dynamics measured with a novel low-Ca($^{2+}$) affinity aequorin probe. *Biochemical Journal, 445*(3), 371–376.

de la Fuente, S., Fonteriz, R. I., Montero, M., & Alvarez, J. (2013). Ca^{2+} homeostasis in the endoplasmic reticulum measured with a new low-Ca^{2+}-affinity targeted aequorin. *Cell Calcium, 54*(1), 37–45.

De Stefani, D., Raffaello, A., Teardo, E., Szabo, I., & Rizzuto, R. (2011). A forty-kilodalton protein of the inner membrane is the mitochondrial calcium uniporter. *Nature, 476*(7360), 336–340.

Ferri, K. F., & Kroemer, G. (2001). Organelle-specific initiation of cell death pathways. *Nature Cell Biology, 3*(11), E255–E263.

Garcia-Cao, I., Song, M. S., Hobbs, R. M., Laurent, G., Giorgi, C., de Boer, V. C., et al. (2012). Systemic elevation of PTEN induces a tumor-suppressive metabolic state. *Cell, 149*(1), 49–62.

George, C. H., Kendall, J. M., Campbell, A. K., & Evans, W. H. (1998). Connexin-aequorin chimerae report cytoplasmic calcium environments along trafficking pathways leading to gap junction biogenesis in living COS-7 cells. *Journal of Biological Chemistry, 273*(45), 29822–29829.

Giorgi, C., Baldassari, F., Bononi, A., Bonora, M., De Marchi, E., Marchi, S., et al. (2012). Mitochondrial Ca($^{2+}$) and apoptosis. *Cell Calcium, 52*(1), 36–43.

Giorgi, C., Ito, K., Lin, H. K., Santangelo, C., Wieckowski, M. R., Lebiedzinska, M., et al. (2010). PML regulates apoptosis at endoplasmic reticulum by modulating calcium release. *Science, 330*(6008), 1247–1251.

Giorgi, C., Romagnoli, A., Agnoletto, C., Bergamelli, L., Sorrentino, G., Brini, M., et al. (2011). Translocation of signalling proteins to the plasma membrane revealed by a new bioluminescent procedure. *BMC Cell Biology, 12*, 27.

Giorgi, C., Romagnoli, A., Pinton, P., & Rizzuto, R. (2008). Ca^{2+} signaling, mitochondria and cell death. *Current Molecular Medicine, 8*(2), 119–130.

Inouye, S., Noguchi, M., Sakaki, Y., Takagi, Y., Miyata, T., Iwanaga, S., et al. (1985). Cloning and sequence analysis of cDNA for the luminescent protein aequorin. *Proceedings of the National Academy of Sciences of the United States of America, 82*(10), 3154–3158.

Kendall, J. M., Dormer, R. L., & Campbell, A. K. (1992). Targeting aequorin to the endoplasmic reticulum of living cells. *Biochemical and Biophysical Research Communications, 189*(2), 1008–1016.

Kendall, J. M., Sala-Newby, G., Ghalaut, V., Dormer, R. L., & Campbell, A. K. (1992). Engineering the CA($^{2+}$)-activated photoprotein aequorin with reduced affinity for calcium. *Biochemical and Biophysical Research Communications, 187*(2), 1091–1097.

Lasorsa, F. M., Pinton, P., Palmieri, L., Scarcia, P., Rottensteiner, H., Rizzuto, R., et al. (2008). Peroxisomes as novel players in cell calcium homeostasis. *Journal of Biological Chemistry, 283*(22), 15300–15308.

Manjarres, I. M., Chamero, P., Domingo, B., Molina, F., Llopis, J., Alonso, M. T., et al. (2008). Red and green aequorins for simultaneous monitoring of Ca^{2+} signals from two different organelles. *Pflügers Archiv*, *455*(5), 961–970.

Marchi, S., Lupini, L., Patergnani, S., Rimessi, A., Missiroli, S., Bonora, M., et al. (2013). Downregulation of the mitochondrial calcium uniporter by cancer-related miR-25. *Current Biology*, *23*(1), 58–63.

Marchi, S., Marinello, M., Bononi, A., Bonora, M., Giorgi, C., Rimessi, A., et al. (2012). Selective modulation of subtype III IP(3)R by Akt regulates ER Ca(2)(+) release and apoptosis. *Cell Death & Disease*, *3*, e304.

Marchi, S., Rimessi, A., Giorgi, C., Baldini, C., Ferroni, L., Rizzuto, R., et al. (2008). Akt kinase reducing endoplasmic reticulum Ca^{2+} release protects cells from Ca^{2+}-dependent apoptotic stimuli. *Biochemical and Biophysical Research Communications*, *375*(4), 501–505.

Marsault, R., Murgia, M., Pozzan, T., & Rizzuto, R. (1997). Domains of high Ca^{2+} beneath the plasma membrane of living A7r5 cells. *EMBO Journal*, *16*(7), 1575–1581.

Martin, J. R. (2008). In vivo brain imaging: Fluorescence or bioluminescence, which to choose? *Journal of Neurogenetics*, *22*(3), 285–307.

Martin, P. E., George, C. H., Castro, C., Kendall, J. M., Capel, J., Campbell, A. K., et al. (1998). Assembly of chimeric connexin-aequorin proteins into functional gap junction channels. Reporting intracellular and plasma membrane calcium environments. *Journal of Biological Chemistry*, *273*(3), 1719–1726.

Martin, J. R., Rogers, K. L., Chagneau, C., & Brulet, P. (2007). In vivo bioluminescence imaging of Ca signalling in the brain of *Drosophila*. *PLoS One*, *2*(3), e275.

McCombs, J. E., & Palmer, A. E. (2008). Measuring calcium dynamics in living cells with genetically encodable calcium indicators. *Methods*, *46*(3), 152–159.

Mitchell, K. J., Pinton, P., Varadi, A., Tacchetti, C., Ainscow, E. K., Pozzan, T., et al. (2001). Dense core secretory vesicles revealed as a dynamic $Ca(^{2+})$ store in neuroendocrine cells with a vesicle-associated membrane protein aequorin chimaera. *Journal of Cell Biology*, *155*(1), 41–51.

Miyawaki, A., Llopis, J., Heim, R., McCaffery, J. M., Adams, J. A., Ikura, M., et al. (1997). Fluorescent indicators for Ca^{2+} based on green fluorescent proteins and calmodulin. *Nature*, *388*(6645), 882–887.

Montero, M., Brini, M., Marsault, R., Alvarez, J., Sitia, R., Pozzan, T., et al. (1995). Monitoring dynamic changes in free Ca^{2+} concentration in the endoplasmic reticulum of intact cells. *EMBO Journal*, *14*(22), 5467–5475.

Nomura, M., Inouye, S., Ohmiya, Y., & Tsuji, F. I. (1991). A C-terminal proline is required for bioluminescence of the $Ca(^{2+})$-binding photoprotein, aequorin. *FEBS Letters*, *295*(1–3), 63–66.

Ottolini, D., Cali, T., & Brini, M. (2013). Measurements of Ca(2)(+) concentration with recombinant targeted luminescent probes. *Methods in Molecular Biology*, *937*, 273–291.

Ottolini, D., Cali, T., Negro, A., & Brini, M. (2013). The Parkinson disease-related protein DJ-1 counteracts mitochondrial impairment induced by the tumour suppressor protein p53 by enhancing endoplasmic reticulum-mitochondria tethering. *Human Molecular Genetics*, *22*(11), 2152–2168.

Perocchi, F., Gohil, V. M., Girgis, H. S., Bao, X. R., McCombs, J. E., Palmer, A. E., et al. (2010). MICU1 encodes a mitochondrial EF hand protein required for $Ca(^{2+})$ uptake. *Nature*, *467*(7313), 291–296.

Pinton, P., Ferrari, D., Magalhaes, P., Schulze-Osthoff, K., Di Virgilio, F., Pozzan, T., et al. (2000). Reduced loading of intracellular $Ca(^{2+})$ stores and downregulation of capacitative $Ca(^{2+})$ influx in Bcl-2-overexpressing cells. *Journal of Cell Biology*, *148*(5), 857–862.

Pinton, P., Pozzan, T., & Rizzuto, R. (1998). The Golgi apparatus is an inositol 1,4,5-trisphosphate-sensitive Ca^{2+} store, with functional properties distinct from those of the endoplasmic reticulum. *EMBO Journal*, *17*(18), 5298–5308.

Pitter, J. G., Maechler, P., Wollheim, C. B., & Spat, A. (2002). Mitochondria respond to Ca^{2+} already in the submicromolar range: Correlation with redox state. *Cell Calcium*, *31*(2), 97–104.

Plovanich, M., Bogorad, R. L., Sancak, Y., Kamer, K. J., Strittmatter, L., Li, A. A., et al. (2013). MICU2, a paralog of MICU1, resides within the mitochondrial uniporter complex to regulate calcium handling. *PLoS One*, *8*(2), e55785.

Rimessi, A., Marchi, S., Patergnani, S., & Pinton, P. (2013). H-Ras-driven tumoral maintenance is sustained through caveolin-1-dependent alterations in calcium signaling. *Oncogene*. http://dx.doi.org/10.1038/onc.2013.192.

Rizzuto, R., Pinton, P., Carrington, W., Fay, F. S., Fogarty, K. E., Lifshitz, L. M., et al. (1998). Close contacts with the endoplasmic reticulum as determinants of mitochondrial Ca^{2+} responses. *Science*, *280*(5370), 1763–1766.

Rizzuto, R., Simpson, A. W., Brini, M., & Pozzan, T. (1992). Rapid changes of mitochondrial Ca^{2+} revealed by specifically targeted recombinant aequorin. *Nature*, *358*(6384), 325–327.

Robert, V., De Giorgi, F., Massimino, M. L., Cantini, M., & Pozzan, T. (1998). Direct monitoring of the calcium concentration in the sarcoplasmic and endoplasmic reticulum of skeletal muscle myotubes. *Journal of Biological Chemistry*, *273*(46), 30372–30378.

Rogers, K. L., Picaud, S., Roncali, E., Boisgard, R., Colasante, C., Stinnakre, J., et al. (2007). Non-invasive in vivo imaging of calcium signaling in mice. *PLoS One*, *2*(10), e974.

Rogers, K. L., Stinnakre, J., Agulhon, C., Jublot, D., Shorte, S. L., Kremer, E. J., et al. (2005). Visualization of local Ca^{2+} dynamics with genetically encoded bioluminescent reporters. *European Journal of Neuroscience*, *21*(3), 597–610.

Romoser, V. A., Hinkle, P. M., & Persechini, A. (1997). Detection in living cells of Ca^{2+}-dependent changes in the fluorescence emission of an indicator composed of two green fluorescent protein variants linked by a calmodulin-binding sequence. A new class of fluorescent indicators. *Journal of Biological Chemistry*, *272*(20), 13270–13274.

Roncali, E., Savinaud, M., Levrey, O., Rogers, K. L., Maitrejean, S., & Tavitian, B. (2008). New device for real-time bioluminescence imaging in moving rodents. *Journal of Biomedical Optics*, *13*(5), 054035.

Scorrano, L., Oakes, S. A., Opferman, J. T., Cheng, E. H., Sorcinelli, M. D., Pozzan, T., et al. (2003). BAX and BAK regulation of endoplasmic reticulum Ca^{2+}: A control point for apoptosis. *Science*, *300*(5616), 135–139.

Tsien, R. Y., Pozzan, T., & Rink, T. J. (1982). T-cell mitogens cause early changes in cytoplasmic free Ca^{2+} and membrane potential in lymphocytes. *Nature*, *295*(5844), 68–71.

Watkins, N. J., & Campbell, A. K. (1993). Requirement of the C-terminal proline residue for stability of the Ca($^{2+}$)-activated photoprotein aequorin. *Biochemical Journal*, *293*(Pt. 1), 181–185.

Webb, S. E., & Miller, A. L. (2012). Aequorin-based genetic approaches to visualize Ca^{2+} signaling in developing animal systems. *Biochimica et Biophysica Acta*, *1820*(8), 1160–1168.

CHAPTER THREE

Measuring Baseline Ca^{2+} Levels in Subcellular Compartments Using Genetically Engineered Fluorescent Indicators

Julia M. Hill[*], Diego De Stefani[†], Aleck W.E. Jones[*], Asier Ruiz[‡], Rosario Rizzuto[†], Gyorgy Szabadkai[*,†,1]

[*]Department of Cell and Developmental Biology, Consortium for Mitochondrial Research, University College London, London, United Kingdom
[†]Department of Biomedical Sciences, CNR Neuroscience Institute, University of Padua, Padua, Italy
[‡]Department of Neurosciences, University of the Basque Country (UPV/EHU), Achúcarro Basque Center for Neuroscience-UPV/EHU, Leioa, Spain; Instituto de Salud Carlos III, Centro de Investigación Biomédica en Red de Enfermedades Neurodegenerativas, (CIBERNED), Madrid, Spain
[1]Corresponding author: e-mail address: g.szabadkai@ucl.ac.uk

Contents

Abstract

Intracellular Ca^{2+} signaling is involved in a series of physiological and pathological processes. In particular, an intimate crosstalk between bioenergetic metabolism and Ca^{2+} homeostasis has been shown to determine cell fate in resting conditions as well as in response to stress. The endoplasmic reticulum and mitochondria represent key hubs of cellular metabolism and Ca^{2+} signaling. However, it has been challenging to specifically detect highly localized Ca^{2+} fluxes such as those bridging these two organelles. To circumvent this issue, various recombinant Ca^{2+} indicators that can be targeted to specific

Methods in Enzymology, Volume 543
ISSN 0076-6879
http://dx.doi.org/10.1016/B978-0-12-801329-8.00003-9

subcellular compartments have been developed over the past two decades. While the use of these probes for measuring agonist-induced Ca^{2+} signals in various organelles has been extensively described, the assessment of basal Ca^{2+} concentrations within specific organelles is often disregarded, in spite of the fact that this parameter is vital for several metabolic functions, including the enzymatic activity of mitochondrial dehydrogenases of the Krebs cycle and protein folding in the endoplasmic reticulum. Here, we provide an overview on genetically engineered, organelle-targeted fluorescent Ca^{2+} probes and outline their evolution. Moreover, we describe recently developed protocols to quantify baseline Ca^{2+} concentrations in specific subcellular compartments. Among several applications, this method is suitable for assessing how changes in basal Ca^{2+} levels affect the metabolic profile of cancer cells.

1. INTRODUCTION

Intracellular Ca^{2+} indicators can be either fluorescent or luminescent, or use a combination of both in bioluminescence resonance energy transfer-based probes. Luminescent methods of Ca^{2+} measurement rely on the apoprotein aequorin and its cofactor coelenterazine. Light emission by aequorin does not require potentially phototoxic excitation, as it has very low-calcium-buffering activity and high signal-to-noise ratio, but it emits comparatively very few photons of light per molecule, requires the presence of a cofactor, and undergoes an irreversible reaction leading to depletion of the probe itself with time (Bonora et al., 2013; Chiesa et al., 2001; Váradi & Rutter, 2002). Thus, while aequorin-based probes are mostly used for thoroughly quantified, short-term population measurement of Ca^{2+} signals (see chapter 2 in this volume), fluorescent indicators, due to their relatively high brightness and reversible binding of Ca^{2+} are more suitable for use on a single-cell level.

Fluorescent indicators can be broadly defined as undergoing a change in their fluorescent properties when Ca^{2+} bound, and there are two main classes—the chemically engineered fluorescent indicator molecules (Grynkiewicz, Poenie, & Tsien, 1985) and recombinant fluorescent probes, also called genetically encoded Ca^{2+} indicator (GECI) proteins. Both types enable the visualization of free Ca^{2+} concentration ($[Ca^{2+}]$); however, specific targeting of chemically engineered fluorophores is historically more problematical, they are prone to nonspecific compartmentalization into a series of intracellular organelles and can be extruded throughout the course of an experiment (Bianchi, Rimessi, Prandini, Szabadkai, & Rizzuto, 2004; Malgaroli, Milani, Meldolesi, & Pozzan, 1987; Thomas et al., 2000). The directed evolution of recombinant Ca^{2+} probes was aimed to overcome these issues, enabling specific visualization of intraorganellar Ca^{2+} signals at high spatial and temporal resolution, while trying to achieve the sensitivity (dynamic range) and brightness of the best chemical probes. In addition, the creation of GECI-containing transgenic organisms enables cell type-specific targeting and study of tissues not compatible with dye loading. To our knowledge, only very few organelle-targeted recombinant Ca^{2+} probe has yet been utilized to generate *in vivo* models (Jiménez-Moreno, Wang, Messi, & Delbono, 2010; Mizuno, Sassa, Higashijima, Okamoto, & Miyawaki, 2013), in contrast to the widespread use of transgenic animals expressing cytoplasmic GECIs, particularly driven by the development of methods to measure neural activity *in vivo* (see, e.g., Akerboom et al., 2012; Hasan et al., 2004; Wallace et al., 2008; Yu, Baird, Tsien, & Davis, 2003).

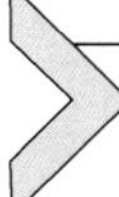

2. EVOLUTION OF RECOMBINANT Ca^{2+} PROBES AND ORGANELLE TARGETING

Recombinant Ca^{2+} probes are derivatives of fluorescent proteins (FPs) engineered to contain Ca^{2+} responsive elements (CREs). Initially, almost exclusively the *Aequora*-derived green fluorescent protein (GFP) mutants were used due to their somewhat higher quantum yield, but red probes have been introduced recently, derived from the *Anthozoan* mApple (Zhao et al., 2011), mKO (Waldeck-Weiermair et al., 2012), mRuby (Akerboom et al., 2013), and mCherry (Carlson & Campbell, 2013). The existing probes can be grouped into single or two fluorescent protein-containing probes. CREs can be inserted into single FPs thus Ca^{2+} binding causes conformational changes altering the fluorescent properties of the probes. A large variety of these probes have been developed where the Ca^{2+}-induced changes include variations in their intensity and/or spectrum (either in excitation or emission). The two fluorescent protein probes are invariably based on the fluorescence resonance energy transfer (FRET) principle: Ca^{2+} binding to the CRE inserted between the donor–acceptor pair modulates their interaction resulting in changes in FRET efficiency, measured as the acceptor/donor emission ratio following the donor excitation. CREs are based either on calmodulin (CaM), in combination with its substrate (smooth or skeletal muscle myosin light chain kinase (MLCK), M13) or troponin C, which were genetically altered to adjust sensitivity to Ca^{2+} or prevent binding to other cellular proteins. The generation of the first (FRET-based) Ca^{2+} sensor more than 15 years ago (Miyawaki et al., 1997; Romoser, Hinkle, & Persechini, 1997) was followed by extensive directed evolution of a large number of probes that has been reviewed comprehensively in the literature (see, e.g., Mank & Griesbeck, 2008; Miyawaki, Nagai, & Mizuno, 2013; Palmer, Qin, Park, & McCombs, 2011; Palmer & Tsien, 2006; Pérez Koldenkova & Nagai, 2013; Rudolf, Mongillo, Rizzuto, & Pozzan, 2003; Tian, Hires, & Looger, 2012; Whitaker, 2010). Here, we focus on the properties of organelle-targeted recombinant Ca^{2+} probes (we compiled and summarize all targeted probes published so far to our knowledge in Table 3.1). In brief, genetic engineering of organelle-targeted recombinant Ca^{2+}-sensitive probes focused on the following three fundamental issues.

3. ENHANCING THE FLUOROPHORES

Mutations in the FPs (GFP and variants) were introduced to enhance brightness and the rate of conformational maturation of the probes. These include point mutations (Griesbeck et al., 2001; Miyawaki et al., 1997, 1999; Nagai et al., 2002), insertion of different linkers and circular permutation (i.e., domain swapping) (Baird, Zacharias, & Tsien, 1999; Nagai et al., 2001, 2004; Nakai et al., 2001; Souslova et al., 2007; Tian et al., 2009; Zhao et al., 2011). These random or rationally designed mutations were also aimed to improve the originally relatively poor dynamic range of fluorescent changes upon Ca^{2+}-induced conformational switch. This was achieved by redesigning protein surfaces to modify interactions between domains and to rewire internal protonation networks. Indeed, only the latest generation of probes was able to reproduce the properties of the best available chemically engineered probes, for example, Oregon-Green BAPTA-1AM (Chen et al., 2013). Finally, by the variations introduced in the FPs, the inherent pH and chloride sensitivity of the probes has been tuned, in order to render the probe independent of the environment, at least in a certain working pH and $[Cl^-]$ range (Griesbeck et al., 2001; Miyawaki et al., 1999; Nagai et al., 2002). Overall, pioneered by a couple of research groups, these efforts led to the development of a series of recombinant design families, such as the cameleons (CFP/YFP FRET-based probes CaM/M13 pair insertion) (Csordás et al., 2010; Giacomello et al., 2010; Liu et al., 2011; Miyawaki et al., 1997; Palmer et al., 2006; Romoser et al., 1997; Truong et al., 2001; Waldeck-Weiermair et al., 2012); camgaroos (single YFP probe with CaM insertion) (Baird et al., 1999; Griesbeck et al., 2001; Hasan et al., 2004); pericams (circularly permuted single YFP probes with CaM/M13 pair as CRE) (Baird et al., 1999; Ishii et al., 2006; Kettlewell et al., 2009; Lu et al., 2013; Nagai et al., 2001); and GCaMP, GECO (Chen et al., 2013; Nakai et al., 2001; Tallini et al., 2006; Tian et al., 2009; Zhao et al., 2011),

Table 3.1 Recombinant fluorescent probes targeted to various subcellular compartments

Compartment	Probe	Targeting sequence (number of tandem repeats)	Single FP	Two FP donor/acceptor	Ca^{2+} sensor	Ratio-metric	Single Wave-length	K_d (μM)	Excitation (nm)	Emission (nm)	Ref.
Mitochondria	$YC2_{mit}$	COX-VIII (1) (pCMV/*myc*/mito; Invitrogen)		ECFP/EYFP[a]	CaM/M13	Yes, Em		1.26 (monophasic, *in situ*)	430	475/535	Arnaudeau, Kelley, Walsh, and Demaurex (2001)
	$YC2.1_{mit}$	COX-VIII (1) (pCMV/*myc*/mito; Invitrogen)		ECFP/EYFP-V68L/Q69K[b]	CaM/M13	Yes, Em		Not determined similar to $YC2_{mit}$	430	475/535	Arnaudeau et al. (2001)
	$YC3.1_{mit}$	COX-VIII (1) (pCMV/*myc*/mito; Invitrogen)		ECFP/EYFP-V68L/Q69K[b]	E104QCaM/M13	Yes, Em		3.98 (monophasic, *in situ*)	430	475/535	Arnaudeau et al. (2001)
	$YC4.1_{mit}$	COX-VIII (1) (pCMV/*myc*/mito, Invitrogen)		ECFP/EYFP-V68L/Q69K[b]	E31QCaM/M13	Yes, Em		0.105/104 (biphasic, *in situ*)	430	475/535	Arnaudeau et al. (2001)
	2mt8YC2.1	COX-VIII (2)		ECFP/EYFP-V68L/Q69K[b] see $YC2.1_{mit}$	CaM/M13	Yes, Em		Not determined, similar to $YC2_{mit}$	440	480/545	Filippin et al. (2005)
	2mt8YC2.12	COX-VIII (2)		ECFP/Venus[c]	CaM/M13	Yes, Em		Not determined, similar to $YC2_{mit}$	440	480/545	Filippin et al. (2005)
	2mt8YC2.3	COX-VIII (2)		ECFP/citrine[d]	CaM/M13	Yes, Em		Not determined, similar to $YC2_{mit}$	440	480/545	Filippin et al. (2005)
	2mtD1cpv	**COX-VIII (2)**		**ECFP/cpVenus[e,f]**	**Design 1: mutant CaM/mutant M13[g]**	**Yes, Em**		**Not determined, probably similar to D1ER: 0.81/60 (*in vitro*) or 4mtD1GO-Cam: 1.53**	**425**	**480/535**	Giacomello et al. (2010)

Continued

Table 3.1 Recombinant fluorescent probes targeted to various subcellular compartments—cont'd

Compartment	Probe	Targeting sequence (number of tandem repeats)	Single FP	Two FP donor/acceptor	Ca^{2+} sensor	Ratio-metric	Single Wave-length	K_d (μM)	Excitation (nm)	Emission (nm)	Ref.
	2mtD2cpv	**COX-VIII (2; up to 8)[h]**		**ECFP/cpVenus[e,f]**	**Design 2: mutant CaM/mutant smooth muscle MLCK segment[g]**	**Yes, Em**		**0.03/3 (*in vitro*)**	**436**	**475/535**	Palmer et al. (2006)
	2mtD3cpv	**COX-VIII (2; up to 8)[h]**		**ECFP/cpVenus[e,f]**	**Design 3: mutant CaM/mutant smooth muscle MLCK segment[g]**	**Yes, Em**		**0.6 (*in vitro*)**	**436**	**475/535**	Palmer et al. (2006)
	2mtD4cpv	**COX-VIII (2; up to 8)[h]**		**ECFP/cpVenus[e,f]**	**Design 4: mutant CaM/mutant smooth muscle MLCK segment[g]**	**Yes, Em**		**64 (*in vitro*)**	**436**	**475/535**	Palmer et al. (2006)
	4mtD1GO-Cam	COX-VIII (4)		cp173-mEGFP/mKO_K (Orange)	Design 1: mutant CaM/mutant M13[g]	Yes, Em		1.53 (25 °C, pH 7.25)	477	510/560	Waldeck-Weiermair et al. (2012)
	RP_{mt}, $RP3.1_{mt}$, mt4PR (ratiometric pericam)	**COX-IV (1)**	**cpEYFP-V68L/Q69K[b,i]**		**E104QCaM/M13**	**Yes, Ex (excitation at 480 nm is more sensitive to ΔpH than to $\Delta[Ca^{2+}]$)**	**Due to high pH sensitivity at 480 nm excitation, RP in mito-chondria is rather used at single wavelength (ex 410 nm)**	**1.7 (*in vitro*) 11 (*in situ*)**	**480/410**	**535**	Filippin et al. (2005), Filippin, Magalhães, Di Benedetto, Colella, and Pozzan (2003), Frieden et al. (2004), Nagai, Sawano, Park, and Miyawaki (2001)

mt8PR, 2mt8PR (ratiometric pericam)	**COX-VIII (1,2)**	**cpEYFP-V68L/Q69K[b,e]**	**E104QCaM/M13**	**Yes, Ex (but see RP3.1$_{mt}$ above)**		**2.5 (*in situ*)**	**480/410**	**535**	Filippin et al. (2005)
mitycam (inverse pericam)	COX-VIII (1)	cpEYFP-V68L/Q69K[b,j]	E67Q, E104QCaM/M13		Yes	0.047 (method not reported)	488	505–530	Kettlewell et al. (2009), Terhzaz et al. (2006)
mt-inverse pericam2	**COX-VIII (1) (pCMV/*myc*/mito, Invitrogen)**	**cpCitrine**	**E104QCaM/M13**		**Yes**	**0.08 (*in situ*)**	**490**	**535**	Ishii, Hirose, and Iino (2006)
mito-Case12	**COX-VIII (1)**	**GCaMP1 mutant[k]**	**CaM/M13**		**Yes**	**1 (*in situ*)**	**491**	**516**	Souslova et al. (2007) commercial: Evrogen; BioCat
GCaMP2-mt	COX-VIII (1)	GCaMP2[k]	CaM/M13		Yes	0.136 (pH 7.5; *in situ*)	488	500–530	(Iguchi et al., 2012)
mitGC3 (GCaMP3)	COX-VIII (1)	GCaMP3[k]	N60DCaM/M13		Yes	0.66 (*in vitro*; Tian et al., 2009)	488	500–530	Isshiki, Nishimoto, Mizuno, and Fujita (2013)
2mt8-GEM-GECO1	**COX-VIII (2)**	**GCAMP3 mutant[k,l]**	**Mutant CaM/M13[l]**	**Yes, Em**		**0.34 (*in vitro*)**	**377**	**520/447**	Llorente-Folch et al. (2013), Zhao et al. (2011)
2mt8-R-GECO1	COX-VIII (2)	mApple-derived GCAMP3 analog[k,l]	Mutant CaM/M13[l]		Yes	0.482 (*in vitro*)	562	624	Lynes et al. (2013), Zhao et al. (2011)

Continued

Table 3.1 Recombinant fluorescent probes targeted to various subcellular compartments—cont'd

Compartment	Probe	Targeting sequence (number of tandem repeats)	Single FP	Two FP donor/ acceptor	Ca^{2+} sensor	Ratio-metric	Single Wave-length	K_d (μM)	Excitation (nm)	Emission (nm)	Ref.
	2mtGCaMP6m	COX-VIII (2)	GCaMP6m[k]		Mutant CaM/ M13		Yes	0.167 (*in vitro*)	474	515	Logan et al. (2014) this chapter
	mt-camgaroo-2	COX-VIII (1) pMITO (Clontech) (2) (Filippin et al., 2005)	Citrine[d]		CaM		Yes	5.3 (*in vitro*)	480	535	Griesbeck, Baird, Campbell, Zacharias, and Tsien (2001)
Mitochondrial outer surface; MAM; mitochondria–ER interface	N33D1cpv	TOM20 N-terminal 33aa		ECFP/cpVenus[e,f]	Design 1: mutant CaM/mutant M13[g]	Yes, Em		Not determined, probably similar to D1ER: 0.81/60 (*in vitro*) or 4mtD1GO-Cam: 1.53	425	480/535	Giacomello et al. (2010)
	OMM-pcm-ER	AKAP1 (N-terminal 1–43) + FKBP	cpEYFP-V68L/Q69K[b,i]		E104QCaM/ M13	Yes, Ex		0.46 (*in situ*)	485/420	535	Csordás et al. (2010)
	OMM-pcm-D2-ER	AKAP1 (N-terminal 1–43) + FKBP	cpEYFP-V68L/Q69K[b,i]		Design 2: mutant CaM/mutant M13[m]	Yes, Ex		3.14 (*in situ*)	485/420	535	Csordás et al. (2010)
Endoplasmic reticulum	ER-cameleon-3	Calreticulin signal sequence: MLLSVPLLL GLLGLAAAD; ER retention signal KDEL		EBFP/EGFP[a]	E104QcaM/ M13	Yes, Em		4.4 (*in vitro*)	381	445/510	Miyawaki et al. (1997)
	ER-yellow cameleon-3	Calreticulin signal sequence: MLLSVPLLL GLLGLAAAD; ER retention signal KDEL		ECFP/EYFP[a]	E104QcaM/ M13	Yes, Em		4.4 (*in vitro*)	443	476/528	Miyawaki et al. (1997)

	ER-yellow cameleon-4	Calreticulin signal sequence: MLLSVPLLL GLLGLAAAD; ER retention signal KDEL		ECFP/EYFP[a]	E31QcaM/M13	Yes, Em		0.083/700 (*in vitro*) 0.039/292 (*in situ*: Arnaudeau et al., 2001)	443	476/528	Miyawaki et al. (1997)
	YC4.3ER	Calreticulin signal sequence: MLLSVPLLL GLLGLAAAD; ER retention signal KDEL		ECFP/citrine[d]	E31QcaM/M13	Yes, Em		Probably as ER-yellow cameleon-4	436	475/535	Palmer, Jin, Reed, and Tsien (2004)
	D1ER	**Calreticulin signal sequence: MLLSVPLLL GLLGLAAAD; ER retention signal KDEL**		**ECFP/citrine[d]**	**Design 1: mutant CaM/ mutant M13[g]**	**Yes, Em**		**0.81/60 (*in vitro*)**	**436**	**475/535**	Palmer et al. (2004)
	split-YC7.3er	Calreticulin signal sequence: MLLSVPLLL GLLGLAAAD; ER retention signal KDEL		ECFP/citrine[d]	ECFP-E31Dcalmodulin and M13–Citrine			130 (*in situ*)	434	480/535	Ishii et al. (2006)
	catchER	Calreticulin signal sequence: MLLSVPLLL GLLGLAAAD; ER retention signal KDEL	EGFP mutantS147E, S202D, Q204E, F223E, and T225E,		Ca^{2+}-binding site generated by the S147E, S202D, Q204E, F223E, and T225E mutations		Yes	180 (*in situ*)	488	535	Tang et al. (2011)

Continued

Table 3.1 Recombinant fluorescent probes targeted to various subcellular compartments—cont'd

Compartment	Probe	Targeting sequence (number of tandem repeats)	Single FP	Two FP donor/acceptor	Ca^{2+} sensor	Ratio-metric	Single Wave-length	K_d (μM)	Excitation (nm)	Emission (nm)	Ref.
	apoK1-er	Calreticulin signal sequence: MLLSVPLLL GLLGLAAAD; ER retention signal KDEL		ECFP/EYFP[a]	Kringle IV domain of apolipoproteinA, mediates binding to calreticulin or ERp57	Yes, Em		No direct Ca^{2+} binding, apparent affinity is similar to ER-YC4	488	535	Osibow, Malli, Kostner, and Graier (2006)
Nucleus	Nu-cameleon-2	Nuclear localizaiton signal: PKKKRKVEDA (C-terminal)		EBFP/EGFP[a]	CaM/M13	Yes, Em		0.07/11 (*in vitro*)	381	445/510	Miyawaki et al. (1997)
	Nu-pericam	**Nuclear localizaiton signal: PKKKRKVEDA (C-terminal)**	**cpEYFP-V68L/Q69K[b,e]**		**E104QCaM/M13**	**Yes, Ex (but see RP3.1$_{mt}$ above)**		**1.7 (*in vitro*)**	**480/410**	**535**	Nagai et al. (2001)
	Nu-R-GECO1	**Nuclear localization signal: DPKKKRKV (C-terminal) (2)**	**mApple-derived GCAMP3 analogue[k,l]**		**Mutant CaM/M13[l]**		**Yes**	**0.482 (*in vitro*)**	**562**	**624**	Zhao et al. (2011)
	Nu-CaYang1/CaYin1	(NLS) DPKKKRKV C-terminus of Ying1 (NES) LALKLAGL-DIGS C-terminus of Yang1		CaYang1: mTFP1/cpv173 CaYin1: tdTomato/mAmetrine1.2	CaYang1: Design1 (D1)[g] CaYin1: Desing3 (D3)[g]	Yes, Em		CaYang1: 1.04/66.8 (*in vitro*) CaYin1: 0.87 (*in vitro*)	CaYang1:455 CaYin1:377	CaYang1 545/495 CaYin1:585/535	Ding, Ai, Hoi, and Campbell (2011)
Golgi	GT-YC3.3	81 aa N-terminal galactosyltransferase type II		ECFP/citrine[d]	E104QcaM/M13	Yes, Em		See ER-YC3	443	476/528	Griesbeck et al. (2001)

	Go-D1cpv	69 aa N-terminal of sialyl-transferase I	ECFP/cpVenus[e,f]	Design 1: mutant CaM/mutant M13[g]	Yes, Em		Not determined, probably similar to D1ER: 0.81/**60** (*in vitro*) or 4mtD1GO-Cam: 1.53	425	480/535	Lissandron, Podini, Pizzo, and Pozzan (2010)
Caveolae	LOXD1	lectin-like oxidized low-density lipoprotein receptor 1	ECFP/citrine[d]	D1	Yes, Em		Not determined, probably similar to D1ER: 0.81/60 (*in vitro*) or 4mtD1GO-Cam: 1.53	425	480/535	Isshiki et al. (2013)
Lysosome	LAMP1-YC3.6	LAMP1 Targets YC3.6 to the cytosolic surface of the lysosome	ECFP/cp173Venus[e]	E104QcaM/ M13	Yes, Em		0.25 (*in vitro*, Nagai, Yamada, Tominaga, Ichikawa, & Miyawaki, 2004)	405	450–490/ 520–570	McCue, Wardyn, Burgoyne, and Haynes (2013)
Peroxisomes	D3cpv-KVK-SKL	C-terminal KVK-SKL	ECFP/cpVenus[e,f]	Design 3: mutant CaM/mutant smooth muscle MLCK segment[g]	Yes, Em		1 (*in situ*)	436	475/535	Drago, Giacomello, Pizzo, and Pozzan (2008)

[a]The significant mutations in the four GFPs are: EBFP, F64L/Y66H/Y145[b];EGFP, F64L/S65T[f];ECFP, F64L/S65T/Y66W/N146I/M153T/V163A/N212K[b];and EYFP, S65G/S72A/T203Y[g];for references, see Miyawaki et al. (1997).
[b]Relative to wild-type GFP, the significant mutations in the GFP variants are: ECFP, F64L/S65T/Y66W/N146I/M153T/V163A/N164H; EYFP, S65G/S72A/T203Y; and EYFP-V68L/Q69K, S65G/V68L/Q69K/S72A/T203Y. Reduced pH sensitivity of cameleons by introducing mutations V68L and Q69K into the acceptor yellow green fluorescent protein. See Miyawaki, Griesbeck, Heim, and Tsien (1999).
[c]Venus: relative to EYFP the following mutations were introduced to further reduce Cl-sensitivity, folding, and oxidation/maturation: F46L/F64L/M153T/V163A/S175G. See Nagai et al. (2002).
[d]Citrine: brighter and acid resistant yellow mutant of GFP with mutations S65G/V68L/Q69M/S72A/T203Y. See Griesbeck et al. (2001).
[e]Circularly permuted Venus at position 173. Introduced to increase dynamic range of YCs of previous generations such as YCx.1[b], YCx.3[d], and YCx.12[c]. See Nagai et al. (2004).
[f]Citrine[d] containing constructs (2mtD2, 2mtD3, 2mtD4) with similar spectral characteristics but different affinities are also available, but still suffer from relatively low dynamic range. See Palmer et al. (2006).
[g]Designs 1–4 (D1–D4) were created to reduce interaction of the CaM/substrate pair with endogenous CaM. D1 was created from M13 (skeletal muscle MLCK; Palmer et al., 2004) while D2–D4 were designed starting from smooth muscle MLCK (Palmer et al., 2006).
[h]Mitochondria-targeted constructs with two to eight tandem COX-VIII N-terminal segments have been created, with increasing efficiency in specificity to the mitochondrion. However, higher (six to eight) repeats also caused mitochondrial damage. Four repeats might represent the optimal compromise. See Palmer et al. (2006).
[i]Ratiometric pericam was created from cpEYFP-V68L/Q69K by introducing H203F, H148D, and F46L, deleting a glycine before CaM, and replacing the GGSGG linker with VDGGSGGTG, between the original N- and C-termini (Nagai et al., 2001).
[j]Inverse pericam was made from ratiometric pericam by a D148T substitution (Nagai et al., 2001). In mitycam, the CaM domain has been further mutated (E67Q) to increase Ca^{2+} affinity of the probe (Terhzaz et al., 2006).
[k]GCaMP probes were generated from circularly permuted (144/149) EGFP (GCaMP1, Nakai, Ohkura, & Imoto, 2001) and further evolved into a series of probes with modified Ca^{2+} sensitivity, intensity, and color, of which GCaMP2 (Tallini et al., 2006), GCaMP3 (Tian et al., 2009), GCaMP6 (Chen et al., 2013), Case12 (Souslova et al., 2007), and different GECOs (Zhao et al., 2011) have been so far targeted to mitochondria or the nucleus.
[l]List of substitutions relative to GCaMP3: GEM-GECO1: 3L60P/K69E/N77Y/D86G/N98I/K119I/L173Q/T223S/N302S/R377P/K380Q/S404G/E430. These mutations led to the development of a unique single FP probe, brighter than GCaMP3, and ratiometric in emission. R-GECO1: Substitutions relative to the mApple-derived analog of GCaMP3: T47A/L60P/E61V/S63V/E64S/R81G/K83R/Y134C/M158L/N164aD/V228A/S290P/I366F/K380N/S404G/N414D/E430V (Zhao et al., 2011).
[m]D2 in ratiometric pericam was created from M13 (skeletal muscle MLCK) by introducing the mutations F68L, M72L, and Q104E (Csordás et al., 2010) analogous to mutations in the original D2 cameleons starting from smooth muscle MLCK (Palmer et al., 2006).
The table intended to summarize most available targeted recombinant Ca^{2+} probes, as they evolved during the last 15 years. Many of the probes are obsolete for various reasons, most frequently because of low basal intensity, slow or incomplete maturation at 37 °C and particularly due to small dynamic range. Where several probes exist, we highlighted in bold the currently best available probes, recommended for use or further development.
YC, yellow cameleon; PR, RP, pcm: ratiometric pericam.

and Case1 × (Souslova et al., 2007) generation of probes (circularly permuted single GFP probes with CaM/M13 CRE).

4. MODULATING Ca^{2+}-BINDING PROPERTIES

Variations in the CRE Ca^{2+}-binding domains allowed the adaptation of Ca^{2+}-binding affinities to enable the detection of Ca^{2+} concentrations varying between nanomolar to near millimolar ranges in different compartments. The original CREs based on (i) concatenated CaM and either skeletal (M13) or smooth muscle MLCK peptide pairs (Miyawaki et al., 1997) or (ii) skeletal or cardiac muscle troponin C (Heim & Griesbeck, 2004; Mank & Griesbeck, 2008; Mank et al., 2006) were used as a starting point for point mutations in Ca^{2+}-binding domains to modify Ca^{2+}-binding affinity. In addition, the interacting surfaces of the CaM/MLCK pairs have been extensively redesigned to limit interaction with endogenous CaM or other substrates (Palmer et al., 2004, 2006), leading to the D series of Cameleon probes.

5. DEVELOPING ORGANELLE TARGETING

Finally, targeting to cellular organelles was achieved by the addition of signal sequences, or by fusion to a protein which localizes to the organelle of interest. As listed and extensively referenced in Table 3.1., the recombinant Ca^{2+} probes have been targeted to the nucleus, mitochondria, endoplasmic reticulum (ER), Golgi apparatus, peroximes, and caveolae. In most cases (e.g., the ER, nucleus), a standard targeting strategy appears efficient, while in other cases, such as the mitochondria, dedicated studies analyzed the optimal strategy for enhancing the localization of the probe (Filippin et al., 2005; Palmer et al., 2006), and different strategies were adopted for targeting to subdomains of the organelle (Csordás et al., 2010; Giacomello et al., 2010).

6. QUANTITATIVE CONSIDERATIONS OF MEASUREMENTS OF ORGANELLE-TARGETED PROBES

Several parameters have to be taken into consideration when planning to quantitatively measure [Ca^{2+}] in organelles. Unambiguous localization and Ca^{2+} specificity are the most fundamental issues for any kind of estimation of organelle Ca^{2+} handling. Specific targeting should be controlled using high-resolution imaging and colocalization of known organelle markers. As detailed above and in Table 3.1, probes were iteratively optimized to control pH, chloride sensitivity, and interaction with other cellular components, these being the most frequent issues distorting Ca^{2+} sensitivity of the probe. Fluorescent proteins of the same design but lacking Ca^{2+}-binding domains can be used as controls to avoid this problem. A series of protocols and reviews have been recently published on the theoretic and practical issues regarding the quantitative use of recombinant Ca^{2+} probes (Grynkiewicz et al., 1985; Miyawaki et al., 2013; Palmer & Tsien, 2006; Pérez Koldenkova & Nagai, 2013; Tian et al., 2012); therefore, here we will refer only to some considerations regarding the use of organelle-targeted probes. First, probe affinities and dynamic ranges can be strikingly different when targeted into different compartments, thus when any of the calibration methods are used to determine absolute $[Ca^{2+}]$ values, we suggest the determination of K_d values *in situ*. In addition, calibration requires the determination of maximal (Ca^{2+} saturated) and minimal (Ca^{2+} free) intensities or ratios of the different probes, which is usually more problematic in case of organelles. For instance, Ca^{2+} ionophores (e.g., ionomycin) exchange Ca^{2+} to H^+, thus proton gradients can affect efficient equilibration of Ca^{2+} between the extracellular, cytoplasm, and organelle $[Ca^{2+}]$. Protocols for efficient Ca^{2+} equilibration of mitochondria (Csordás et al., 2010; Szabadkai, Pitter, & Spät, 2001), ER (Arnaudeau et al., 2001) and other organelles (Drago et al., 2008; Lissandron et al., 2010) have been developed and should be followed. In particular, equilibration of extra- and intra-organellar $[Ca^{2+}]$ requires the dissipation of transmembrane potentials (e.g., mitochondria) and active transport mechanisms (e.g., ER and Golgi) to avoid the formation of concentration gradients. Sometimes, it is particularly difficult to equilibrate organelles with Ca^{2+}-free, EGTA-buffered medium, thus calibration protocols relying partially on the *in vitro* specifications of the probes have to be applied (see, e.g., Maravall, Mainen, Sabatini, & Svoboda, 2000). However, in many cases, no exact determination of $[Ca^{2+}]$ is required to answer a specific experimental question. For examining kinetics of organelle Ca^{2+} release/uptake, relative changes normalized to background corrected basal intensities are sufficient for both single wavelength (Szabadkai et al., 2004) and ratiometric probes (Fülöp, Szanda, Enyedi, Várnai, & Spät, 2011). Similarly, here we present three protocols which can be used for such semiquantitative analysis of basal organelle $[Ca^{2+}]$ in the mitochondria and ER.

7. EXPERIMENTAL PROTOCOLS

The essential requirements for an experiment to measure $[Ca^{2+}]$ in intracellular organelles using fluorescent-targeted recombinant probes are (i) an appropriate probe (ii) a suitable cellular model, and (iii) an apposite fluorescence imaging system. The available probes and their critical parameters (such as Ca^{2+} affinity, suitability for ratiometric imaging, and spectral properties) are listed in Table 3.1. Any cellular model in which the expression of the probes can be achieved (by transfection, transduction) can be used, given the expression constructs are available. Finally, while any modern confocal or wide-field imaging system will be capable of generating valuable data using these probes, clearly the speed, sensitivity, and spatial resolution of imaging systems will determine whether the experimenter will be able to fully explore the kinetic, quantitative, and spatial parameters of the often complex localized organelle Ca^{2+} signals. Since it is not the focus of the chapter, for details of the systems used, we refer the readers to the references used in the relevant items in Table 3.1.

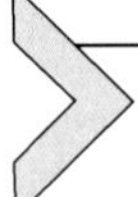

8. SEMIQUANTITATIVE IMAGING OF BASAL MITOCHONDRIAL $[Ca^{2+}]$ WITH A NOVEL MITOCHONDRIA-TARGETED GCaMP6m (2mtGCaMP6m)

Mitochondrial Ca^{2+} uptake is a thoroughly regulated process following complex kinetics (Nicholls, 2005; Rizzuto, De Stefani, Raffaello, & Mammucari, 2012; Spät, Szanda, Csordás, & Hajnóczky, 2008; Szabadkai & Duchen, 2008), due to the complex interactions of the recently described components of the Ca^{2+} uptake machinery (for recent reviews, see Csordás, Várnai, Golenár, Sheu, & Hajnóczky, 2012; Marchi & Pinton, 2014; Patron et al., 2013; Pizzo, Drago, Filadi, & Pozzan, 2012; Raffaello, De Stefani, & Rizzuto, 2012; Williams, Boyman, Chikando, Khairallah, & Lederer, 2013). In order to quantitatively measure small differences in resting $[Ca^{2+}]_m$ with high sensitivity, we developed a new probe based on the last-generation GCaMP probes (Chen et al., 2013) targeted to the mitochondrial matrix. We chose the GCaMP6m version, due to its high Ca^{2+} affinity, with its K_d (167 n*M*, measured *in vitro*) in the range of the predicted normal resting $[Ca^{2+}]_m$.

- *Construct design*: 2mtGCaMP6m was obtained by the addition of two tandem mitochondrial targeting sequences (derived from COX8A of human origin, see Filippin et al., 2005) to the GCaMP6m sequence (Addgene plasmid 40754) and was created by custom gene synthesis (Life Technologies). The construct and map are available on request.
- *Cellular models*: We have so far tested the probe in HeLa and HEK293 cell lines and human primary fibroblasts. Cells were grown on 24-mm coverslips and transfected with 2mtGCaMP6m encoding plasmid using Lipofectamine LTX according to the manufacturer's instruction (Life Technologies). After 36 h, coverslips were placed in 1 mL of KRB/ Ca^{2+} buffer for imaging.
- *Cellular imaging and analysis*: Imaging was performed on a Zeiss Axiovert 200 microscope equipped with a 63×/1.4 NA Plan Apochromat objective. Excitation was performed with a Deltaram V high speed monochromator (Photon Technology International) equipped with a 75-W xenon arc lamp. Images were captured with a high-sensitivity Evolve 512 Delta EMCCD (Photometrics). The system is controlled by Metamorph 7.5 and was assembled by Crisel Instruments. We noted that the GCaMP fluorophore has an isosbestic point (~410 nm) in its excitation spectrum (Chen et al., 2013). Thus next, we have experimentally confirmed in living cells that exciting GCaMP6m at 410 nm leads to fluorescence emission which is not Ca^{2+} dependent. As a consequence, the ratio between 474 and 410 nm excitation wavelengths is proportional to $[Ca^{2+}]$ while independent on probe expression levels (Fig. 3.1). By using appropriate equilibration protocols (see below, and Szabadkai et al., 2001; Vay et al., 2009), this will enable the determination of the exact *in situ* K_d of the probe and the calibration of the signal into $[Ca^{2+}]$. However, even without calibration, the method allows the relative assessment of basal mitochondrial $[Ca^{2+}]$ in different cell lines or conditions. Cells were thus alternatively illuminated at 474 and 410 nm, and fluorescence was collected through a 515/30-nm bandpass filter (Semrock). Exposure time was set to 300 ms. At least five fields were collected per coverslip, and each field was acquire for 20 s (0.5 frame/s). Analyses were performed with the Fiji distribution of ImageJ (Schindelin et al., 2012).

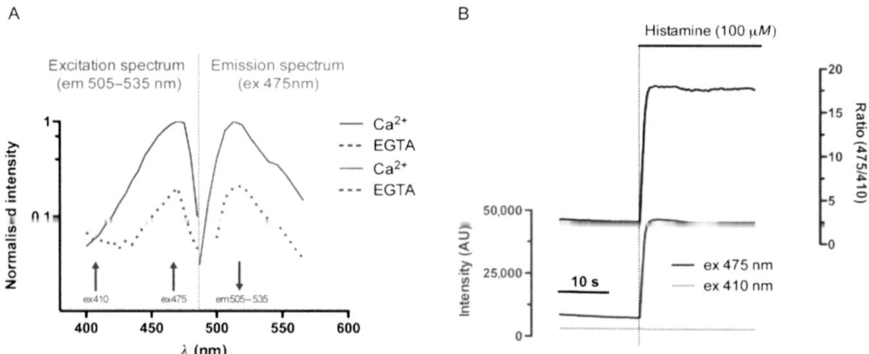

Figure 3.1 *Mitochondrial [Ca^{2+}] measured with the 2mtGCaMP6m circularly permuted GFP-based high-affinity Ca^{2+} probe.* (A) Excitation and emission spectra of the probe expressed in HeLa cells, obtained at 475 nm excitation and 505–535 nm emission, respectively. Spectra were measured either in the presence of Ca^{2+} or in intracellular buffer in the presence of EGTA following digitonin permeabilization to determine the Ca^{2+}-free traces, as previously described (Szabadkai et al., 2001). Arrows indicate the excitation (410 nm: isosbestic point; 475 nm) and emission wavelengths used in the experiments. (B) Basal and histamine (100 μ*M*) stimulated mitochondrial Ca^{2+} signals in HeLa cells transiently transfected with 2mtGCaMP6m. The two single wavelengths at 410 and 475 nm excitations as well as the ratio are shown. The experiments are representative of at least three different preparations. (See the color plate.)

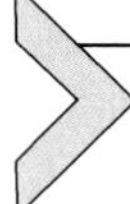

9. SEMIQUANTITATIVE IMAGING OF AGONIST-INDUCED MITOCHONDRIAL Ca^{2+} SIGNALS IN PRIMARY NEURONS USING THE LOW-AFFINITY 2mtD4cpv PROBE

One of the principal challenges of quantitative [Ca^{2+}] imaging in many subcellular organelles is to apply a probe with the appropriate affinity and dynamic range to cover rapid and sometimes vast changes in the actual [Ca^{2+}]. Accurately capturing agonist-induced mitochondrial Ca^{2+} signals is particularly challenging since the change in [Ca^{2+}] ranges from ~100 n*M* to ~1 m*M*, depending on the cell type; the latter being measured in neuroendocrine cells (Montero et al., 2000). The currently available targeted fluorescent probes with the widest range of Ca^{2+} sensitivity are from the design family of cameleons (Palmer et al., 2006). The 2mtD1cpv (K_{d1} ~ 0.8 μ*M*; K_{d2} ~ 60 μ*M*) (Giacomello et al., 2010) and 2mtD2cpv (K_{d1} ~ 0.3 μ*M*; K_{d1} ~ 3 μ*M*) (Palmer et al., 2006) probes both show biphasic sensitivity *in vitro* to Ca^{2+}, covering a wide concentration range, as shown and discussed previously (Palmer et al., 2011; Palmer & Tsien, 2006). Here, we demonstrate a less frequently used protocol designed specifically to measure high Ca^{2+} load in the mitochondria of neurons during conditions modeling glutamate-induced excitotoxicity with the monophasic low-affinity probe 2mtD4cpv.

- *Cellular model*: Cortical neurons were obtained from the cortical lobes of E18 Sprague–Dawley rat embryos according to previously described procedures (Cheung, Pascoe, Giardina, John, & Beart, 1998; Larm, Cheung, & Beart, 1996) and resuspended in Neurobasal® medium (Life Techonologies) supplemented with B27® (Life Technologies), 10% FBS, 0.5 m*M* L-glutamine and antibiotic–antimycotic (Life Technologies). For transfection of cells, 4 × 106 rat neurons were transfected in suspension with 3 μg of cDNA using Rat Neuron Nucleofector® Kit and Nucleofector® II device (Lonza, Switzerland), according to the manufacturer's instructions, plated onto poly-L-ornithine-coated glass-bottom μ-dishes (Ibidi GmbH, Germany) and maintained at 37 °C and 5% CO_2. Medium was replaced by supplemented FBS-free Neurobasal® medium 24 h later, and cultures were used at 8–9 days *in vitro*.
- *Cellular imaging and analysis*: For single-cell mitochondrial Ca^{2+} imaging, neurons transfected with mitochondria-targeted 2mtD4cpv Ca^{2+} indicator (Palmer et al., 2006) were transferred to incubation buffer (Ca^{2+}- and Mg^{2+}-free HBSS containing 20 m*M* HEPES, pH 7.4, 10 m*M* glucose, 10 μ*M* glycine, and 2.6 m*M* $CaCl_2$) and imaged by a TCS SP8 × confocal microscope (Leica, Germany). Cells were excited at 458 nm (Argon laser line), and CFP and YFP emission were acquired by two HyD™ detectors for FRET ratio quantification at an acquisition rate of 1 frame/15 s. Pinhole was set at 300 μm, and images were taken with a 63 × oil objective (1.4 NA). For data analysis, a homogeneous population of 5–12 cells was selected in the field of view and neuronal somata selected as Region of interest (ROI). Background values were subtracted and data were expressed as $\Delta R/R$ in which R represents the YFP/CFP fluorescence ratio for a given time point and R_0 represents the mean of the resting FRET ratios of the population imaged. Representative traces are shown in Fig. 3.2.

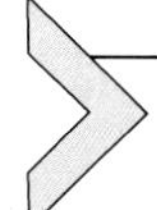

10. SEMIQUANTITATIVE IMAGING OF BASAL, STEADY-STATE [Ca^{2+}] IN THE ER

Although techniques to equilibrate [Ca^{2+}] in intracellular organelles have been applied, these are usually technically challenging and prone to artifacts (see above). Thus, quantitation of organellar [Ca^{2+}] without disturbing the existing steady-state would be desirable. Here, we show a technique, based on the quantitation of FRET efficiency (a measure of

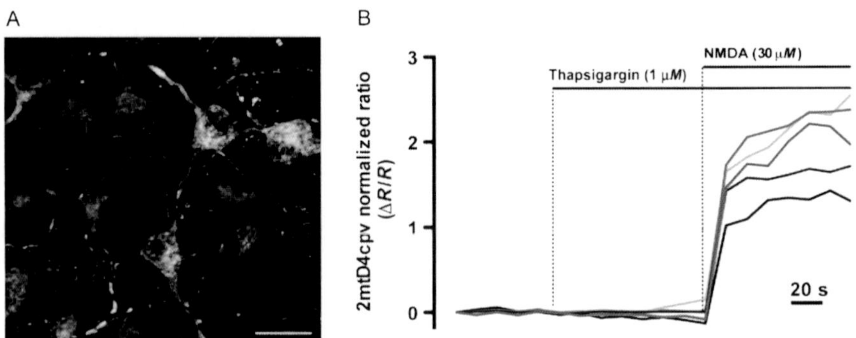

Figure 3.2 *Mitochondrial Ca^{2+} dynamics in neurons measured with the 2mtD4cpv cameleon probe.* (A) Micrograph shows the subcellular localization of the probe. Images were acquired in the cpv channel. Scale bar: 20 μm. (B) Neurons in the left panel were exposed to thapsigargin (1 μ*M*) and NMDA (30 μ*M*) as indicated, and the changes in the 535/485 ratio of the 2mtD4cpv probe were measured and normalized to the control period (Δ*R*/*R*). Traces show no increase in the FRET ratio after the thapsigargin-induced slow and low concentration cytosolic Ca^{2+} rise due to the low affinity of the probe. However, mitochondrial Ca^{2+} uptake upon NMDA receptor activation causes a fast, two- to fourfold increase in the ratio, indicating strong Ca^{2+} load. (See the color plate.)

steady-state [Ca^{2+}]) of the ER-targeted ECFP/citrine FRET-based probe ERD1. A short imaging protocol exploiting acceptor photobleaching measures FRET efficiency. The protocol can be employed for any FRET-based probe where the steady-state [Ca^{2+}] is at least in the one order of magnitude range of the K_d of the Ca^{2+}-sensing element (e.g., ~60 μ*M* for D1).

- *Cellular model*: We have used this technique to explore differences between cisplatin (CDDP) sensitive and resistant clones of the human lung cancer cell line A549 (Yao et al., 2013). Cells were grown on 22-mm coverslips and transfected with ERD1 encoding plasmid (Palmer et al., 2004) using Lipofectamine 2000 according to the manufacturer's instruction (Life Technologies). After 48 h, coverslips were placed in 1 mL of KRB/Ca^{2+} buffer for imaging.
- *Cellular imaging and analysis*: Measurements were carried out on a Zeiss LSM510 META confocal system, the probe was excited by a 405-nm laser diode, emission spectra were acquired from 420 to 600 nm, and the YFP and CFP signals were obtained by unmixing the spectrum based on previously registered spectra of separate CFP and YFP proteins, as well as the autofluorescence of nontransfected cells. FRET efficiency, which is the function of the ER luminal [Ca^{2+}] was quantified using the acceptor bleaching method (Gu, Di, Kelsell, & Zicha, 2004). Briefly, after five acquisitions, YFP was bleached (using maximal laser power at both

488 and 514 nm excitation wavelengths, with the time of illumination set to reach typically about 80–90% of the original intensity), followed by acquisition of further five image spectra. Reduction of the YFP signal leads to an increase in the CFP signal, which was normalized to the decrease in YFP intensity during bleaching. The normalized increase in CFP intensity is presented as FRET efficiency (see Fig. 3.3).

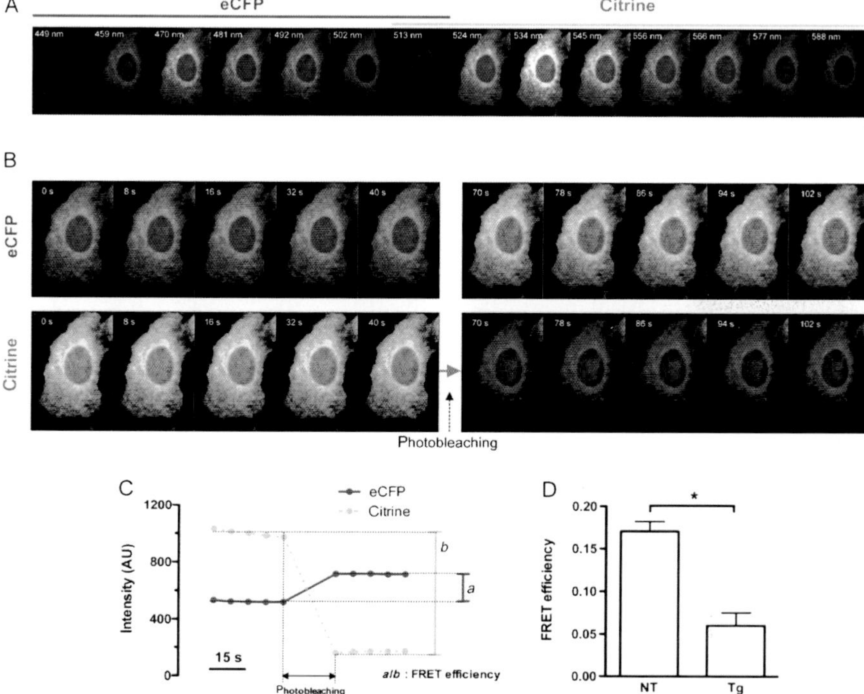

Figure 3.3 *Quantification of steady-state [Ca^{2+}] in the ER by acceptor photobleaching of the ERD1 FRET-based Ca^{2+} sensor.* (A) Emission spectra of the eCFP and citrine signals of the FRET probe in the 450–588 nm range following excitation at 405 nm. (B) Emission spectra were acquired during a short time lapse, composed of five image series, followed by photobleaching of the citrine probe (using 488 and 514 nm laser lines) and further five image series. The spectra were used for spectral deconvolution and separation of the specific eCFP and citrine signals using the Zeiss Meta system. (C) Quantification of FRET efficiency using the fluorescent intensities measured in (B). FRET efficiency was defined as the ratio between the increase in CFP and decrease in YFP following bleaching. (D) Steady-state ERD1 FRET efficiency in control, nontreated (NT) and thapsigargin (1 μ*M*, 30 min, in Ca^{2+}-free KRB) treated (Tg) A549 cells. The SERCA inhibitor Tg causes depletion of the ER Ca^{2+} content, reflected by the reduction in the FRET efficiency. Mean ± SEM of three independent experiments are shown. $^{*}p < 0.01$ (*t*-test). (See the color plate.)

11. CONCLUDING REMARKS

Organelle-targeted recombinant Ca^{2+}-sensitive probes proved as powerful research tools, helped explore the kinetics of agonist-induced Ca^{2+} signals, and estimated basal intraorganelle Ca^{2+} concentrations in several physiological and pathological settings. Altogether the use of these probes sparked the establishment of a novel research field, that is, compartmental signaling by Ca^{2+}, reflected also in other signaling systems (e.g., cAMP, cGMP, ATP). However, despite extensive redesign and improvement over the past two decades, the available probes still suffer some limitations to their use, and future developments are likely to focus on addressing these, such as dealing with pH sensitivity and other artifacts related to calibration and quantification; increasing quantum yield, fluorescence intensity, and dynamic range; improving selective targeting; reducing probe size to avoid perturbing the experimental model; further developing multiwavelength excitation and emission in order to parallelly observe multiple processes; and finally improvements in allowing the *in vivo* use of the probes.

REFERENCES

Akerboom, J., Carreras Calderón, N., Tian, L., Wabnig, S., Prigge, M., Tolö, J., et al. (2013). Genetically encoded calcium indicators for multi-color neural activity imaging and combination with optogenetics. *Frontiers in Molecular Neuroscience*, *6*, 2. http://dx.doi.org/10.3389/fnmol.2013.00002.

Akerboom, J., Chen, T., Wardill, T. J., Tian, L., Marvin, J. S., Mutlu, S., et al. (2012). Optimization of a GCaMP calcium indicator for neural activity imaging. *The Journal of Neuroscience: The Official Journal of the Society for Neuroscience*, *32*(40), 13819–13840. http://dx.doi.org/10.1523/JNEUROSCI.2601-12.2012.

Arnaudeau, S., Kelley, W. L., Walsh, J. V., & Demaurex, N. (2001). Mitochondria recycle Ca(2+) to the endoplasmic reticulum and prevent the depletion of neighboring endoplasmic reticulum regions. *The Journal of Biological Chemistry*, *276*(31), 29430–29439. http://dx.doi.org/10.1074/jbc.M103274200.

Baird, G. S., Zacharias, D. A., & Tsien, R. Y. (1999). Circular permutation and receptor insertion within green fluorescent proteins. *Proceedings of the National Academy of Sciences of the United States of America*, *96*(20), 11241–11246. http://dx.doi.org/10.1073/pnas.97.22.11984.

Bianchi, K., Rimessi, A., Prandini, A., Szabadkai, G., & Rizzuto, R. (2004). Calcium and mitochondria: Mechanisms and functions of a troubled relationship. *Biochimica et Biophysica Acta*, *1742*(1–3), 119–131. http://dx.doi.org/10.1016/j.bbamcr.2004.09.015.

Bonora, M., Giorgi, C., Bononi, A., Marchi, S., Patergnani, S., Rimessi, A., et al. (2013). Subcellular calcium measurements in mammalian cells using jellyfish photoprotein aequorin-based probes. *Nature Protocols*, *8*(11), 2105–2118. http://dx.doi.org/10.1038/nprot.2013.127.

Carlson, H. J., & Campbell, R. E. (2013). Circular permutated red fluorescent proteins and calcium ion indicators based on mCherry. *Protein Engineering, Design & Selection*, *26*(12), 763–772. http://dx.doi.org/10.1093/protein/gzt052.

Chen, T., Wardill, T. J., Sun, Y., Pulver, S. R., Renninger, S. L., Baohan, A., et al. (2013). Ultrasensitive fluorescent proteins for imaging neuronal activity. *Nature*, *499*(7458), 295–300. http://dx.doi.org/10.1038/nature12354.

Cheung, N. S., Pascoe, C. J., Giardina, S. F., John, C. A., & Beart, P. M. (1998). Micromolar L-glutamate induces extensive apoptosis in an apoptotic-necrotic continuum of insult-dependent, excitotoxic injury in cultured cortical neurones. *Neuropharmacology*, *37*(10–11), 1419–1429.

Chiesa, A., Rapizzi, E., Tosello, V., Pinton, P., de Virgilio, M., Fogarty, K. E., et al. (2001). Recombinant aequorin and green fluorescent protein as valuable tools in the study of cell signalling. *The Biochemical Journal*, *355*(Pt. 1), 1–12.

Csordás, G., Várnai, P., Golenár, T., Roy, S., Purkins, G., Schneider, T. G., et al. (2010). Imaging interorganelle contacts and local calcium dynamics at the ER-mitochondrial interface. *Molecular Cell*, *39*(1), 121–132. http://dx.doi.org/10.1016/j.molcel.2010.06.029.

Csordás, G., Várnai, P., Golenár, T., Sheu, S.-S., & Hajnóczky, G. (2012). Calcium transport across the inner mitochondrial membrane: Molecular mechanisms and pharmacology. *Molecular and Cellular Endocrinology*, *353*(1–2), 109–113. http://dx.doi.org/10.1016/j.mce.2011.11.011.

Ding, Y., Ai, H., Hoi, H., & Campbell, R. E. (2011). Förster resonance energy transfer-based biosensors for multiparameter ratiometric imaging of Ca^{2+} dynamics and caspase-3 activity in single cells. *Analytical Chemistry*, *83*(24), 9687–9693. http://dx.doi.org/10.1021/ac202595g.

Drago, I., Giacomello, M., Pizzo, P., & Pozzan, T. (2008). Calcium dynamics in the peroxisomal lumen of living cells. *The Journal of Biological Chemistry*, *283*(21), 14384–14390. http://dx.doi.org/10.1074/jbc.M800600200.

Filippin, L., Abad, M. C., Gastaldello, S., Magalhães, P. J., Sandonà, D., Pozzan, T., et al. (2005). Improved strategies for the delivery of GFP-based Ca^{2+} sensors into the mitochondrial matrix. *Cell Calcium*, *37*(2), 129–136. http://dx.doi.org/10.1016/j.ceca.2004.08.002.

Filippin, L., Magalhães, P. J., Di Benedetto, G., Colella, M., & Pozzan, T. (2003). Stable interactions between mitochondria and endoplasmic reticulum allow rapid accumulation of calcium in a subpopulation of mitochondria. *The Journal of Biological Chemistry*, *278*(40), 39224–39234. http://dx.doi.org/10.1074/jbc.M302301200.

Frieden, M., James, D., Castelbou, C., Danckaert, A., Martinou, J.-C., & Demaurex, N. (2004). Ca(2+) homeostasis during mitochondrial fragmentation and perinuclear clustering induced by hFis1. *The Journal of Biological Chemistry*, *279*(21), 22704–22714. http://dx.doi.org/10.1074/jbc.M312366200.

Fülöp, L., Szanda, G., Enyedi, B., Várnai, P., & Spät, A. (2011). The effect of OPA1 on mitochondrial Ca^{2+} signaling. *PloS One*, *6*(9), e25199. http://dx.doi.org/10.1371/journal.pone.0025199.

Giacomello, M., Drago, I., Bortolozzi, M., Scorzeto, M., Gianelle, A., Pizzo, P., et al. (2010). Ca^{2+} hot spots on the mitochondrial surface are generated by Ca^{2+} mobilization from stores, but not by activation of store-operated Ca^{2+} channels. *Molecular Cell*, *38*(2), 280–290. http://dx.doi.org/10.1016/j.molcel.2010.04.003.

Griesbeck, O., Baird, G. S., Campbell, R. E., Zacharias, D. A., & Tsien, R. Y. (2001). Reducing the environmental sensitivity of yellow fluorescent protein. Mechanism and applications. *Journal of Biological Chemistry*, *276*(31), 29188–29194. http://dx.doi.org/10.1074/jbc.M102815200.

Grynkiewicz, G., Poenie, M., & Tsien, R. Y. (1985). A new generation of Ca^{2+} indicators with greatly improved fluorescence properties. *Journal of Biological Chemistry*, *260*(6), 3440–3450.

Gu, Y., Di, W. L., Kelsell, D. P., & Zicha, D. (2004). Quantitative fluorescence resonance energy transfer (FRET) measurement with acceptor photobleaching and spectral unmixing. *Journal of Microscopy*, *215*(Pt. 2), 162–173. http://dx.doi.org/10.1111/j.0022-2720.2004.01365.x.

Hasan, M. T., Friedrich, R. W., Euler, T., Larkum, M. E., Giese, G., Both, M., et al. (2004). Functional fluorescent Ca^{2+} indicator proteins in transgenic mice under TET control. *PLoS Biology*, *2*(6), e163. http://dx.doi.org/10.1371/journal.pbio.0020163.

Heim, N., & Griesbeck, O. (2004). Genetically encoded indicators of cellular calcium dynamics based on troponin C and green fluorescent protein. *Journal of Biological Chemistry*, *279*(14), 14280–14286. http://dx.doi.org/10.1074/jbc.M312751200.

Iguchi, M., Kato, M., Nakai, J., Takeda, T., Matsumoto-Ida, M., Kita, T., et al. (2012). Direct monitoring of mitochondrial calcium levels in cultured cardiac myocytes using a novel fluorescent indicator protein, GCaMP2-mt. *International Journal of Cardiology*, *158*(2), 225–234. http://dx.doi.org/10.1016/j.ijcard.2011.01.034.

Ishii, K., Hirose, K., & Iino, M. (2006). Ca^{2+} shuttling between endoplasmic reticulum and mitochondria underlying Ca^{2+} oscillations. *EMBO Reports*, 7(4), 390–396. http://dx.doi.org/10.1038/sj.embor.7400620.

Isshiki, M., Nishimoto, M., Mizuno, R., & Fujita, T. (2013). FRET-based sensor analysis reveals caveolae are spatially distinct Ca(2+) stores in endothelial cells. *Cell Calcium*, *54*, 395–403. http://dx.doi.org/10.1016/j.ceca.2013.09.002.

Jiménez-Moreno, R., Wang, Z.-M., Messi, M. L., & Delbono, O. (2010). Sarcoplasmic reticulum Ca^{2+} depletion in adult skeletal muscle fibres measured with the biosensor D1ER. *Pflügers Archiv: European Journal of Physiology*, *459*(5), 725–735. http://dx.doi.org/10.1007/s00424-009-0778-4.

Kettlewell, S., Cabrero, P., Nicklin, S. A., Dow, J. A. T., Davies, S., & Smith, G. L. (2009). Changes of intra-mitochondrial Ca^{2+} in adult ventricular cardiomyocytes examined using a novel fluorescent Ca^{2+} indicator targeted to mitochondria. *Journal of Molecular and Cellular Cardiology*, *46*(6), 891–901. http://dx.doi.org/10.1016/j.yjmcc.2009.02.016.

Larm, J. A., Cheung, N. S., & Beart, P. M. (1996). (S)-5-fluorowillardiine-mediated neurotoxicity in cultured murine cortical neurones occurs via AMPA and kainate receptors. *European Journal of Pharmacology*, *314*(1–2), 249–254.

Lissandron, V., Podini, P., Pizzo, P., & Pozzan, T. (2010). Unique characteristics of Ca^{2+} homeostasis of the trans-Golgi compartment. *Proceedings of the National Academy of Sciences of the United States of America*, *107*(20), 9198–9203. http://dx.doi.org/10.1073/pnas.1004702107.

Liu, S., He, J., Jin, H., Yang, F., Lu, J., & Yang, J. (2011). Enhanced dynamic range in a genetically encoded Ca^{2+} sensor. *Biochemical and Biophysical Research Communications*, *412*(1), 155–159. http://dx.doi.org/10.1016/j.bbrc.2011.07.065.

Llorente-Folch, I., Rueda, C. B., Amigo, I., del Arco, A., Saheki, T., Pardo, B., et al. (2013). Calcium-regulation of mitochondrial respiration maintains ATP homeostasis and requires ARALAR/AGC1-malate aspartate shuttle in intact cortical neurons. *The Journal of Neuroscience: The Official Journal of the Society for Neuroscience*, *33*(35), 13957–13971. http://dx.doi.org/10.1523/JNEUROSCI.0929-13.2013, 13971a.

Logan, C. V., Szabadkai, G., Sharpe, J. A., Parry, D. A., Torelli, S., Childs, A. M., et al. (2014). Loss-of-function mutations in MICU1 cause a brain and muscle disorder linked to primary alterations in mitochondrial calcium signaling. *Nature Genetics*, *46*(2), 188–193. http://dx.doi.org/10.1038/ng.2851.

Lu, X., Ginsburg, K. S., Kettlewell, S., Bossuyt, J., Smith, G. L., & Bers, D. M. (2013). Measuring local gradients of intramitochondrial [Ca(2+)] in cardiac myocytes during sarcoplasmic reticulum Ca(2+) release. *Circulation Research*, *112*(3), 424–431. http://dx.doi.org/10.1161/CIRCRESAHA.111.300501.

Lynes, E. M., Raturi, A., Shenkman, M., Ortiz Sandoval, C., Yap, M. C., Wu, J., et al. (2013). Palmitoylation is the switch that assigns calnexin to quality control or ER Ca^{2+} signaling. *Journal of Cell Science, 126*(Pt. 17), 3893–3903. http://dx.doi.org/10.1242/jcs.125856.

Malgaroli, A., Milani, D., Meldolesi, J., & Pozzan, T. (1987). Fura-2 measurement of cytosolic free Ca^{2+} in monolayers and suspensions of various types of animal cells. *The Journal of Cell Biology, 105*(5), 2145–2155. http://dx.doi.org/10.1083/jcb.105.5.2145.

Mank, M., & Griesbeck, O. (2008). Genetically encoded calcium indicators. *Chemical Reviews, 108*(5), 1550–1564. http://dx.doi.org/10.1021/cr078213v.

Mank, M., Reiff, D. F., Heim, N., Friedrich, M. W., Borst, A., & Griesbeck, O. (2006). A FRET-based calcium biosensor with fast signal kinetics and high fluorescence change. *Biophysical Journal, 90*(5), 1790–1796. http://dx.doi.org/10.1529/biophysj.105.073536.

Maravall, M., Mainen, Z. F., Sabatini, B. L., & Svoboda, K. (2000). Estimating intracellular calcium concentrations and buffering without wavelength ratioing. *Biophysical Journal, 78*(5), 2655–2667. http://dx.doi.org/10.1016/S0006-3495(00)76809-3.

Marchi, S., & Pinton, P. (2014). The mitochondrial calcium uniporter complex: molecular components, structure and physiopathological implications. *Journal of Physiology, 592*, 829–839. http://dx.doi.org/10.1113/jphysiol.2013.268235.

McCue, H. V., Wardyn, J. D., Burgoyne, R. D., & Haynes, L. P. (2013). Generation and characterization of a lysosomally targeted, genetically encoded Ca(2+)-sensor. *Biochemical Journal, 449*(2), 449–457. http://dx.doi.org/10.1042/BJ20120898.

Miyawaki, A., Griesbeck, O., Heim, R., & Tsien, R. Y. (1999). Dynamic and quantitative Ca^{2+} measurements using improved cameleons. *Proceedings of the National Academy of Sciences of the United States of America, 96*(5), 2135–2140.

Miyawaki, A., Llopis, J., Heim, R., McCaffery, J. M., Adams, J. A., Ikura, M., et al. (1997). Fluorescent indicators for Ca^{2+} based on green fluorescent proteins and calmodulin. *Nature, 388*(6645), 882–887. http://dx.doi.org/10.1038/42264.

Miyawaki, A., Nagai, T., & Mizuno, H. (2013). Imaging intracellular free Ca^{2+} concentration using yellow cameleons. *Cold Spring Harbor Protocols. 2013*(11). http://dx.doi.org/10.1101/pdb.prot078642.

Mizuno, H., Sassa, T., Higashijima, S.-I., Okamoto, H., & Miyawaki, A. (2013). Transgenic zebrafish for ratiometric imaging of cytosolic and mitochondrial Ca^{2+} response in teleost embryo. *Cell Calcium, 54*(3), 236–245. http://dx.doi.org/10.1016/j.ceca.2013.06.007.

Montero, M., Alonso, M. T., Carnicero, E., Cuchillo-Ibáñez, I., Albillos, A., García, A. G., et al. (2000). Chromaffin-cell stimulation triggers fast millimolar mitochondrial Ca^{2+} transients that modulate secretion. *Nature Cell Biology, 2*(2), 57–61. http://dx.doi.org/10.1038/35000001.

Nagai, T., Ibata, K., Park, E. S., Kubota, M., Mikoshiba, K., & Miyawaki, A. (2002). A variant of yellow fluorescent protein with fast and efficient maturation for cell-biological applications. *Nature Biotechnology, 20*(1), 87–90. http://dx.doi.org/10.1038/nbt0102-87.

Nagai, T., Sawano, A., Park, E. S., & Miyawaki, A. (2001). Circularly permuted green fluorescent proteins engineered to sense Ca^{2+}. *Proceedings of the National Academy of Sciences of the United States of America, 98*(6), 3197–3202. http://dx.doi.org/10.1073/pnas.051636098.

Nagai, T., Yamada, S., Tominaga, T., Ichikawa, M., & Miyawaki, A. (2004). Expanded dynamic range of fluorescent indicators for Ca(2+) by circularly permuted yellow fluorescent proteins. *Proceedings of the National Academy of Sciences of the United States of America, 101*(29), 10554–10559. http://dx.doi.org/10.1073/pnas.0400417101.

Nakai, J., Ohkura, M., & Imoto, K. (2001). A high signal-to-noise Ca(2+) probe composed of a single green fluorescent protein. *Nature Biotechnology*, *19*(2), 137–141. http://dx.doi.org/10.1038/84397.

Nicholls, D. G. (2005). Mitochondria and calcium signaling. *Cell Calcium*, *38*(3–4), 311–317. http://dx.doi.org/10.1016/j.ceca.2005.06.011.

Osibow, K., Malli, R., Kostner, G. M., & Graier, W. F. (2006). A new type of non-Ca^{2+}−buffering Apo(a)-based fluorescent indicator for intraluminal Ca^{2+} in the endoplasmic reticulum. *Journal of Biological Chemistry*, *281*(8), 5017–5025. http://dx.doi.org/10.1074/jbc.M508583200.

Palmer, A. E., Giacomello, M., Kortemme, T., Hires, S. A., Lev-Ram, V., Baker, D., et al. (2006). Ca^{2+} indicators based on computationally redesigned calmodulin-peptide pairs. *Chemistry & Biology*, *13*(5), 521–530. http://dx.doi.org/10.1016/j.chembiol.2006.03.007.

Palmer, A. E., Jin, C., Reed, J. C., & Tsien, R. Y. (2004). Bcl-2-mediated alterations in endoplasmic reticulum Ca^{2+} analyzed with an improved genetically encoded fluorescent sensor. *Proceedings of the National Academy of Sciences of the United States of America*, *101*(50), 17404–17409. http://dx.doi.org/10.1073/pnas.0408030101.

Palmer, A. E., Qin, Y., Park, J. G., & McCombs, J. E. (2011). Design and application of genetically encoded biosensors. *Trends in Biotechnology*, *29*(3), 144–152. http://dx.doi.org/10.1016/j.tibtech.2010.12.004.

Palmer, A. E., & Tsien, R. Y. (2006). Measuring calcium signaling using genetically targetable fluorescent indicators. *Nature Protocols*, *1*(3), 1057–1065. http://dx.doi.org/10.1038/nprot.2006.172.

Patron, M., Raffaello, A., Granatiero, V., Tosatto, A., Merli, G., De Stefani, D., et al. (2013). The mitochondrial calcium uniporter (MCU): Molecular identity and physiological roles. *Journal of Biological Chemistry*, *288*(15), 10750–10758. http://dx.doi.org/10.1074/jbc.R112.420752.

Pérez Koldenkova, V., & Nagai, T. (2013). Genetically encoded Ca(2+) indicators: Properties and evaluation. *Biochimica et Biophysica Acta*, *1833*(7), 1787–1797. http://dx.doi.org/10.1016/j.bbamcr.2013.01.011.

Pizzo, P., Drago, I., Filadi, R., & Pozzan, T. (2012). Mitochondrial Ca^{2+} homeostasis: Mechanism, role, and tissue specificities. *Pflügers Archiv: European Journal of Physiology*, *464*(1), 3–17. http://dx.doi.org/10.1007/s00424-012-1122-y.

Raffaello, A., De Stefani, D., & Rizzuto, R. (2012). The mitochondrial Ca(2+) uniporter. *Cell Calcium*, *52*(1), 16–21. http://dx.doi.org/10.1016/j.ceca.2012.04.006.

Rizzuto, R., De Stefani, D., Raffaello, A., & Mammucari, C. (2012). Mitochondria as sensors and regulators of calcium signalling. *Nature Reviews Molecular Cell Biology*, *13*(9), 566–578. http://dx.doi.org/10.1038/nrm3412.

Romoser, V. A., Hinkle, P. M., & Persechini, A. (1997). Detection in living cells of Ca^{2+}−dependent changes in the fluorescence emission of an indicator composed of two green fluorescent protein variants linked by a calmodulin-binding sequence. A new class of fluorescent indicators. *The Journal of Biological Chemistry*, *272*(20), 13270–13274. http://dx.doi.org/10.1074/jbc.272.20.13270.

Rudolf, R., Mongillo, M., Rizzuto, R., & Pozzan, T. (2003). Looking forward to seeing calcium. *Nature Reviews Molecular Cell Biology*, *4*(7), 579–586. http://dx.doi.org/10.1038/nrm1153.

Schindelin, J., Arganda-Carreras, I., Frise, E., Kaynig, V., Longair, M., Pietzsch, T., et al. (2012). Fiji: An open-source platform for biological-image analysis. *Nature Methods*, *9*(7), 676–682. http://dx.doi.org/10.1038/nmeth.2019.

Souslova, E. A., Belousov, V. V., Lock, J. G., Strömblad, S., Kasparov, S., Bolshakov, A. P., et al. (2007). Single fluorescent protein-based Ca^{2+} sensors with increased dynamic range. *BMC Biotechnology*, 7, 37. http://dx.doi.org/10.1186/1472-6750-7-37.

Spät, A., Szanda, G., Csordás, G., & Hajnóczky, G. (2008). High- and low-calcium-dependent mechanisms of mitochondrial calcium signalling. *Cell Calcium, 44*(1), 51–63. http://dx.doi.org/10.1016/j.ceca.2007.11.015.

Szabadkai, G., & Duchen, M. R. (2008). Mitochondria: The hub of cellular Ca^{2+} signaling. *Physiology (Bethesda, MD), 23*, 84–94. http://dx.doi.org/10.1152/physiol.00046.2007.

Szabadkai, G., Pitter, J. G., & Spät, A. (2001). Cytoplasmic Ca^{2+} at low submicromolar concentration stimulates mitochondrial metabolism in rat luteal cells. *Pflügers Archiv: European Journal of Physiology, 441*(5), 678–685. http://dx.doi.org/10.1007/s004240000466.

Szabadkai, G., Simoni, A. M., Chami, M., Wieckowski, M. R., Youle, R. J., & Rizzuto, R. (2004). Drp-1-dependent division of the mitochondrial network blocks intraorganellar Ca^{2+} waves and protects against Ca^{2+}–mediated apoptosis. *Molecular Cell, 16*(1), 59–68. http://dx.doi.org/10.1016/j.molcel.2004.09.026.

Tallini, Y. N., Ohkura, M., Choi, B.-R., Ji, G., Imoto, K., Doran, R., et al. (2006). Imaging cellular signals in the heart in vivo: Cardiac expression of the high-signal Ca^{2+} indicator GCaMP2. *Proceedings of the National Academy of Sciences of the United States of America, 103*(12), 4753–4758. http://dx.doi.org/10.1073/pnas.0509378103.

Tang, S., Wong, H.-C., Wang, Z.-M., Huang, Y., Zou, J., Zhuo, Y., et al. (2011). Design and application of a class of sensors to monitor Ca^{2+} dynamics in high Ca^{2+} concentration cellular compartments. *Proceedings of the National Academy of Sciences of the United States of America, 108*(39), 16265–16270. http://dx.doi.org/10.1073/pnas.1103015108.

Terhzaz, S., Southall, T. D., Lilley, K. S., Kean, L., Allan, A. K., Davies, S. A., et al. (2006). Differential gel electrophoresis and transgenic mitochondrial calcium reporters demonstrate spatiotemporal filtering in calcium control of mitochondria. *Journal of Biological Chemistry, 281*(27), 18849–18858. http://dx.doi.org/10.1074/jbc.M603002200.

Thomas, D., Tovey, S. C., Collins, T. J., Bootman, M. D., Berridge, M. J., & Lipp, P. (2000). A comparison of fluorescent Ca^{2+} indicator properties and their use in measuring elementary and global Ca^{2+} signals. *Cell Calcium, 28*(4), 213–223. http://dx.doi.org/10.1054/ceca.2000.0152.

Tian, L., Hires, S. A., & Looger, L. L. (2012). Imaging neuronal activity with genetically encoded calcium indicators. *Cold Spring Harbor Protocols, 2012*(6), 647–656. http://dx.doi.org/10.1101/pdb.top069609.

Tian, L., Hires, S. A., Mao, T., Huber, D., Chiappe, M. E., Chalasani, S. H., et al. (2009). Imaging neural activity in worms, flies and mice with improved GCaMP calcium indicators. *Nature Methods, 6*(12), 875–881. http://dx.doi.org/10.1038/nmeth.1398.

Truong, K., Sawano, A., Mizuno, H., Hama, H., Tong, K. I., Mal, T. K., et al. (2001). FRET-based in vivo Ca^{2+} imaging by a new calmodulin-GFP fusion molecule. *Nature Structural Biology, 8*(12), 1069–1073. http://dx.doi.org/10.1038/nsb728.

Váradi, A., & Rutter, G. A. (2002). Green fluorescent protein calcium biosensors. Calcium imaging with GFP cameleons. *Methods in Molecular Biology, 183*, 255–264. http://dx.doi.org/10.1385/1-59259-280-5:255.

Vay, L., Hernández-SanMiguel, E., Lobatón, C. D., Moreno, A., Montero, M., & Alvarez, J. (2009). Mitochondrial free [Ca2+] levels and the permeability transition. *Cell Calcium, 45*(3), 243–250. http://dx.doi.org/10.1016/j.ceca.2008.10.007.

Waldeck-Weiermair, M., Alam, M. R., Khan, M. J., Deak, A. T., Vishnu, N., Karsten, F., et al. (2012). Spatiotemporal correlations between cytosolic and mitochondrial Ca(2+) signals using a novel red-shifted mitochondrial targeted cameleon. *PloS One*, 7(9), e45917. http://dx.doi.org/10.1371/journal.pone.0045917.

Wallace, D. J., Meyer zum Alten Borgloh, S., Astori, S., Yang, Y., Bausen, M., Kügler, S., et al. (2008). Single-spike detection in vitro and in vivo with a genetic Ca^{2+} sensor. *Nature Methods, 5*(9), 797–804. http://dx.doi.org/10.1038/nmeth.1242.

Whitaker, M. (2010). Genetically encoded probes for measurement of intracellular calcium. *Methods in Cell Biology*, *99*, 153–182. http://dx.doi.org/10.1016/B978-0-12-374841-6.00006-2.

Williams, G. S. B., Boyman, L., Chikando, A. C., Khairallah, R. J., & Lederer, W. J. (2013). Mitochondrial calcium uptake. *Proceedings of the National Academy of Sciences of the United States of America*, *110*(26), 10479–10486. http://dx.doi.org/10.1073/pnas.1300410110.

Yao, Z., Jones, A. W. E., Fassone, E., Sweeney, M. G., Lebiedzinska, M., Suski, J. M., et al. (2013). PGC-1β mediates adaptive chemoresistance associated with mitochondrial DNA mutations. *Oncogene*, *32*(20), 2592–2600. http://dx.doi.org/10.1038/onc.2012.259.

Yu, D., Baird, G. S., Tsien, R. Y., & Davis, R. L. (2003). Detection of calcium transients in Drosophila mushroom body neurons with camgaroo reporters. *The Journal of Neuroscience: The Official Journal of the Society for Neuroscience*, *23*(1), 64–72.

Zhao, Y., Araki, S., Wu, J., Teramoto, T., Chang, Y.-F., Nakano, M., et al. (2011). An expanded palette of genetically encoded Ca^{2+} indicators. *Science (New York, NY)*, *333*(6051), 1888–1891. http://dx.doi.org/10.1126/science.1208592.

CHAPTER FOUR

Autophagy and Autophagic Flux in Tumor Cells

Nicolas Dupont, Idil Orhon, Chantal Bauvy, Patrice Codogno[1]
INSERM U1151-CNRS UMR 8253, Institut Necker Enfants-Malades, University Paris Descartes, Paris cedex 14, France
[1]Corresponding author: e-mail address: patrice.codogno@inserm.fr

Contents

Abstract

Macroautophagy (hereafter referred to as autophagy), a central mechanism mediating the lysosomal degradation of cytoplasmic components, can be stimulated by a wide panel of adverse stimuli, including a panoply of anticancer agents. The central autophagic organelle is the autophagosome, a double membrane-bound vacuole that sequesters the cytoplasmic material destined to disposal. The ultimate destiny of the autophagosome is to fuse with a lysosome, resulting in the degradation of the autophagic cargo. In this setting, it is important to discriminate whether a particular stimulus actually promotes autophagy or it simply blocks the fusion of autophagosomes with lysosomes. To this aim, the methods that assess autophagy should assess not only the number of autophagosomes but also the so-called autophagic flux, that is, the clearance of the autophagy cargo from the lysosomal compartment. Here, we present a compendium of methods to assess the autophagic flux in cultured malignant cells. This approach should allow for the study of the intimate link between autophagy and oncometabolism in several experimental paradigms.

Methods in Enzymology, Volume 543
ISSN 0076-6879
http://dx.doi.org/10.1016/B978-0-12-801329-8.00004-0

1. INTRODUCTION

Macroautophagy (referred to below as autophagy) is a form of autophagy that degrades cellular macromolecules and cytoplasmic structures in eukaryotic cells (Boya, Reggiori, & Codogno, 2013; Yang & Klionsky, 2010). Autophagy begins with the formation of a phagophore or isolation membrane that subsequently elongates to produce double membrane-bound vacuoles known as autophagosomes, which sequester the cytoplasmic material in bulk or in a selective manner via autophagy adapters such as SQSTM1/p62 (Lamb, Yoshimori, & Tooze, 2013). The discovery of autophagy-related (Atg) proteins (Mizushima, Yoshimori, & Ohsumi, 2011), which are involved in autophagosome formation, was a major breakthrough in the understanding of autophagy and its importance in physiology and pathology (Mizushima & Komatsu, 2011; Ravikumar et al., 2010; Rubinsztein, Codogno, & Levine, 2012). In mammalian cells, once the autophagosome has been formed, it acquires acidic and degradative capacities by merging with endocytic compartments to form a compartment known as the amphisome (Stromhaug & Seglen, 1993). The final stage of autophagy is the fusion of autophagic vacuoles (amphisomes or autophagosomes) with lysosomes to form autolysosomes, where the autophagy cargo is totally degraded. The role of autophagy in cancer is complex and context dependent (Liu & Ryan, 2012; Lorin, Hamaï, Mehrpour, & Codogno, 2013; White, 2012). Defects in macroautophagy promote DNA damage and genomic instability. In contrast, macroautophagy also protects cancer cells from metabolic stress by providing substances that allow them to maintain their metabolism. Moreover, several drugs used in cancer therapy stimulate autophagy, which in many instances constitutes a form of resistance to therapy (Lorin et al., 2013; Rubinsztein et al., 2012). However, in some cases, cancer treatments can trigger cell death involving autophagy (Lorin et al., 2013; Rubinsztein et al., 2012). Whatever the outcome of autophagy, the mere visualization of autophagosomes is not sufficient to conclude that autophagy has been stimulated (Klionsky et al., 2012; Mizushima, Yoshimori, & Levine, 2010). Here, two sets of information are of fundamental importance before any conclusion can be reached about the stimulation of autophagy: (1) Determination of the increase in the number of autophagosomes formed (induction of autophagy), and (2) the assay of the autophagic flux, that is, the clearance of the autophagy cargo by the lysosomal compartment. An increase in the number of autophagosomes can be the consequence of an increase in the induction of autophagy and in autophagic flux, but it can also be the consequence of a blockade of the autophagic flux without any stimulation of the induction step (Klionsky et al., 2012; Mizushima et al., 2010). In this chapter, we describe some of the methods most commonly used for determining the induction of autophagy and autophagic flux. Many of the methods described are based on the detection of protein LC3 (a member of the Atg8 family) (Mizushima et al., 2011). During autophagy, LC3-I is converted into LC3-II by the formation of a covalent link between the C-terminus of LC3-I and the polar head of phosphatidylethanol amine. LC3-II and other members of the Atg8 family are involved in the elongation of the preautophagosomal membrane and in the sealing of the membrane (Mizushima & Komatsu, 2011; Mizushima et al., 2011; Weidberg et al., 2010). The fraction of LC3-II associated with the inner face of the autophagosome is transported into the lysosomal compartment where it is degraded. We will therefore describe LC3-based methods used to analyze the induction of autophagy and autophagic flux. We will also describe methods based on the degradation of the autophagy cargo, such as SQSTM1/p62, which is transported to the lysosome via the interaction between its LC3-Interaction Region (LIR) with LC3-II (Birgisdottir, Lamark, & Johansen, 2013) and the degradation of radiolabeled, long-lived proteins used to analyze the autophagic flux.

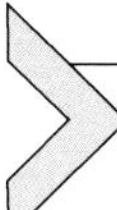

2. AUTOPHAGY INITIATION AND AUTOPHAGIC FLUX

2.1. LC3 Western blot

2.1.1 *Materials and reagents*

1. Dulbecco's Modified Eagle's Medium (DMEM, 4.5 g/l glucose, glutamax) (Invitrogen, Life Technologies, ref. 31966) supplemented with 10% fetal bovine serum (Invitrogen, Life Technologies).
2. Hanks' balanced salt solution (HBSS) without sodium bicarbonate (Invitrogen, Life Technologies, ref. 14025) or Earle's balanced salt solution (EBSS) (Invitrogen, Life Technologies, ref. 24010-043).
3. Protease Inhibitors: Bafilomycin A1 (Sigma B1793, final concentration of 100 μ*M*)/chloroquine (Sigma C6628, final concentration of 50 μ*M*).
4. Phosphate-buffered saline (PBS): 137 m*M* NaCl, 2.7 m*M* KCl, 4.3 m*M* Na_2HPO_4, 1.4 m*M* KH_2PO_4, pH 7.3.
5. Laemmli (2×) sample buffer (3.75% SDS, 25% glycerol, 25% β-mercaptoethanol, 0.125% bromophenol blue, and 31.5% of 1.5 *M* buffer Tris–HCl (containing 0.4% SDS), pH 8.8).

6. RIPA buffer (150 m*M* NaCl, 1% NP-40, 0.5% sodium deoxycholate, 0.1% SDS, 50 m*M* Tris, pH 8).
7. PBS–Tween 20 (0.1%) (Sigma-Aldrich, ref. P7949).
8. Blocking solution (5% milk PBS–Tween 20).
9. Antibody solution (2% milk PBS–Tween 20).
10. LC3B antibody (Sigma-Aldrich, ref. L7543).

2.1.2 Protocol

The day before the experiment, cells are seeded in six-well plates to be used near confluence:

1. In most cases, leave the cells for 24 h to adhere before applying any stimuli. Several starvation conditions can be used to stimulate autophagy, and these include Earle's Balanced Salt Solution (EBSS), serum deprivation (medium without FBS), amino acid starvation, or glucose starvation (see Notes A and B).
2. In order to observe the autophagic flux, Bafilomycin A1 (100 or 200 n*M*) or hydroxychloroquine (50 μ*M*) is added directly to the medium for 2–4 h (see Note C) before the experiment is stopped.
3. Discard the medium and wash the cells with ice-cold PBS, while the plates are kept on ice. Aspirate all the PBS and then store the plates at −80 °C for Western blotting (see Note D).
4. On the day of the Western blot experiment, thaw the stored cells on ice. Add around 100 μl (the volume depends on the cell type) of Laemmli 2 × directly to the wells at room temperature. It is important to note that cell lysis must be carried out on the same day as the Western blotting in order to avoid degradation of the LC3 (see Note E).
5. Scrape the cells using a cell scraper and collect the cell suspension in a microcentrifuge tube. Boil the samples at 100 °C for 5 min.
6. Load the samples onto a 12.5% SDS-PAGE. It is important to note that a gel at less than 12.5% would not be sufficient to separate LC3-I (16 kDa) and LC3-II (14 kDa).
7. Carry out a wet transfer using a transfer buffer consisting of 24 m*M* Tris and 190 m*M* glycine, for 1 h at 120 V.

Flowchart for LC3 Western blot sample preparation

Cells are seeded in the usual medium in six-well plates and used near confluence.

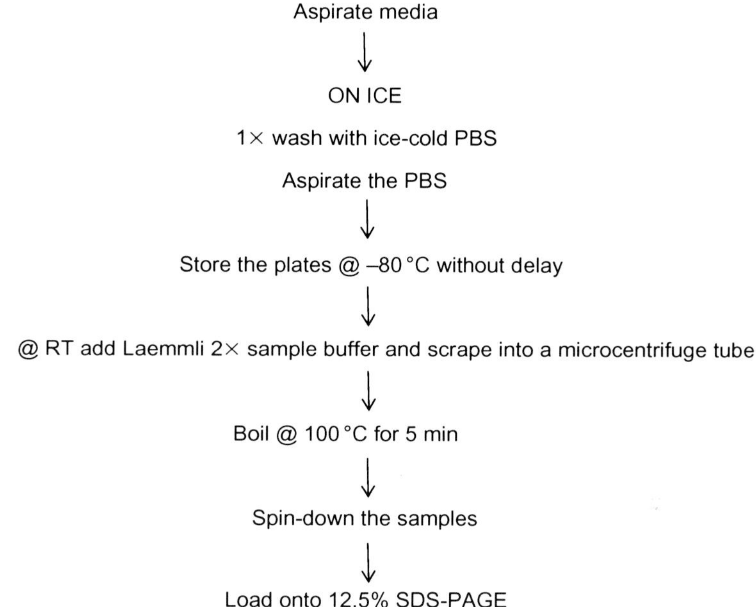

The relative band intensity of the LC3-II levels is quantified using ImageJ or some other densitometry software. In order to quantify the changes in autophagy initiation, the levels of LC3-II are calculated over the expression of actin (or some other housekeeping protein). The increase or decrease in the expression of LC3-II, the autophagosome-bound form of LC3, serves to monitor the formation of mature autophagosomes triggered by different stimuli. The total accumulation of LC3-II is observed by blocking the fusion of autophagosome with lysosomes by adding lysosomal inhibitors (e.g., Baf A1) (Tanida, Minematsu-Ikeguchi, Ueno, & Kominami, 2005). In response to autophagic stimuli, such as starvation, autophagosome synthesis would be expected to increase, which is indeed observed by the increase in lipidated LC3-II levels (LC3-II normalized versus actin) under these conditions (Fig. 4.1) (see Note F).

Autophagic flux is quantified from the ratio of LC3-II levels in the presence or absence of lysosomal inhibitors. Another method of monitoring autophagic activity is to use two different time points of inhibition. By using two time-points, the turnover of the LC3-II protein can be observed over time. This method is useful to determine the rate of autophagosome biosynthesis over time (Rubinsztein et al., 2009).

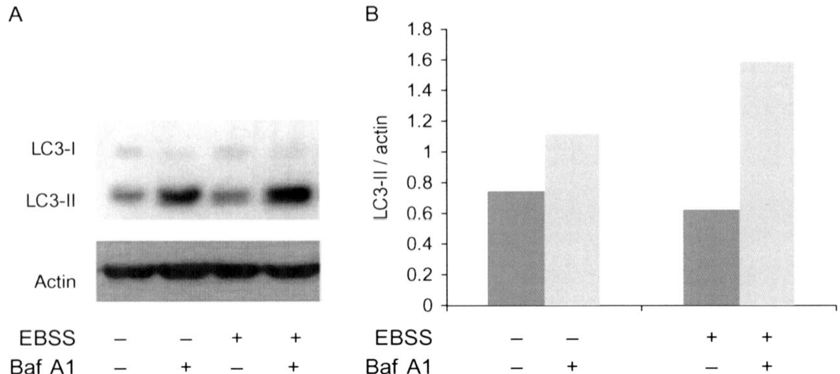

Figure 4.1 *LC3-II turnover during autophagy.* LC3 expression levels in response to 2 h of EBSS treatment, with or without Bafilomycin A1 (100 n*M*). HeLa cells were cultured in DMEM for 24 h and then starved for 2 h in EBSS medium with or without the lysosomal inhibitor Bafilomycin A1 (Baf A1) in the medium. (A) Cells were lysed and proteins resolved by SDS-PAGE. LC3 was detected by immunoblotting. (B) LC3-II expression is normalized versus actin expression. Increased levels of LC3-II lipidation are observed in response to starvation, in the presence of Baf A1, which shows the induction of autophagosome synthesis in response to EBSS treatment.

2.1.3 Notes and potential pitfalls

A. HBSS is used when the cells are incubated in a humidified chamber at 37 °C in the absence of CO_2; otherwise EBSS is used. The reason is that EBSS, unlike HBSS, contains bicarbonate, and the pH of the medium is governed by the Henderson–Hasselbalch equation: $\mathrm{pH} = \mathrm{p}K_a + \log[\mathrm{HCO_3}^-]/[\mathrm{CO_2}] = 6.1 + \log[\mathrm{HCO_3}^-]/[\mathrm{CO_2}]$.

B. A necessary precaution to ensure the correct stimulus: during the experiment, it is important to wash the cells with preheated medium containing the desired stimulus (EBSS, HBSS, or other) three times, incubating for at least 2 min during each wash in order to induce autophagic activation.

C. Bafilomycin A1, a specific inhibitor of vacuolar type H^+-ATPase (V-ATPase), inhibits the acidification of lysosomes and therefore blocks the last step of autophagy, that is, the fusion of the autophagosome with a lysosome, leading to an accumulation of early autophagosome vacuoles (Yamamoto et al., 1998). The time window for the incubation of cells with lysosomal inhibitors is critical and depends on the cell type involved. It is advisable to use a saturating concentration of Bafilomycin A1 (Klionsky, Elazar, Seglen, & Rubinsztein, 2008). The inhibitor treatment time is important for monitoring the blockade of the autophagic flux because at early time-points (2–4 h) the drug seems to have a greater blocking effect. However, for treatments lasting more than 6 h, impairment of the fusion of autophagosomes with late endosomes and lysosomes is observed (Klionsky et al., 2008). It is advisable to use two different time-points for the inhibitor treatment in order to observe the change in autophagic flux. Hydroxychloroquine can be used for longer incubations if needed.

D. Some treatments such as cell death inducers can lead to unequal loading with Laemmli buffer. In this step, other lysis buffers, such as RIPA buffer, can be used on the samples in order to quantify protein concentration prior to Western blot experiment. Collect the cells in cold PBS on ice in microcentrifuge tubes, centrifuge, and discard the supernatant. Add 50 μl of RIPA buffer to lyse the cells, centrifuge, and collect the supernatant. On the day of the experiment, add 50 μg of protein with sample buffer and boil the samples at 100 °C for 5 min. The protein concentration can be measured using a standard protein assay kit (Pierce BCA Protein Assay Kit #23335).

E. LC3-II is more sensitive to repeated freeze–thaw cycles of the samples, so it is advisable to use fresh samples in order to avoid any degradation of the protein. It is known that a third band of LC3 can be observed in some experiments after the repeated use of lysates (Klionsky et al., 2012). It is therefore advisable to use fresh cell lysates to analyze LC3 turnover.

F. LC3 turnover is cell type and stress dependent. It should be noted that the turnover of LC3 protein from LC3-I to LC3-II can also be used as a marker for autophagosome formation by determining the LC3-II/LC3-I ratio; however, it is preferable to avoid this determination. Depending on the tissue or cell lines used, the abundance of LC3-I may change. The choice of LC3 antibody is also important, and the detection of the two forms of the LC3 protein may change depending on the affinity of the antibody. This means that it is preferable to quantify the turnover of LC3 from the ratio of the intensity of LC3-II over a housekeeping gene rather from than LC3-II/LC3-I (Klionsky et al., 2012).

2.2. Assaying the autophagic flux in cultured cell lines using the tandem probe

An mRFP-GFP tandem fluorescent-tagged LC3 has been specially designed to monitor the autophagic flux. This method is based on differential quenching of the fluorescence emitted by the red and green probes in the

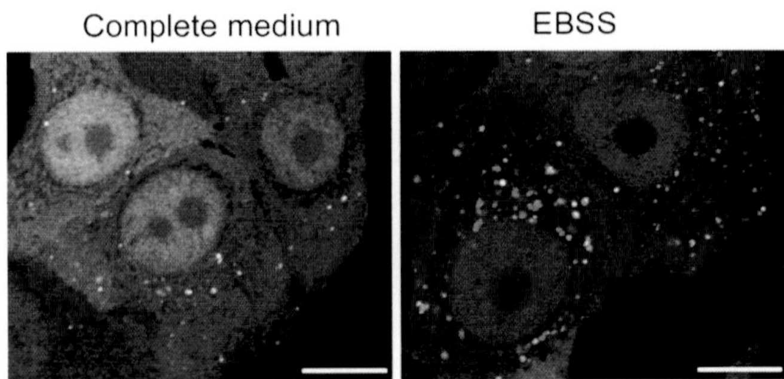

Figure 4.2 *Monitoring the autophagic flux by the tandem mRFP-GFP LC3 probe.* HeLa cells stably expressing mRFP-GFP LC3 were or were not subjected to starvation (EBSS) for 2 hours, fixed, and then examined by confocal microscopy. Bar indicates 10 μm. (See the color plate.)

lysosomal acidic compartment (for details, see Kimura, Noda, & Yoshimori, 2007). Briefly, autophagosomes appear yellow (GFP+RFP+) and autolysosomes red (GFP−RFP+), because of the quenching of the GFP-LC3 signal at an acidic pH (Fig. 4.2). An alternative method to monitor autophagic flux by confocal microscopy is to quantify lysosomal proteins in autophagolysosomes (see Note A).

2.2.1 Materials and reagents

1. DMEM (4.5 g/l glucose, glutamax) (Invitrogen, Life Technologies, ref. 31966) supplemented with 10% fetal bovine serum (Invitrogen, Life Technologies).
2. PBS: 137 mM NaCl, 2.7 mM KCl, 4.3 mM Na_2HPO_4, 1.4 mM KH_2PO_4, pH 7.3.
3. HBSS (Invitrogen, Life Technologies, ref. 14025) or EBSS (Invitrogen, Life Technologies, ref. 24010-043).
4. 3-Methyladenine (3-MA, Sigma-Aldrich, ref. M9281).
5. Paraformaldehyde 32% (PAF, Electron Microscopy Science, ref. 15714-S).
6. Dako fluorescent mounting medium (Dako, ref. S3023).

2.2.2 Protocol

The protocol described below was originally validated for use in HeLa cells (Kimura, Fujita, Noda, & Yoshimori, 2009) and can be optimized as appropriate for the cell system used. Stable or transiently transfected mRFP-GFP LC3 cells can be used (the plasmid expressing mRFP-GFP LC3 is available on addgene) (see Notes B–). This protocol is designed for stably transfected mRFP-GFP LC3 HeLa cells. The day before the experiment, the cells are seeded in six-well plates with a cover slip (previously sterilized) and used near confluence (usually the next day):

1. On the day of the experiment, the cells are washed three times with HBSS (or EBSS) (see Note E) and then incubated for at least 2 h (for some cell lines, this period can be extended to 4 or 6 h) with fresh HBSS (or EBSS). The negative control consists of cells incubated with full medium for the same periods of time. Throughout the period of autophagy stimulation, 3-MA can be added at a final concentration of 10 m*M* to inhibit *de novo* formation of autophagic vacuoles (Seglen & Gordon, 1982; also see Note F).
2. Next, remove the medium by aspiration, wash the cells three times with PBS, fixed with 4% PAF (previously diluted from 32% PAF in PBS) for 10 min at room temperature, and then wash again with PBS.
3. To quench the fluorescence signal from free aldehyde groups, incubate the cells with NH_4Cl 50 m*M* in PBS for 10 min at room temperature, and then wash again with PBS.
4. After mounting cover slips with mounting medium on microscope slides, pictures are obtained using a confocal microscope (see Note G and Fig. 4.2); count the yellow (GFP+RFP+) and red (GFP−RFP+) dots using appropriate computer software, such as ImageJ.

2.2.3 Notes and potential pitfalls

A. The quantification of the colocalization between a lysosomal protein such as LAMP2 and LC3 could be used via a standard immunofluorescence protocol (fixation using 4% PFA (10 min) and permeabilization using 0.1% Triton X-100 (5 min)) as an alternative method to monitor autophagic flux. The primary antibodies used for this alternative method are from MBL (PM036) and from Santa Cruz (clone H4B4) for LC3 and LAMP2, respectively.

B. It has been reported that weak fluorescence of EGFP is still present in acidic lysosomes (pH between 4 and 5), leading to some yellow signals (GFP+RFP+) from autolysosomes, which can result in misinterpretation of the autophagic flux results. It is therefore preferable to choose a monomeric green fluorescent protein that is more acid sensitive than EGFP, such as the mTagRFP-mWasabi-LC3 reporter (Zhou et al., 2012), in which mWasabi is more acid sensitive than EGFP.

C. Another necessary precaution in interpreting the tandem fluorescent marker is that colocalization of GFP and mRFP can also be seen in the context of impaired proteolytic degradation within autolysosomes or altered lysosomal pH. The accumulation of yellow dots does not therefore always mean that the fusion of the lysosome with the autophagosome has been inhibited.

D. A limitation on the use of the tandem fluorescent marker is that green dots can also be observed (GFP+RFP−). If they are, this is due to unfolding of the RFP proteins. These green puncta should not be taken into consideration.

E. HBSS is used when the cells are incubated in a humidified chamber at 37 °C in the absence of CO_2; otherwise, EBSS is used. The reason is that EBSS, unlike HBSS, contains bicarbonate, and the pH of the medium is governed by the Henderson–Hasselbalch equation: $pH = pK_a + \log[HCO_3^-]/[CO_2] = 6.1 + \log[HCO_3^-]/[CO_2]$.

F. 3-MA blocks autophagy by inhibiting class-III phosphatidylinositol 3-kinase (Petiot, Ogier-Denis, Blommaart, Meijer, & Codogno, 2000). However, it should be kept in mind that 3-MA is a phosphatidylinositol 3-kinase inhibitor (Blommaart, Krause, Schellens, Vreeling-Sindelarova, & Meijer, 1997), which interferes with other intracellular trafficking pathways that are dependent on phosphatidylinositol 3-kinase. 3-MA also affects some other intracellular events (Tolkovsky, Xue, Fletcher, & Borutaite, 2002). It is a good idea to prepare a more concentrated stock solution, for example, 50 m*M* in water, culture medium, or balanced salt solution. It may be necessary to heat the solution under a warm water faucet to dissolve such a high concentration of 3-MA. Once prepared, the solution can be stored frozen, but the 3-MA may precipitate on thawing.

G. The quantification of "yellow" and "red-only" dots can be automated using an automatic microscope (such as Cellomics from Thermoscientific) that can be used to assess a huge population of cells (1000 or more) and provide more accurate data SQSTM1/p62 than can be achieved by manual assessment of a few selected cells.

3. AUTOPHAGIC FLUX AND DEGRADATION

SQSTM1/p62 is another protein marker that can be used to monitor autophagic flux. The interaction of SQSTM1/p62 with LC3 via the LIR domain means that the degradation of SQSTM1/p62 is specific to autophagy. Increased levels of SQSTM1/p62 are associated with the inhibition of autophagy, whereas decreased SQSTM1/p62 levels are associated with the activation of autophagy. The expression and localization of endogenous SQSTM1/p62 or ectopically expressed SQSTM1/p62 can also be observed using immunofluorescence with commercially available antibodies (for details see Bjorkoy et al., 2009 and see Notes A–C).

3.1. SQSTM1/p62 Western blot

3.1.1 Materials and reagents

Materials and reagents used to perform SQSTM1/p62 Western blot are identical to the materials and reagents used to perform LC3 Western blot (cf. Section 2.1.1). The SQSTM1/p62 antibody used is the BD science ref. 610830.

3.1.2 Protocol

The protocol for SQSTM1/p62 Western blot is identical to the protocol for LC3 Western blot (cf. Section 2.1.2) with the exception of the way to prepare the SDS-PAGE. We recommend to use 8% or 10% SDS-PAGE for high molecular weigh proteins such as SQSTM1/p62. In response to autophagic stimuli, such as starvation, SQSTM1/p62 is degraded as a result of the activation of autophagy, which is observed at decreased levels of the protein (Fig. 4.3).

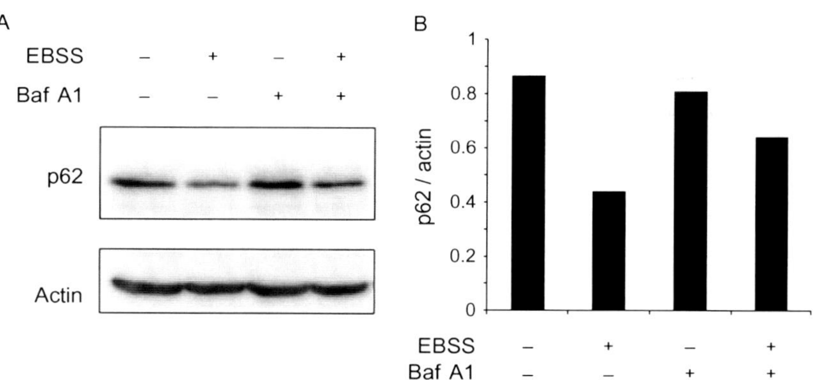

Figure 4.3 *SQSTM1/p62 degradation during autophagy.* SQSTM/p62 expression levels in response to 2 h of EBSS treatment, with or without Bafilomycin A1 (100 n*M*). HeLa cells were cultured in DMEM for 24 h and then starved for 2 h in EBSS medium with or without the lysosomal inhibitor Bafilomycin A1 (Baf A1) in the medium. (A) Cells were lysed and proteins resolved by SDS-PAGE. SQSTM1/p62 was detected by immunoblotting. (B) SQSTM1/p62 levels are normalized versus actin expression. Decreasing levels of the SQSTM1/p62 protein are observed in response to starvation in the presence of Baf A1. This decrease indicates the total degradation of SQSTM1/p62 by autophagolysosomes when autophagy is activated.

3.1.3 Notes and potential pitfalls

Most of the notes and pitfalls for LC3 Western blot can be applied for SQSTM1/p62 Western blot also (cf. Section 2.1.3); however, there are additional specific notes to look out for this assay:

A. Transcriptional activation of SQSTM1/p62 may result in increased accumulation of SQSTM1/p62, which can be misleading for interpreting autophagy. Transcriptional regulation of SQSTM1/p62 can be misleading for interpreting autophagic activity depending on protein levels. SQSTM1/p62 can be upregulated under different conditions, thus increasing the Western blot levels independently of any autophagic degradation. In some cases, autophagic stimuli can induce the expression of the SQSTM1/p62 gene and protein, as in the case of muscle atrophy induced by cancer. One way to avoid misinterpretations is to use a stable cell line expression EGFP-tagged SQSTM1/p62 under the control of an inducible promoter to monitor SQSTM1/p62 degradation. Other solutions are suggested in detail in the guidelines for monitoring autophagy (Klionsky et al., 2012), including a radioactive pulse-chase measurement (Bjorkoy et al., 2009).

B. SQSTM1/p62 degradation is a good readout for monitoring autophagic activity; however, because of the context- and stimulus-dependent behavior of the protein, it is recommended that an SQSTM1/p62 assay be used with different autophagy methods, as well as different SQSTM1/p62 measurements, such as pulse chase or immunofluorescence (for details see Bjorkoy et al., 2009).

C. SQSTM1/p62 aggregates are insoluble in NP-40 or Triton X-100 solutions, and therefore cell lysates prepared with lysis buffers containing these detergents may show lower SQSTM1/p62 levels in Western blots. In order to monitor the correct level of SQSTM1/p62 expression, it is recommended to avoid using lysis buffers containing these detergents, but rather to use one that contain SDS, as described in the protocol (Lim et al., 2011).

3.2. Proteolysis

3.2.1 Materials and reagents

1. DMEM (4.5 g/l glucose, glutamax) (Invitrogen, Life Technologies, ref. 31966) supplemented with 10% fetal bovine serum (Invitrogen, Life Technologies).

2. PBS: 137 mM NaCl, 2.7 mM KCl, 4.3 mM Na_2HPO_4, 1.4 mM KH_2PO_4, pH 7.3.
3. HBSS (Invitrogen, Life Technologies, ref. 14025) or EBSS (Invitrogen, Life Technologies, ref. 24010-043).
4. L-[U-^{14}C] valine (266 mCi/mmol, 9.84 GBq/mmol, Amersham Biosciences, ref. CFB75).
5. "Cold" valine (Merck, ref. 1.08495.0100).
6. Trichloroacetic acid (TCA) (Sigma-Aldrich, ref. T4885).
7. 0.2 M NaOH.
8. 3-MA (Sigma-Aldrich, ref. M9281).

3.2.2 Protocol

This protocol described below was originally validated for use with various different cancer cells (see Bauvy, Meijer, & Codogno, 2009). Cells are seeded in the usual medium in six-well plates and used near confluence:

1. Intracellular proteins are labeled for 18 h at 37 °C with 0.2 μCi/ml of L-[U-^{14}C] valine (spec. activity: 266 mCi/mmol) in complete medium (Pulse).
2. Any unincorporated radioactivity is eliminated by rinsing the cells three times with PBS.
3. The cells are then incubated with fresh complete medium containing 10 mM cold valine (see Note A) for 1 h to degrade short-lived proteins (in some cell lines, this period can be extended to 24 h).
4. After this period, the medium is removed and replaced with the appropriate fresh complete medium supplemented with 10 mM cold valine, and incubated for a further 4 h (longer incubations in the chase medium are also possible, and in some cases it may be desirable to use multiple points). Throughout the chase period, 3-MA (see Note B) can be added at a final concentration of 10 mM to inhibit the *de novo* formation of autophagic vacuoles (Seglen & Gordon, 1982). To stimulate autophagy HBSS (or EBSS) (see Note C) containing 10 mM cold valine and 0.1% of bovine serum albumin can be added.
5. The medium is then precipitated overnight after adding TCA at a final concentration of 10%.
6. After centrifuging the culture medium for 10 min at 470 × g at 4 °C, the acid-soluble radioactivity is measured by liquid scintillation counting.
7. The cells are washed twice with cold TCA (w/v), plus 10 mM cold valine to make sure that no radioactivity has remained adsorbed to the

denatured proteins. The cell pellet is then dissolved at 37 °C in 0.2 *M* NaOH for 2 h.

8. The radioactivity is then measured by liquid scintillation counting. The rate of degradation of longer-lived proteins is calculated from the ratio of the acid-soluble radioactivity in the medium to that in the acid-precipitable cell fraction.

3.2.3 Notes and potential pitfalls

A. Valine (amino acids) is used because this amino acid does not interfere with autophagy in most cell types. Since amino acids such as leucine are physiological inhibitors of autophagy (Meijer & Dubbelhuis, 2004), the choice of the amino acid used during the pulse chase is important.

B. 3-MA blocks autophagy by inhibiting class-III phosphatidylinositol 3-kinase (Petiot et al., 2000). However, it should be kept in mind that 3-MA is a phosphatidylinositol 3-kinase inhibitor (Blommaart et al., 1997), which also interferes with other intracellular trafficking pathways dependent on the phosphatidylinositol 3-kinases. 3-MA also affects some other intracellular events (Tolkovsky et al., 2002). 3-MA is routinely used at a final concentration of 10 m*M*. It is a good idea to prepare a stronger stock solution, for example,100 m*M* in water. It may be necessary to heat the solution under a warm water faucet to dissolve such a high concentration of 3-MA.The prepared solution can be stored frozen, but the 3-MA may precipitate on thawing.

C. HBSS is used when the cells are incubated in a humidified chamber at 37 °C in the absence of CO_2; otherwise, EBSS is used. The reason is that EBSS, unlike HBSS, contains bicarbonate, and the pH of the medium is governed by the Henderson–Hasselbalch equation: $pH = pK_a + \log[HCO_3^-]/[CO_2] = 6.1 + \log[HCO_3^-]/[CO_2]$.

4. CONCLUSION

The assays presented here have been widely used. However, the gold standard assay remains the detection of autophagic structures (autophagosomes, amphisomes, and autolysosomes) by electron microscopy (Eskelinen, Reggiori, Baba, Kovacs, & Seglen, 2011). Other methods for measuring autophagic flux have also been developed, such as fluorescence-activated cell sorter-based methods (Shvets, Fass, & Elazar,

2008). Researchers should perform several autophagic assays before reaching any final conclusions about whether the initiation of autophagy or/and autophagy flux is affected under their experimental conditions.

ACKNOWLEDGMENTS

N. D. is recipient of a Fondation pour la Recherche Médicale (FRM) fellowship. Studies in P. C.'s laboratory are supported by institutional funding from INSERM, University Paris Descartes, and grants from ANR and INCa.

REFERENCES

Bauvy, C., Meijer, A. J., & Codogno, P. (2009). Assaying of autophagic protein degradation. *Methods in Enzymology*, *452*, 47–61.

Birgisdottir, A. B., Lamark, T., & Johansen, T. (2013). The LIR motif—Crucial for selective autophagy. *Journal of Cell Science*, *126*(Pt 15), 3237–3247.

Bjorkoy, G., Lamark, T., Pankiv, S., Overvatn, A., Brech, A., & Johansen, T. (2009). Monitoring autophagic degradation of p62/SQSTM1. *Methods in Enzymology*, *452*, 181–197.

Blommaart, E. F., Krause, U., Schellens, J. P., Vreeling-Sindelarova, H., & Meijer, A. J. (1997). The phosphatidylinositol 3-kinase inhibitors wortmannin and LY294002 inhibit autophagy in isolated rat hepatocytes. *European Journal of Biochemistry*, *243*(1–2), 240–246.

Boya, P., Reggiori, F., & Codogno, P. (2013). Emerging regulation and functions of autophagy. *Nature Cell Biology*, *15*(7), 713–720.

Eskelinen, E. L., Reggiori, F., Baba, M., Kovacs, A. L., & Seglen, P. O. (2011). Seeing is believing: The impact of electron microscopy on autophagy research. *Autophagy*, 7(9), 935–956.

Kimura, S., Fujita, N., Noda, T., & Yoshimori, T. (2009). Monitoring autophagy in mammalian cultured cells through the dynamics of LC3. *Methods in Enzymology*, *452*, 1–12.

Kimura, S., Noda, T., & Yoshimori, T. (2007). Dissection of the autophagosome maturation process by a novel reporter protein, tandem fluorescent-tagged LC3. *Autophagy*, *3*(5), 452–460.

Klionsky, D. J., Abdalla, F. C., Abeliovich, H., Abraham, R. T., Acevedo-Arozena, A., Adeli, K., et al. (2012). Guidelines for the use and interpretation of assays for monitoring autophagy. *Autophagy*, *8*(4), 445–544.

Klionsky, D. J., Elazar, Z., Seglen, P. O., & Rubinsztein, D. C. (2008). Does bafilomycin A1 block the fusion of autophagosomes with lysosomes? *Autophagy*, *4*(7), 849–950.

Lamb, C. A., Yoshimori, T., & Tooze, S. A. (2013). The autophagosome: Origins unknown, biogenesis complex. *Nature Reviews. Molecular Cell Biology*, *14*(12), 759–774.

Lim, J., Kim, H. W., Youdim, M. B., Rhyu, I. J., Choe, K. M., & Oh, Y. J. (2011). Binding preference of p62 towards LC3-ll during dopaminergic neurotoxin-induced impairment of autophagic flux. *Autophagy*, 7(1), 51–60.

Liu, E. Y., & Ryan, K. M. (2012). Autophagy and cancer—Issues we need to digest. *Journal of Cell Science*, *125*(Pt 10), 2349–2358.

Lorin, S., Hamaï, A., Mehrpour, M., & Codogno, P. (2013). Autophagy regulation and its role in cancer. *Seminars in Cancer Biology*, *23*(5), 361–379.

Meijer, A. J., & Dubbelhuis, P. F. (2004). Amino acid signalling and the integration of metabolism. *Biochemical and Biophysical Research Communications*, *313*(2), 397–403.

Mizushima, N., & Komatsu, M. (2011). Autophagy: Renovation of cells and tissues. *Cell*, *147*(4), 728–741.

Mizushima, N., Yoshimori, T., & Levine, B. (2010). Methods in mammalian autophagy research. *Cell*, *140*(3), 313–326.

Mizushima, N., Yoshimori, T., & Ohsumi, Y. (2011). The role of Atg proteins in autophagosome formation. *Annual Review of Cell and Developmental Biology*, *27*, 107–132.

Petiot, A., Ogier-Denis, E., Blommaart, E. F., Meijer, A. J., & Codogno, P. (2000). Distinct classes of phosphatidylinositol 3′-kinases are involved in signaling pathways that control macroautophagy in HT-29 cells. *Journal of Biological Chemistry*, *275*(2), 992–998.

Ravikumar, B., Sarkar, S., Davies, J. E., Futter, M., Garcia-Arencibia, M., Green-Thompson, Z. W., et al. (2010). Regulation of mammalian autophagy in physiology and pathophysiology. *Physiological Reviews*, *90*(4), 1383–1435.

Rubinsztein, D. C., Codogno, P., & Levine, B. (2012). Autophagy modulation as a potential therapeutic target for diverse diseases. *Nature Reviews. Drug Discovery*, *11*(9), 709–730.

Rubinsztein, D. C., Cuervo, A. M., Ravikumar, B., Sarkar, S., Korolchuk, V., Kaushik, S., et al. (2009). In search of an "autophagomometer". *Autophagy*, *5*(5), 585–589.

Seglen, P. O., & Gordon, P. B. (1982). 3-Methyladenine: Specific inhibitor of autophagic/lysosomal protein degradation in isolated rat hepatocytes. *Proceedings of the National Academy of Sciences of the United States of America*, *79*(6), 1889–1892.

Shvets, E., Fass, E., & Elazar, Z. (2008). Utilizing flow cytometry to monitor autophagy in living mammalian cells. *Autophagy*, *4*(5), 621–628.

Stromhaug, P. E., & Seglen, P. O. (1993). Evidence for acidity of prelysosomal autophagic/endocytic vacuoles (amphisomes). *Biochemical Journal*, *291*(Pt 1), 115–121.

Tanida, I., Minematsu-Ikeguchi, N., Ueno, T., & Kominami, E. (2005). Lysosomal turnover, but not a cellular level, of endogenous LC3 is a marker for autophagy. *Autophagy*, *1*(2), 84–91.

Tolkovsky, A. M., Xue, L., Fletcher, G. C., & Borutaite, V. (2002). Mitochondrial disappearance from cells: A clue to the role of autophagy in programmed cell death and disease? *Biochimie*, *84*(2–3), 233–240.

Weidberg, H., Shvets, E., Shpilka, T., Shimron, F., Shinder, V., & Elazar, Z. (2010). LC3 and GATE-16/GABARAP subfamilies are both essential yet act differently in autophagosome biogenesis. *EMBO Journal*, *29*(11), 1792–1802.

White, E. (2012). Deconvoluting the context-dependent role for autophagy in cancer. *Nature Reviews Cancer*, *12*(6), 401–410.

Yamamoto, A., Tagawa, Y., Yoshimori, T., Moriyama, Y., Masaki, R., & Tashiro, Y. (1998). Bafilomycin A1 prevents maturation of autophagic vacuoles by inhibiting fusion between autophagosomes and lysosomes in rat hepatoma cell line, H-4-II-E cells. *Cell Structure and Function*, *23*(1), 33–42.

Yang, Z., & Klionsky, D. J. (2010). Eaten alive: A history of macroautophagy. *Nature Cell Biology*, *12*(9), 814–822.

Zhou, C., Zhong, W., Zhou, J., Sheng, F., Fang, Z., Wei, Y., et al. (2012). Monitoring autophagic flux by an improved tandem fluorescent-tagged LC3 (mTagRFP-mWasabi-LC3) reveals that high-dose rapamycin impairs autophagic flux in cancer cells. *Autophagy*, *8*(8), 1215–1226.

CHAPTER FIVE

Methods to Assess Autophagy *In Situ*—Transmission Electron Microscopy Versus Immunohistochemistry

Wim Martinet[*,1], **Jean-Pierre Timmermans**[†], **Guido R. Y. De Meyer**[*]
[*]Laboratory of Physiopharmacology, University of Antwerp, Antwerp, Belgium
[†]Laboratory of Cell Biology and Histology, University of Antwerp, Antwerp, Belgium
[1]Corresponding author: e-mail address: wim.martinet@uantwerpen.be

Contents

Abstract

Autophagy is a well-conserved lysosomal degradation pathway that plays a major role in both oncogenesis and tumor progression. Transmission electron microscopy (TEM) as well as immunohistochemistry are indispensable tools for the evaluation of autophagy *in situ*. Here, we describe an optimized protocol for the study of autophagic vacuoles by TEM and elaborate on the immunohistochemical detection of microtubule-associated protein 1 light chain (MAP1LC3, best known as LC3), which is currently considered as one of the most reliable markers of the autophagic process. The advantages, potential pitfalls, and limitations of these methods, as well as their value in the field of autophagy and oncometabolism research are discussed. Overall, we recommend a combined use of different techniques including TEM, immunohistochemistry, and molecular approaches (such as immunoblotting) for the unambiguous detection of autophagy in malignant as well as in normal tissues.

Methods in Enzymology, Volume 543
ISSN 0076-6879
http://dx.doi.org/10.1016/B978-0-12-801329-8.00005-2

1. INTRODUCTION

Autophagy is a subcellular process for bulk destruction of cytoplasmic components in lysosomes (Mizushima & Komatsu, 2011). Specific autophagy-related proteins (Atg-proteins) are essential for this process and are responsible for the formation of autophagosomes, which after fusion with a lysosome degrade the cellular content. Autophagy occurs at basal levels in most tissues, but is activated by environmental stimuli or stress-related signals. The autophagic process maintains the balance between biogenesis and production of cell organelles and destroys unwanted or damaged intracellular structures. In this way, cell survival is promoted in an unfavorable environment.

Growing evidence reveals that autophagy is involved in the prevention of a wide range of human pathological conditions including heart and liver disease, neurodegeneration, and gastrointestinal disorders (Choi, Ryter, & Levine, 2013). Furthermore, activation of autophagy could be an effective way of eliminating infectious agents that access the cytosol either directly through the plasma membrane or after being internalized in phagosomes. Also noteworthy is that autophagy plays an essential role in embryogenesis, aging, and lipid metabolism. Recent evidence has revealed that basal autophagy can be intensified by specific drugs (Rubinsztein, Codogno, & Levine, 2012), suggesting that the autophagic machinery can be manipulated to treat human disease.

In cancer, autophagy may exert a dual role depending on the stage of the disease (Choi et al., 2013; White, 2012). Tumor development is suppressed by autophagy in primary cells by safeguarding against metabolic stress through the homeostatic turnover of mitochondria and the clearance of protein aggregates. Defects in autophagy result in the reduced removal of potential sources of genotoxic stress such as reactive oxygen species (ROS) from leaky damaged mitochondria or other organelles (White, 2012). Accumulation of ROS may be an important source of DNA damage, genetic instability, and tumor growth. Recently, mutations in many signaling molecules involved in the PI3K/Akt/mTOR pathway have been identified in cancer cells. These mutations may result in the constitutive activation of mTOR signaling and consequently in the inhibition of autophagy. In established tumors, however, autophagy is frequently upregulated, particularly in regions of the tumor with increased metabolic stress (White, 2012). Therefore, autophagy should also be considered as a survival pathway of cancer cells under stressful conditions.

Despite the development of different methodologies and guidelines for autophagy detection both *in vitro* and *in vivo* (Klionsky et al., 2012), demonstration of autophagy in tissue has not received extensive evaluation. In this chapter, we provide protocols for the *in situ* detection of autophagy via transmission electron microscopy (TEM) or immunohistochemistry. Given that both techniques have their advantages and disadvantages, the pros and cons of each approach and their value in autophagy research will be discussed.

2. TRANSMISSION ELECTRON MICROSCOPY

2.1. Methodology

This protocol is suitable for epoxy embedding of tissue samples and generally allows good identification of both membranes and cytoplasmic content of autophagosomes via TEM. Previously, similar TEM protocols have been reported to analyze autophagy (Swanlund, Kregel, & Oberley, 2010; Yla-Anttila, Vihinen, Jokitalo, & Eskelinen, 2009) and it might be of interest to consult these reports as well to gain full insight in the tips and tricks of electron microscopy.

2.1.1 Preparation of reagents and buffer solutions

1. Fixative and wash buffer: First, a stock solution containing 0.2 *M* sodium cacodylate and 0.1% $CaCl_2 \cdot 2H_2O$ is prepared by dissolving 4.28 g sodium cacodylate and 0.1 g $CaCl_2 \cdot 2H_2O$ in 80 ml ultrapure water. Adjust with HCl to pH 7.3–7.4 and make up to 100 ml with ultrapure water. To prepare TEM fixative, 10 ml glutaraldehyde 25% (electron microscopy grade, such as Merck Cat. No. 1.04239) is added to 50 ml stock solution. The mixture is adjusted with HCl to pH 7.4 and made up to 100 ml with ultrapure water. Wash buffer is prepared by adding 50 ml stock solution to 50 ml water containing 7.5 g sucrose. Fixative should always be prepared as freshly as possible. Wash buffer can be stored for several weeks at 4 °C.
2. Veronal acetate (Michaelis) buffer 0.033 *M*: A veronal acetate stock solution is prepared by dissolving 9.71 g sodium acetate trihydrate and 14.71 g sodium 5,5-diethylbarbiturate in 500 ml ultrapure water. The veronal acetate stock solution can be stored for several weeks at 4 °C. Subsequently, 6 g sucrose is dissolved in a small volume of water. After

the addition of 20 ml veronal acetate stock, the pH is adjusted to 7.2–7.4 using acetic acid. The solution is made up to 100 ml with ultrapure water, yielding 6% sucrose in 0.05 M veronal acetate buffer. Finally, 2 volumes of the latter buffer, 0.5 volume 6% OsO_4, and 0.5 volume water are mixed to obtain 0.033 *M* veronal acetate buffer containing 4% sucrose.

3. Embedding resin mixture: Blend 5 ml EMbed 812 resin (Electron Microscopy Sciences, Cat. No. 14900) with 4 ml dodecyl succinic anhydride (DDSA; Electron Microscopy Sciences, Cat. No. 13710) and 2 ml nadic methyl anhydride (NMA; Electron Microscopy Sciences, Cat. No. 19000). Prior to mixing, the resin and the anhydrides should be warmed (50 °C) to reduce their viscosity.
4. Embedding resin mixture containing accelerator: Mix 20 ml EMbed 812 with 16 ml DDSA, 8 ml NMA, and 0.77 ml 2,4,6-Tris (dimethylaminomethyl)phenol (DMP-30; Electron Microscopy Sciences, Cat. No. 13600). Different formulations of the resin mixture may be used depending on the desired hardness of the resin block to allow various sectioning conditions. The mixture with accelerator as described above will yield a block of medium hardness. Blocks can be made harder by decreasing the proportion of DDSA and increasing the amount of NMA.
5. Reynolds solution: First, slightly heat (and stir) 17.6% sodium citrate. Slowly add an equal volume of 13.3% lead nitrate and leave for 30 min at room temperature. Adjust to pH 10.5 with sodium hydroxide Titrisol®. This stock solution can be stored at room temperature for several weeks. Prior to use, adjust to pH 12.4 with sodium hydroxide Titrisol®. This pH is critical since pH >12.45 will lead to precipitation, whereas pH <12.35 will give poor staining. Reynolds solution should be free of carbon dioxide to prevent formation of $PbCO_3$ deposits. Therefore, always use carbon dioxide-free water and sodium hydroxide (Titrisol®). Carbon dioxide-free water is prepared by boiling ultrapure water. After transfer of the boiling water in a separating funnel, a carbon dioxide trap filled with soda lime is connected to the outlet on top of the separating funnel to prevent re-infiltration of fresh carbon dioxide.

2.1.2 General protocol

1. Isolate a small piece of tissue (approximately 1 mm^3) and add 4 ml 0.1 *M* sodium cacodylate-buffered (pH 7.4) 2.5% glutaraldehyde solution. Allow fixation for 2 h at 4 °C.

2. Remove fixative and rinse tissue (3 × 10 min) with 5 ml 0.1 *M* sodium cacodylate-buffered (pH 7.4) 7.5% sucrose. Samples can be stored in this buffer at 4 °C for several weeks.
3. Remove cacodylate buffer and add 3–4 ml 1% osmium tetroxide (OsO_4) in 0.033 *M* veronal acetate buffer containing 4% sucrose. Allow postfixation for 2 h at 4 °C. Please note that OsO_4 is extremely toxic and should be handled with great care.
4. Remove OsO_4 solution and rinse tissue (3 × 10 min) with 5 ml 0.05 *M* veronal acetate buffer containing 6% sucrose.
5. Dehydrate cell pellet in an ethanol gradient as follows: 50% ethanol (15 min), 70% ethanol (15 min), 90% ethanol (15 min), 96% ethanol (15 min), and 100% ethanol (4 × 20 min).
6. Treat the tissue sample at room temperature with ethanol 100%/EMbed 812 mixture without DMP-30 (1:1) for 2 h, and with ethanol 100%/EMbed 812 mixture without DMP-30 (2:3) overnight. It should be noted that some tissues might need a more powerful solvent (e.g., propylene oxide, acetone) than ethanol for penetration of the embedding medium in the tissue. In that case, we suggest to dehydrate the tissue first with an ethanol gradient (as described in point 5), followed by treatment with 100% propylene oxide (2 × 15 min) and propylene 100%/EMBed 812 mixture (as described above).
7. Treat the tissue sample at room temperature with EMbed 812 resin mixture (without DMP-30) for 2 × 2 h, followed by EMbed 812 resin mixture (containing DMP-30) for 1 h.
8. Transfer tissue into a gelatin capsule, cover with EMbed 812 resin mixture containing DMP-30, and allow polymerization at 60 °C for approximately 36 h.
9. Cut ultrathin sections (±50 nm thick) with an ultramicrotome using a standard diamond knife (Element six).
10. Capture and air-dry sections on 200 mesh copper grids (Gilder, Cat. No. G200).
11. Stain sections with 2% uranyl acetate (in water) for 15 min in the dark. Rinse sections with ultrapure water.
12. Stain sections with Reynolds solution pH 12.4 for 10 min. Rinse sections with 0.05 *M* NaOH Titrisol® and CO_2-free ultrapure water.
13. View samples with a TEM at an accelerating voltage of 80–120 kV.
14. Optional: Perform quantitative morphometric analysis of electron micrographs using previously published protocols (Swanlund et al., 2010; Yla-Anttila et al., 2009).

2.2. TEM as a tool to study autophagy in tissue

Autophagy was first discovered by TEM in the mid-1950s as a result of the pioneering work of several excellent electron microscopists (Eskelinen, Reggiori, Baba, Kovacs, & Seglen, 2011). However, our structural-mechanistic understanding of autophagy only started in the 1960s with the discovery of the lysosome by the late Belgian Nobel prizewinner Christian de Duve, who introduced the terms *autophagy* and *autophagic vacuoles*. He assumed that these vacuolar structures do not necessarily form as a protective response to cytoplasmic damage, and instead argued that they might be normal constituents of healthy cells sequestering and degrading cytoplasm in response to the cell's metabolic demands. Because the presence of autophagic vacuoles is by far the most important morphological feature of cells with autophagic activity, demonstration of these structures by TEM remains the gold standard—even in recent years—to assess autophagy in tissue. It offers high-resolving power (resolution theoretically <0.1 nm) and hence, provides much more detailed information about the cell's morphology in comparison with conventional light microscopy (theoretical resolution ~ 0.17 μm). Even though confocal microscopy and various molecular techniques have become the leading approaches to study autophagy in many different settings, they often fail to provide firm conclusions if not combined with TEM. Indeed, electron microscopy is still vital to confirm and verify results obtained by other methods, and to produce novel knowledge that would not have been obtained by any other experimental approaches (Eskelinen et al., 2011).

The autophagic process starts with the formation of a cup-shaped membranous structure, known as a phagophore or isolation membrane. After elongation and maturation, the phagophore forms a vacuole or autophagosome, which engulfs small portions of the cytosol including protein aggregates, lipid droplets, and complete organelles (e.g., ER membranes, mitochondria). The diameter of autophagosomes in TEM images varies between 300 nm and several micrometers, but is on average 600 nm, at least in cultured cells (Yla-Anttila et al., 2009). Using the above-described TEM protocol, isolation membranes and early autophagic vacuoles can be easily detected in tissue samples (Fig. 5.1). Typical is the presence of a narrow electron-lucent space between the two limiting membranes of the autophagosomes, allowing easy identification, even at low magnification. However, it should be noted that the number and contrast of autophagosome limiting membranes may vary due to limitations in the preservation of lipids during sample preparation

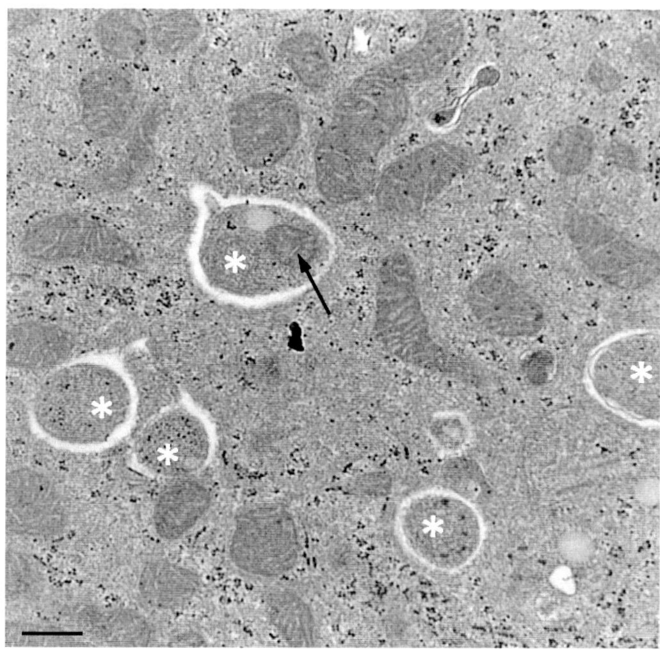

Figure 5.1 *Ultrastructure of early autophagic vacuoles (asterisks) in epoxy embedded liver from mice that were fed a methionine/choline-deficient diet.* Note the presence of glycogen granules (small electron-dense dots) and other cytosolic material inside the autophagosomes. One autophagosome has engulfed a mitochondrion (arrow). Scale bar = 500 nm.

(Eskelinen, 2008). By definition, autophagosomes have a double or (occasionally) multiple membrane. The cytoplasm inside is not degraded and has the same electron density as the cytoplasm outside the autophagosome. Sometimes only one electron-dense membrane is present or the autophagosomes seem to have no limiting membrane at all. This is probably caused by lipid extraction during conventional aldehyde fixation (Eskelinen, 2008). Interestingly, the composition of autophagosome delimiting membranes is rather unique in that they are exceptionally poor in cholesterol and rich in unsaturated lipids (Eskelinen et al., 2011; Punnonen, Pihakaski, Mattila, Lounatmaa, & Hirsimaki, 1989; Yla-Anttila et al., 2009), which is inconsistent with any known organelle membrane. Thus, if the ultrastructure of the autophagosomal membrane is important for further study or the mechanisms involved in trafficking or fusion of these membranes with other organelles need to be investigated in more detail, it is advisable to apply more sophisticated protocols to enhance the contrast of unsaturated lipids (Yla-Anttila et al., 2009). These methods include the use of imidazole-buffered osmium tetroxide, which stains unsaturated lipids, or osmium tetroxide combined with potassium ferrocyanide.

Autophagosomes eventually fuse with early or late endosomes (thereby forming amphisomes) before fusing with lysosomes, or they fuse directly with lysosomal vesicles to become autolysosomes. Incorporation of the outer autophagosomal membrane within the lysosomal membrane allows degradation of the remaining inner single-membrane and the cytoplasmic content of the autophagosome by lysosomal hydrolases. During this step, the cholesterol content in the outer limiting membrane of the autophagic vacuole increases, resulting in less contrast after staining with imidazole-buffered osmium tetroxide as compared with early autophagosomes (Yla-Anttila et al., 2009). Usually, the cytoplasmic content disintegrates before the inner limiting membrane disappears. Particularly noteworthy is that the cytoplasmic content inside autolysosomes becomes electron-dense, often revealing dark granular or amorphous clumps due to partial degradation of ribosomes (Fig. 5.2; Eskelinen, 2008; Yla-Anttila et al., 2009).

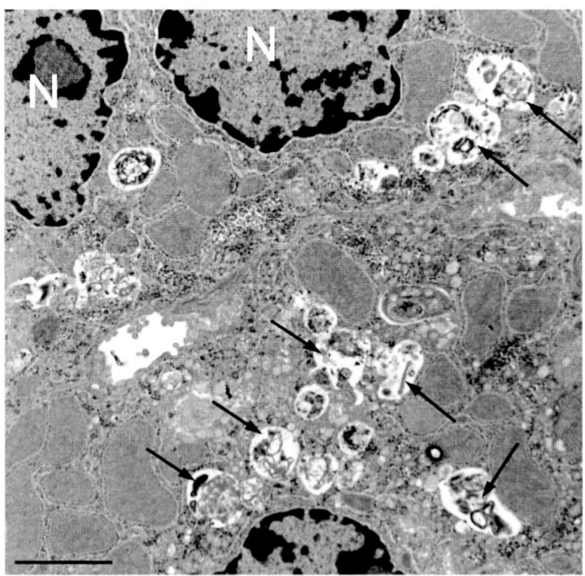

Figure 5.2 *Ultrastructure of late autophagic vacuoles in epoxy embedded liver from starved mice.* C57BL/6 mice underwent starvation for 48 h. The micrographs of starved liver show many autophagic vacuoles (arrows) containing partially degraded electron-dense cytoplasmic content. Scale bar = 2 μm. N, nucleus.

2.3. Guidelines for the correct identification of autophagic vacuoles by TEM

It should be noted that interpreting TEM images is rather challenging. Particularly for inexperienced electron microscopists, it may be difficult to distinguish autophagic compartments from lysosomes, endosomes, or other structures in the cell. As a consequence, too many authors of autophagy-related research articles (and reviewers of such articles) misinterpret autophagic structures. Below we summarize some of the most common pitfalls or mistakes in identifying autophagic vacuoles by TEM in mammalian cells:

1. Autophagic vacuoles in macrophages (or other phagocytes): Because of the strong phagocytic potential, it is difficult to determine via conventional TEM whether the vacuoles in the cytoplasm of macrophages result from autophagy or (hetero)phagocytosis (Fig. 5.3). During heterophagy, the exogenous material is sequestered in a single-membrane-bounded vacuole that originates from the invagination of the plasma membrane. Heterophagic vacuoles can be distinguished from autophagosomes in that the latter structures are surrounded by two (or more) limiting membranes and usually contain intact intracellular organelles targeted for degradation. However, late-stage autophagic vacuoles are recognized as single-membrane vesicles and cannot easily be distinguished from heterophagic vacuoles by morphological criteria.
2. Lipid droplets: Foam cells contain large amounts of lipids in the form of droplets which are oval or round in shape. By definition, these droplets are electron lucent and thus clearly distinguishable from both early and late autophagic vacuoles. However, it should be noted that lipid droplet formation can be induced in certain tissues (e.g., liver) after nutrient deprivation, that is, a potent condition known to stimulate autophagy. Accordingly, numerous lipid droplets are frequently found close to autophagic vacuoles in starved tissue (Fig. 5.4) and might be incorrectly identified as an autophagic compartment.
3. Empty vacuoles: Similar to lipid droplets, empty vacuoles do not have cytoplasmic content, and therefore should not be considered as late-stage autophagic vacuoles. Empty vacuoles in TEM images may occur as a result of a specific cellular process (e.g., formation of secretory granules) or simply through invagination of the plasma membrane. However, it has been the authors' experience that vacuolization may also occur in cells treated with autophagy inducers (Fig. 5.5).

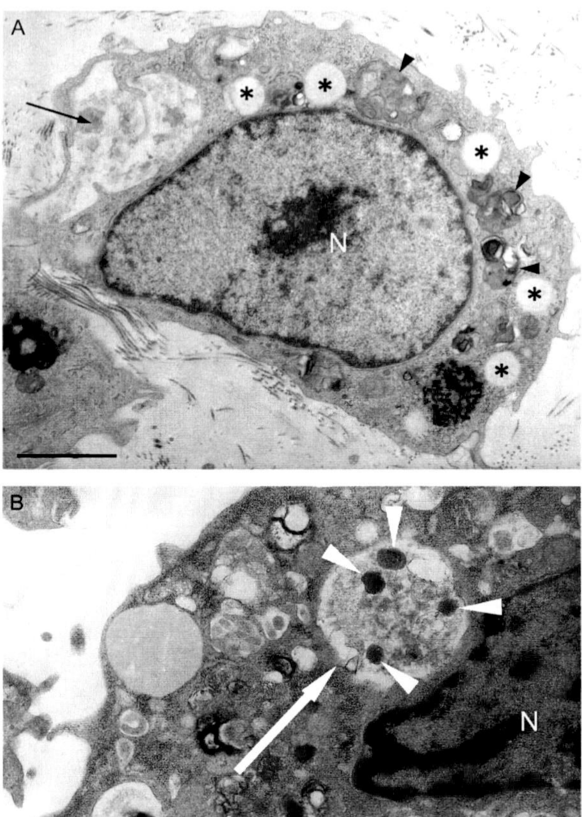

Figure 5.3 *Vesicles originating from heterophagocytosis are sometimes incorrectly identified as autophagic vacuoles.* Panel (A) shows the ultrastructure of a macrophage in an atherosclerotic plaque of a cholesterol-fed rabbit, trying to engulf extracellular material (arrow). Apart from lipid droplets (asterisks), the cytoplasm seems to contain several vesicles. Based on this image, it is hard to identify whether these vesicles are related to autophagy or heterophagy. Panel (B) shows part of a macrophage that has phagocytized a blood platelet (arrow). Several α-granules (arrowheads) can be found inside the phagosome. Scale bar $= 2\ \mu$m. N, nucleus.

4. Swollen organelles: Necrotic cell death or inappropriate fixation and handling of tissue samples may cause swelling of different organelles, which in turn could result in misidentification of intracellular structures. In particular, swollen mitochondria resemble autophagosomes because they also have a double limiting membrane and may contain precipitates (Fig. 5.6). However, mitochondria can be easily identified by the presence of cristae (inward folds of the inner membrane).

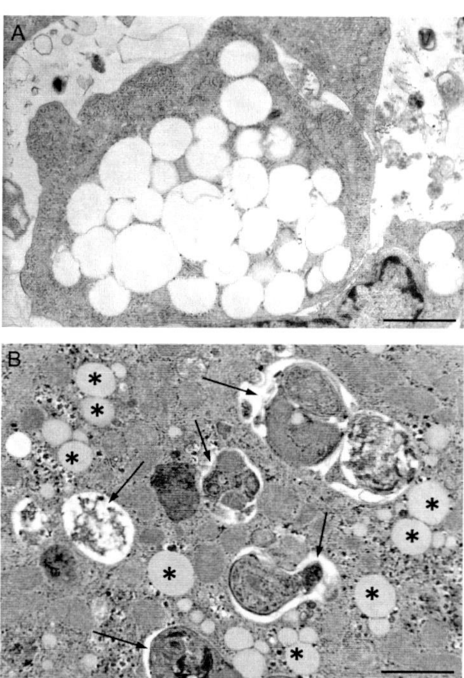

Figure 5.4 *Lipid droplets should not be identified as autophagic vacuoles.* Panel (A) shows the ultrastructure of a macrophage-derived foam cell in an atherosclerotic plaque of a cholesterol-fed rabbit. The cytoplasm of foam cells is often crammed with lipid droplets. Panel (B) illustrates the accumulation of lipid droplets (asterisks) in liver from starved mice. Besides steatosis, starvation stimulates the formation of autophagic vacuoles (arrows) in the liver. Scale bar = 2 μm.

5. Membranous whorls: Some cell types undergoing autophagy in response to specific stimuli reveal vesicles that exclusively contain membrane whorls (Fig. 5.7). These structures, often called myelin figures, represent membranes (or fragments thereof) arranged in concentric rings. It has been claimed by several groups that myelin figures are indicative of autophagic degradation of membranous cellular components. However, membrane vesiculation and the formation of myelin figures could be artifacts of glutaraldehyde fixation (Morgenstern, 1991). Moreover, excess fatty acids (e.g., in adipose tissue) can be visualized as myelin figures if the tissue is exposed (and maintained) above pH 7.4 (Amende, Blanchette-Mackie, Chernick, & Scow, 1985). Therefore, vesicles with myelin figures should not be classified as autophagosomes or autophagic vacuoles if they only contain membrane whorls and not cytoplasm.

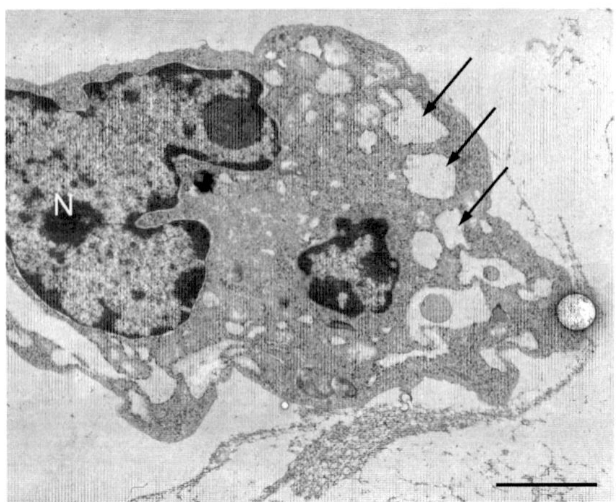

Figure 5.5 *Empty vacuoles may form inside cells, even under autophagy-stimulating conditions, but these vesicles should not be interpreted as standard autophagic vacuoles.* The figure presents the ultrastructure of a mouse macrophage treated with 10 μ*M* everolimus for 24 h. Several electron-lucent vacuoles (arrows) are present in the cytoplasm. Scale bar = 2 μm. N, nucleus.

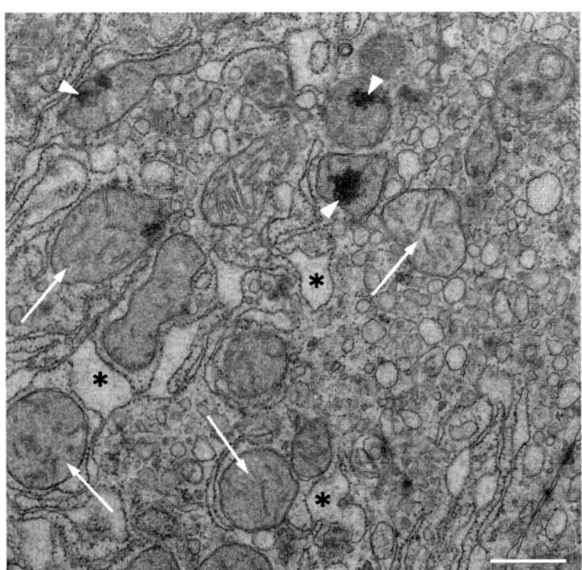

Figure 5.6 *Ultrastructure of liver from a critically ill patient.* From a morphological point of view, livers from critically ill patients reveal an autophagy-deficiency phenotype. Typical is the inadequate removal of damaged intracellular structures. Because mitochondria in these samples (arrows) have a swollen appearance and sometimes contain precipitates (arrow heads), they could be misidentified as autophagosomes. Incorrect identification of swollen mitochondria as autophagic vacuoles is further enhanced by the presence of a double limiting membrane. Importantly, not only mitochondria but also compartments of rough endoplasmic reticulum (asterisks) may show swelling, which may further complicate interpretation of TEM images. Scale bar = 500 nm.

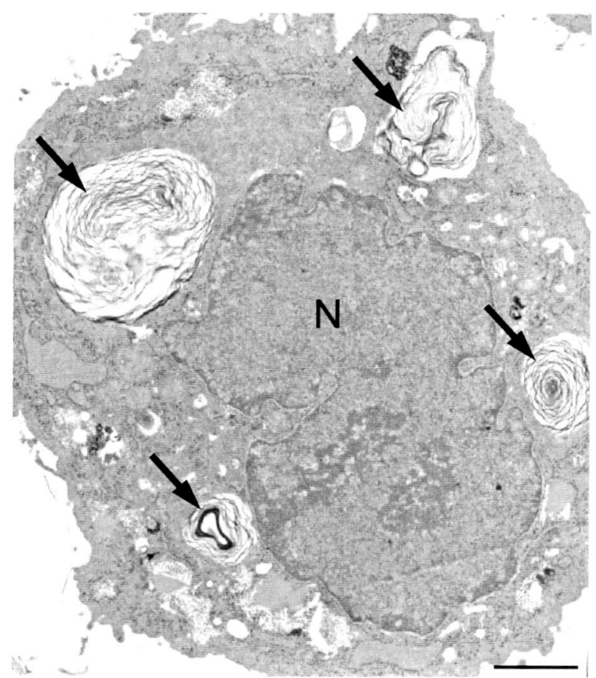

Figure 5.7 *Ultrastructure of smooth muscle cells (SMC) from rabbit aorta exposed to 25 μM 7-ketocholesterol (7-KC) for 18 h.* 7-KC causes severe oxidative damage and autophagy-associated SMC death. Treatment of SMC with 7-KC triggers the formation of large vesicles containing exclusively membrane whorls (arrows). Vesicles with membrane whorls must not be confused with autophagic vacuoles, which as a rule contain cytoplasm. Scale bar = 2 μm. N, nucleus.

6. Other intracellular structures: Sometimes, multivesicular endosomes (MVE) are mistakenly interpreted as autophagosomes. However, the latter structures may fuse with MVE to form amphisomes. In addition, rough endoplasmic reticulum may surround other organelles in the cytoplasm (Eskelinen, 2008) so that it may seem that the formation of an autophagosome is in progress. Ribosomes are not present on the limiting membranes of autophagosomes so that careful examination of the membrane may help to avoid such misinterpretation.

To ensure correct identification of autophagosomal structures, immuno-TEM with gold-labeling is strongly recommended (Klionsky et al., 2012). This approach requires high-quality antibodies, for example, to cargo proteins of cytoplasmic origin and to LC3. Moreover, other EM technologies such as correlative light and electron microscopy (CLEM) could be helpful in confirming the presence of autophagosomes (Klionsky et al., 2012). Due to space limitations, we refer here to previously published protocols for details on immunogold labeling (Swanlund et al., 2010) or CLEM methodology (Razi & Tooze, 2009).

3. IMMUNOHISTOCHEMISTRY

This protocol has been optimized for the detection of the autophagy marker protein LC3B in paraffin-embedded tissue specimens (Martinet, Schrijvers, Timmermans, Bult, & De Meyer, 2013). Other Atg-proteins can also be detected with this method, but minor changes to the protocol might be necessary.

3.1. Methodology

1. Isolate tissue specimens and fix in 4% neutral-buffered formalin for 24 h at room temperature.
2. After fixation, transfer the tissue into 60% isopropanol (in water) for at least 1 h. Samples can be stored in isopropanol for several weeks.
3. Dehydrate tissue samples through a series of graded ethanol baths to displace the water, and then infiltrate with paraffin wax. Because paraffin embedding and sectioning of the wax blocks are standard procedures in most laboratories, please consult protocols in previous literature.
4. Dewax sections (5 μm thick) in toluene (2×5 min).
5. Immerse sections in isopropanol (2×5 min) and rehydrate in distilled water (5 min).
6. Quench endogenous peroxidase by incubation in 3% hydrogen peroxide in water for 5 min.
7. Wash slides in water and incubate for 10 min at 37 °C with trypsin buffer (10 m*M* Tris–HCl, pH 7.8, 0.1% $CaCl_2$, 0.1% NaCl) containing 0.01% trypsin (Sigma, Cat. No. 93613).
8. Rinse in water and perform heat-mediated antigen retrieval in boiling citrate buffer (1.6 m*M* citric acid, 6.9 m*M* sodium citrate-5,5-hydrate, pH 6.0) for 10 min using a microwave.
9. Allow slides to cool down for 30 min at room temperature and then rinse twice with distilled water and once with staining buffer (10 m*M* Tris–HCl, pH 7.6, 0.9% NaCl, 0.1% Triton X-100, 0.004% thimerosal).
10. Block nonspecific protein binding by incubating slides for 20 min with serum (1:5 diluted in staining buffer). Serum is derived from the animal in which the secondary antibody was raised.

11. Apply the primary antibody on the tissue sections overnight in an appropriate dilution (usually 1:100 to 1:1000 in staining buffer). Use a damp box to prevent the sections from drying out.
12. Immerse sections 2 × 5 min in staining buffer.
13. Incubate slides for 30 min with undiluted, ready-to-use Envision+ system (Dako).
14. Rinse slides 2 × 5 min in staining buffer and once in distilled water, prior to incubation with peroxidase substrate solution until the desired stain intensity develops. The peroxidase substrate solution is prepared as follows: 20 mg 3-amino-9-ethylcarbazole is dissolved in 24 ml DMSO, then mixed with 200 ml sodium acetate, pH 5.2, and supplemented with 4 ml 1:100 diluted Perhydrol (hydrogen peroxide 30%) in water.
15. Rinse sections in distilled water and counterstain for 1 min using Carazzi's hematoxylin.
16. Rinse in water and mount sections under glass coverslips.
17. Optional: Measure the immunoreactive area of histological images using ImageJ software (NIH).

3.2. Immunohistochemistry as a tool to study autophagy in tissue

Considering the many pitfalls in correctly identifying autophagic vacuoles by TEM (see Section 2.3), other methods may be required to monitor autophagy in tissue samples. One of the most useful alternatives is the immunohistochemical analysis of autophagy marker proteins. In recent years, ATG8/LC3 (microtubule-associated protein 1 light chain 3) is frequently used as a phagophore or autophagosome marker *in vitro* and *in vivo* because of the apparent changes in its biochemical nature and subcellular localization after autophagy induction (Klionsky et al., 2012). Under nonautophagic conditions, newly synthesized LC3 is cleaved at its C-terminus by the protease ATG4 to generate LC3-I, which distributes diffusely throughout the cytoplasm. However, after induction of autophagy, cytosolic LC3-I is conjugated to phosphatidylethanolamine (PE) to form LC3-II, and recruited via PE to the inner and outer surface of autophagosomal membranes. Accordingly, autophagosomes can be readily identified by immunohistochemistry in tissue sections due to the formation of LC3-positive puncta.

In mammals, the LC3 family includes at least three different isoforms: LC3A, LC3B, and LC3C. Increasing evidence suggests that immunohistochemical staining of both LC3A and LC3B is a valuable technique for the assessment of autophagy in tissue, particularly in the field of cancer

(Ladoire et al., 2012; Rosenfeldt, Nixon, Liu, Mah, & Ryan, 2012; Sivridis et al., 2010). Indeed, many tumor cells tend to express higher LC3 levels than the adjacent, nonmalignant tissue (Ladoire et al., 2012). Unfortunately, immunohistochemical detection of LC3 has several limitations. First, the expression of LC3A and LC3B in different tissues may vary considerably. Expression levels are relatively low in normal liver as well as in many other organs such as heart, spleen, and lung making the immunohistochemical detection of LC3 isoforms not always feasible (Martinet et al., 2013). Moreover, whereas LC3 was found to be moderately to highly elevated in a large number of tumors such as esophageal, gastrointestinal, and breast carcinomas (Ladoire et al., 2012; Yoshioka et al., 2008), decreased expression of LC3 has been reported in lung, brain, and ovary cancer (Jiang, Shao, Wang, Yan, & Liu, 2012; Shen, Li, Wang, Deng, & Zhu, 2008). Also noteworthy is that tumors of the same type tend to be highly heterogeneous in the number and intensity of LC3 puncta (Ladoire et al., 2012), most likely due to differences in tumor stage and/or intake of anticancer drugs, which are often well-known autophagy agonists. As LC3 staining of noncancerous tissue is generally weak, immunohistochemical detection of LC3 depends on a highly sensitive detection method. The protocol in Section 3.1 describes the use of primary antibodies in combination with Envision+ immunodetection. Envision+ is an extremely sensitive system that allows signal amplification via a hydrophilic dextran polymer, conjugated to secondary antibodies and multiple (up to 100) horseradish peroxidase molecules, and is a prerequisite for obtaining maximum sensitivity. However, staining experiments in our laboratory have revealed that even this sensitive detection method requires overexpression of LC3 (e.g., by using GFP-LC3 transgenic mice) to achieve staining (Figs. 5.8 and 5.9). In addition, not all tissues in GFP-LC3 mice undergoing starvation-induced autophagy revealed LC3 puncta. Brain tissue, for example, turned out to be insensitive to food withdrawal, as previously reported (Mizushima, Yamamoto, Matsui, Yoshimori, & Ohsumi, 2004), but also other tissues such as the arterial vessel wall or spleen did not reveal a substantial number of LC3 puncta upon starvation, although autophagy was clearly induced in other organs of the same mouse (Fig. 5.10). This finding suggests that induction of autophagy is not uniform but organ dependent. Interestingly, an optimized method for immunohistochemical detection of endogenous LC3 (quite similar to the protocol in Section 3.1) has recently been published which allows staining of LC3 puncta even in regular (nontransgenic) human and mouse tissue samples (Rosenfeldt et al., 2012). This method recommends the use of 10 m*M*

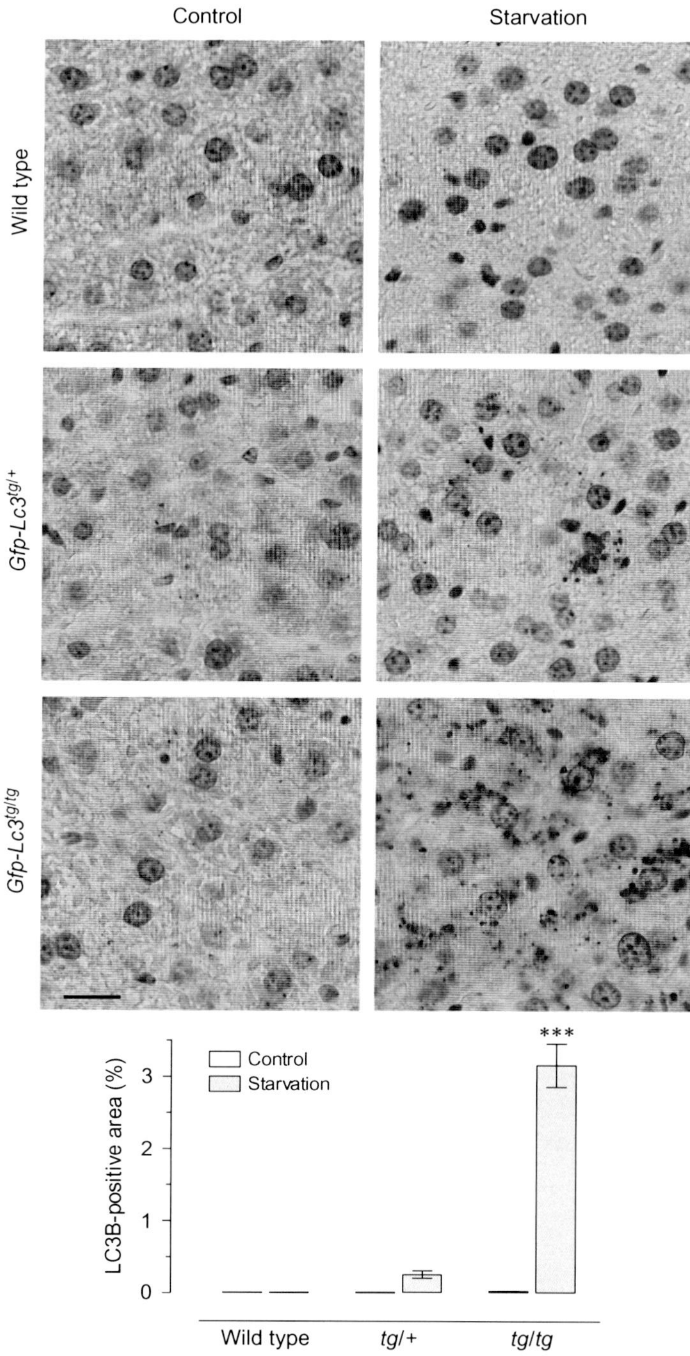

Figure 5.8 *Immunohistochemical detection of LC3B in GFP-LC3 transgenic mice.* Liver samples were isolated from fed wild-type, Gfp-Lc3$^{tg/+}$, or *Gfp-Lc3*$^{tg/tg}$ mice (control) and from mice that underwent starvation for 48 h to induce autophagy. After fixation in neutral buffered formalin, tissues were paraffin-embedded and stained for LC3B using rabbit monoclonal anti-LC3B (clone D11; Cell signaling, 1:100 (wild type); 1:300 (*Gfp-Lc3*$^{tg/+}$) or 1:1000 (*Gfp-Lc3*$^{tg/tg}$)) and Envision+. Scale bar = 20 μm. The LC3B-positive area was quantified. $^{***}P < 0.001$ versus control (two-way ANOVA, followed by Bonferroni's posttest, $n = 10$). (See the color plate.)

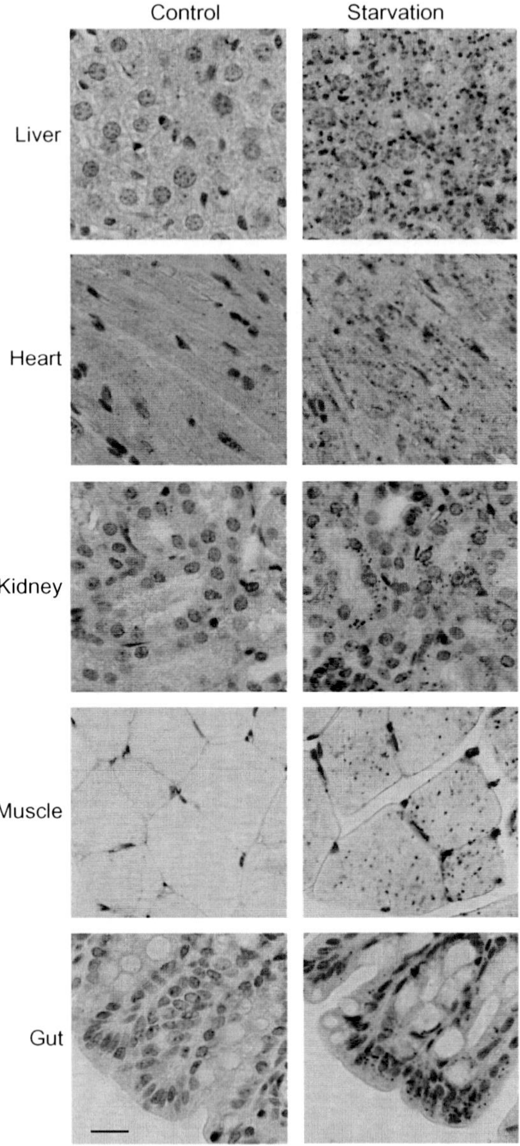

Figure 5.9 *Immunohistochemical detection of LC3B in different organs of starved Gfp-Lc3$^{tg/tg}$ mice.* Gfp-Lc3$^{tg/tg}$ mice were fed regular chow (control) or underwent starvation for 48 h. Tissue samples from different organs were collected and fixed in neutral buffered formalin. Thereafter, tissues were paraffin-embedded and stained for LC3B using mouse monoclonal anti-LC3B (clone 5F10; Nanotools, 1:1000) and Envision+. Scale bar = 20 μm. (See the color plate.)

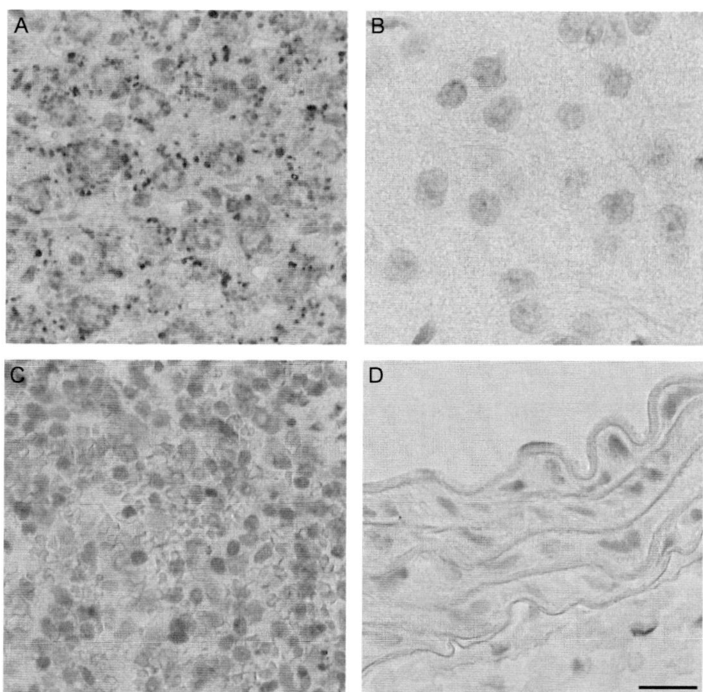

Figure 5.10 *Starvation-induced autophagy in mice is not uniform but tissue-specific. Gfp-Lc3*$^{tg/tg}$ mice underwent starvation for 48 h. Tissue samples from liver (A), brain (B), spleen (C), and aorta (D) were collected and fixed in neutral buffered formalin. Thereafter, tissues were paraffin-embedded and stained for LC3B using rabbit monoclonal anti-LC3B (clone D11; Cell signaling, 1:1000) and Envision+. Scale bar = 20 μm. (See the color plate.)

Tris–EDTA, pH 9.0 (instead of citrate buffer, pH 6.0), during heat-induced epitope retrieval (see Section 3.1). However, EDTA buffer does not provide better results in our hands; on the contrary, it rather inhibits LC3 staining (Martinet et al., 2013).

Next to low LC3 expression levels, the second major problem hampering optimal LC3 staining is the rapid degradation of LC3-II by lysosomal proteases during autophagy, leading us to conclude that LC3 may not be the most practical marker to detect autophagic activity. Indeed, the absence of LC3 staining in spleen and aorta shown in Fig. 5.10 might result from prolonged starvation and degradation of LC3. Administration of drugs that elevate the lysosomal pH (e.g., chloroquine) are instrumental in inhibiting LC3 protein degradation and increasing the number of autophagic compartments (Klionsky et al., 2012). For assessment of chloroquine-enhanced autophagosome accumulation *in vivo*, we would like to refer to excellent, previously published protocols (Iwai-Kanai et al., 2008; Perry et al., 2009).

Finally, it is noteworthy that LC3 stains are optimal in neutral-buffered formalin-fixed tissue. Samples fixed in the precipitant fixative methacarn (60% methanol, 30% 1,1,1-trichloroethane, 10% glacial acetic acid) or Bouin's fixative do not stain for (endogenous) LC3 (Martinet et al., 2013). Possibly, sufficient cross-linking is essential for fixing small proteins such as LC3, which may not be insufficiently large to be made insoluble by precipitant fixatives.

3.3. Guidelines for the correct identification of autophagic vacuoles by immunohistochemistry

Similar to TEM images, immunohistochemical stainings for LC3 (or other autophagy marker proteins) may lead to false conclusions and thus should be interpreted with caution:

1. Large globular intracellular structures may stain positive for LC3A and LC3B in autophagy-deficient cells (Fig. 5.11). Cells with defective autophagy accumulate protein aggregates or cellular structures, in which LC3 can be incorporated in an autophagy-independent manner (Kuma, Matsui, & Mizushima, 2007; Shibata et al., 2009).
2. Many commercialized LC3 antibodies are not isoform-specific. In particular, polyclonal LC3B antibodies (e.g., anti-LC3B [NB100-2331] from Novus Biologicals, anti-LC3B [NB600-1384] and anti-MAPLC3B H-50 [sc-28266] from Santa Cruz, anti-LC3 [L8918] and anti-LC3B [L7543] from Sigma–Aldrich) may show cross-reactivity with LC3A (Martinet et al., 2013; Zois et al., 2011). Examples of isoform-specific LC3 antibodies are rabbit polyclonal anti-LC3A (Abgent, AP1805a), rabbit polyclonal anti-MAP1LC3A (Abcam, ab62720), mouse monoclonal anti-LC3B (clone 5F10; Nanotools, 0231-100/LC3-5F10) and rabbit monoclonal anti-LC3B (clone D11, Cell Signaling, 3868; Martinet et al., 2013; Zois et al., 2011).
3. Granular staining without clear formation of LC3-positive puncta (e.g., see Shintaku, 2011) may not reflect autophagosome formation, but instead represents LC3 protein expression and/or aggregation. Granular staining may also result from nonspecific staining of cytoplasmic components.
4. A growing body of immunohistochemical evidence shows that several tumor types contain large clumps of LC3A-positive amorphous material, often enclosed within cytoplasmic vacuoles (Sivridis et al., 2010). This

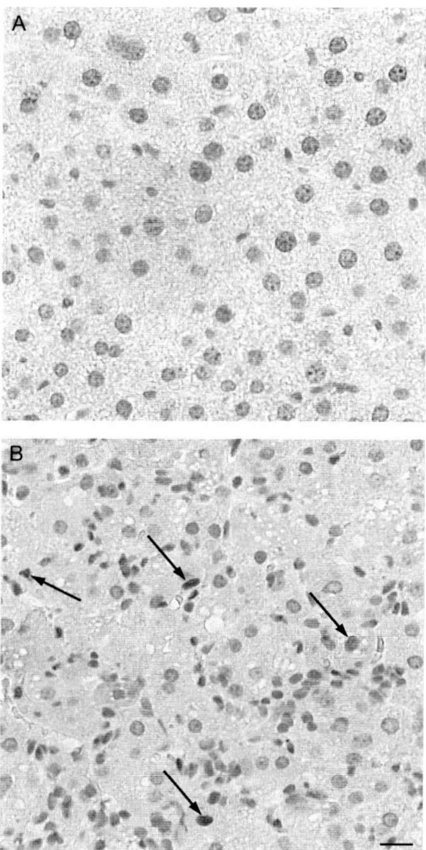

Figure 5.11 *LC3A isoforms accumulate in autophagy-deficient tissues and may cause false positive staining.* Liver samples were isolated from fed autophagy-competent (*Atg7*$^{+/+}$ *Alb-Cre*$^{+}$) mice (A) or liver-specific autophagy-deficient (*Atg7*$^{F/F}$ *Alb-Cre*$^{+}$) mice (B). After fixation in neutral buffered formalin, tissues were paraffin-embedded and stained for LC3A using rabbit polyclonal anti-LC3A (Abgent, 1:100) in combination with Vectastain ABC. Globular structures of different size (arrows) stain positive for LC3A in liver from *Atg7*$^{F/F}$ *Alb-Cre*$^{+}$ mice. Similar results were obtained with antibodies against LC3B. Scale bar = 20 μm. (See the color plate.)

"stone-like" pattern of LC3A expression is associated with malignant phenotypes of human cancer, including distant metastasis, increased tumor size, extensive necrosis, and leads to an unfavorable prognosis. Most likely, "stone-like" LC3A reflects accumulation of autophagic debris that cannot be further degraded, rather than an active autophagic process.

5. We do not recommend the use of frozen tissue specimens for the analysis of autophagic vacuoles because of the poor morphology and resolution of frozen sections. Moreover, frozen sections may show LC3B-positive dots around lipid droplets (Martinet et al., 2013). This finding is consistent with previous reports showing that LC3B is localized on the surface of lipid droplets and critically involved in lipid droplet formation (Shibata et al., 2009). This staining is not autophagy-specific and may lead to incorrect interpretation.
6. Many Atg-proteins, other than LC3, have been used as immunohistochemical markers for autophagy. In the early 2000s, granular cytoplasmic ubiquitin inclusions were considered to be an attractive marker for autophagic degeneration of cardiomyocytes during heart failure (Knaapen et al., 2001). However, more recent evidence indicates that these inclusions may result from defective autophagy as ubiquitin-positive cytoplasmic inclusions colocalize with enhanced levels of SQSTM1/p62, a selective substrate of autophagy that accumulates in cells when autophagy is inhibited (Fig. 5.12). Because p62 accumulates in autophagy-deficient cells, it is not a useful marker for tissue with autophagic activity (Martinet et al., 2013), even though immunohistochemical detection of p62 can be combined with LC3 staining to determine whether LC3 positive signals point to autophagy induction or impaired degradation of the LC3 protein. Besides ubiquitin and p62, ATG5, CTSD/cathepsin D, and BECN1/Beclin 1 have been tested as alternative markers for *in situ* detection of autophagy (Martinet et al., 2013). In starved liver, staining for ATG5 or CTSD is not different from stainings of nonstarved (or autophagy-deficient) controls (Fig. 5.13). Moreover, ATG5 can be found in the nucleus (Fig. 5.13), which does not seem to be autophagy-related and probably reflects unspecific staining. BECN1, which is part of a class III phosphatidylinositol 3-kinase complex that promotes autophagy and may form macroaggregates after induction of autophagy by various stimuli (Klionsky et al., 2012), is used by many researchers as a way to monitor autophagy. However, BECN1 often shows poor upregulation after induction of autophagy (Fig. 5.13). In addition, certain forms of autophagy are induced in a BECN1-independent manner (Klionsky et al., 2012), and consequently cannot be detected via BECN1 immunostains. Also noteworthy is that the expression of BECN1 is downregulated in many types of human cancer, which is predictive of tumor aggressiveness and poor prognosis. Considering all these difficulties, we only recommend monitoring of LC3 puncta for the clear, unambiguous detection of autophagic vacuoles in tissue, if technically possible.

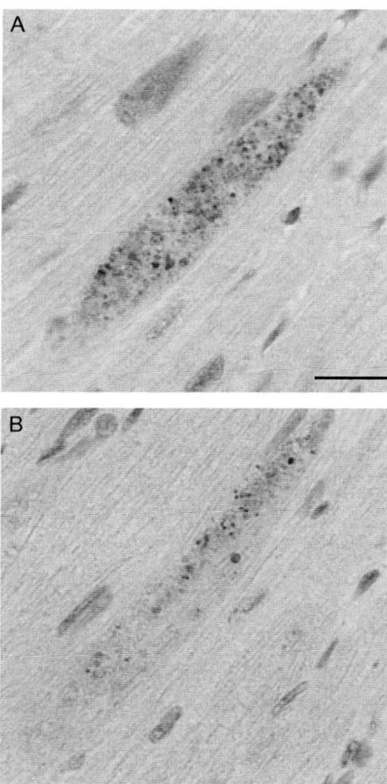

Figure 5.12 *Immunohistochemical detection of ubiquitin (A) and SQSTM1/p62 (B) in cardiomyocytes of failing human heart.* Tissue samples from the heart were isolated from patients with heart failure. After fixation in neutral buffered formalin, tissues were paraffin-embedded and stained for ubiquitin and p62 using mouse monoclonal anti-ubiquitin (clone 6C1; Sigma, 1:300,000) and rabbit polyclonal anti-p62 (Sigma, 1:500), respectively, in combination with Vectastain ABC. Granular ubiquitin inclusions accumulate in the cytosol of degenerating cardiomyocytes during heart failure and colocalize with p62 staining. This finding suggests a malfunction in the autophagic pathway, rather than enhanced autophagic activity mediating cardiomyocyte death. Scale bar = 20 μm. (See the color plate.)

4. CONCLUDING REMARKS

Both TEM and immunohistochemical stains (combined with conventional light microscopy) have benefits and drawbacks. TEM is one of the most accurate and sensitive tools for the detection and quantification of autophagy in tissue (Eskelinen et al., 2011; Swanlund et al., 2010). However, this method requires expensive equipment and is time consuming, rendering it unsuitable for daily routine. Moreover, an experienced electron microscopist is needed

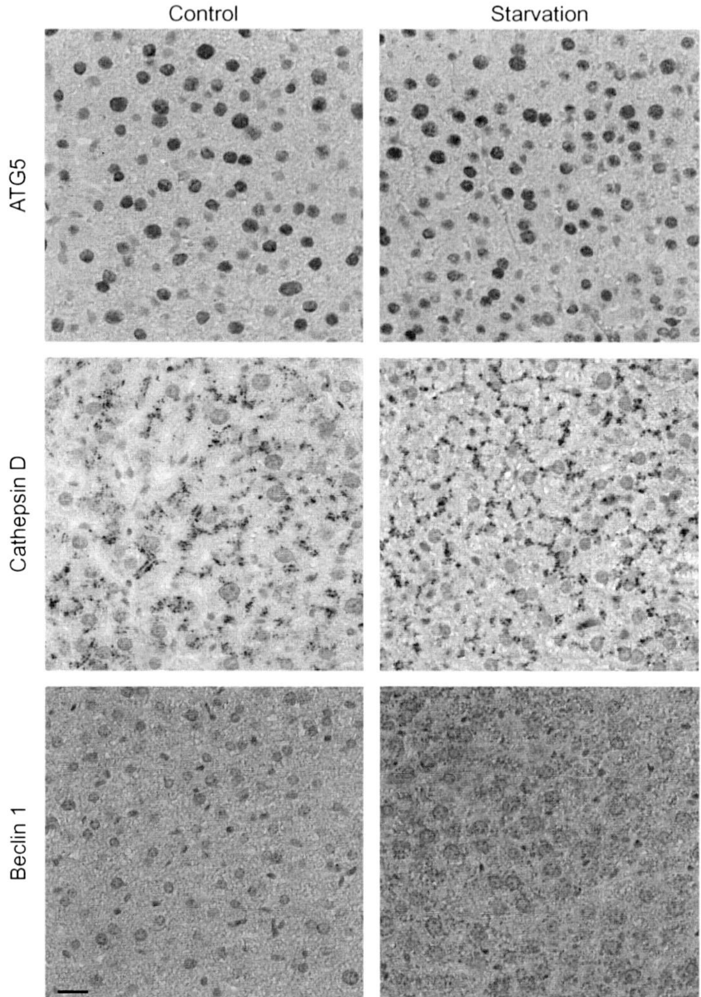

Figure 5.13 *ATG5, CTSD/cathepsin D, and BECN1/Beclin 1 are not ideal targets for the immunohistochemical detection of autophagy.* Liver samples were isolated from control mice and mice that underwent starvation for 48 h to induce autophagy. After fixation in neutral buffered formalin, tissues were paraffin-embedded and stained for ATG5, CTSD, or BECN1 using rabbit polyconal anti-ATG5 (Abcam, ab78073, 1:100), rabbit monoclonal anti-CTSD (clone EPR3057Y; Abcam, ab75852, 1:1000), or rabbit polyclonal anti-BECN1 (Lifespan, LS-B3202, 1:1000) in combination with Vectastain ABC. Scale bar = 20 μm. (See the color plate.)

to correctly interpret the TEM images. Immunoelectron microscopy with antibodies against autophagy-related marker proteins would reduce misidentification of autophagosomal structures (Swanlund et al., 2010), but this technique is particularly cumbersome. Advantages of immunohistochemistry include a fast turnaround time (typically 2 days vs. 5–7 days for TEM), low cost, ease of performance, and widespread familiarity (Martinet et al., 2013). Unfortunately, immunohistochemical detection of LC3 lacks sensitivity, even when using signal amplification systems such as Envision+, so that over-expression of the protein is required. In addition, interpretation of immunohistochemical stains is not error-free. LC3 conversion (LC3-I to LC3-II) by immunoblot analysis is an alternative technique widely used by autophagy researchers to monitor autophagy (Klionsky et al., 2012). However, Western blots detect LC3 in the entire tissue sample (not in single cells), and various caveats regarding detection of LC3 by this method have been described (Mizushima & Yoshimori, 2007). We conclude that ideal methods for monitoring autophagy in tissue do not exist at this moment. Therefore, we recommend the use of multiple techniques (morphological analysis of autophagic structures by TEM combined with standard immunohistochemistry and/or immunoblotting) to study the presence of autophagy *in situ*.

ACKNOWLEDGMENTS

This work was supported by the Fund for Scientific Research (FWO)-Flanders and the University of Antwerp (BOF).

REFERENCES

Amende, L. M., Blanchette-Mackie, E. J., Chernick, S. S., & Scow, R. O. (1985). Effect of pH on visualization of fatty acids as myelin figures in mouse adipose tissue by freeze-fracture electron microscopy. *Biochimica et Biophysica Acta*, *837*, 94–102.

Choi, A. M., Ryter, S. W., & Levine, B. (2013). Autophagy in human health and disease. *New England Journal of Medicine*, *368*, 651–662.

Eskelinen, E. L. (2008). To be or not to be? Examples of incorrect identification of autophagic compartments in conventional transmission electron microscopy of mammalian cells. *Autophagy*, *4*, 257–260.

Eskelinen, E. L., Reggiori, F., Baba, M., Kovacs, A. L., & Seglen, P. O. (2011). Seeing is believing: The impact of electron microscopy on autophagy research. *Autophagy*, *7*, 935–956.

Iwai-Kanai, E., Yuan, H., Huang, C., Sayen, M. R., Perry-Garza, C. N., Kim, L., et al. (2008). A method to measure cardiac autophagic flux in vivo. *Autophagy*, *4*, 322–329.

Jiang, Z. F., Shao, L. J., Wang, W. M., Yan, X. B., & Liu, R. Y. (2012). Decreased expression of Beclin-1 and LC3 in human lung cancer. *Molecular Biology Reports*, *39*, 259–267.

Klionsky, D. J., Abdalla, F. C., Abeliovich, H., Abraham, R. T., Acevedo-Arozena, A., Adeli, K., et al. (2012). Guidelines for the use and interpretation of assays for monitoring autophagy. *Autophagy*, *8*, 445–544.

Knaapen, M. W., Davies, M. J., De Bie, M., Haven, A. J., Martinet, W., & Kockx, M. M. (2001). Apoptotic versus autophagic cell death in heart failure. *Cardiovascular Research*, *51*, 304–312.

Kuma, A., Matsui, M., & Mizushima, N. (2007). LC3, an autophagosome marker, can be incorporated into protein aggregates independent of autophagy: Caution in the interpretation of LC3 localization. *Autophagy*, *3*, 323–328.

Ladoire, S., Chaba, K., Martins, I., Sukkurwala, A. Q., Adjemian, S., Michaud, M., et al. (2012). Immunohistochemical detection of cytoplasmic LC3 puncta in human cancer specimens. *Autophagy*, *8*, 1175–1184.

Martinet, W., Schrijvers, D. M., Timmermans, J. P., Bult, H., & De Meyer, G. R. Y. (2013). Immunohistochemical analysis of macroautophagy: Recommendations and limitations. *Autophagy*, *9*, 386–402.

Mizushima, N., & Komatsu, M. (2011). Autophagy: Renovation of cells and tissues. *Cell*, *147*, 728–741.

Mizushima, N., Yamamoto, A., Matsui, M., Yoshimori, T., & Ohsumi, Y. (2004). In vivo analysis of autophagy in response to nutrient starvation using transgenic mice expressing a fluorescent autophagosome marker. *Molecular Biology of the Cell*, *15*, 1101–1111.

Mizushima, N., & Yoshimori, T. (2007). How to interpret LC3 immunoblotting. *Autophagy*, *3*, 542–545.

Morgenstern, E. (1991). Aldehyde fixation causes membrane vesiculation during platelet exocytosis: A freeze-substitution study. *Scanning Microscopy Supplement*, *5*, S109–S115.

Perry, C. N., Kyoi, S., Hariharan, N., Takagi, H., Sadoshima, J., & Gottlieb, R. A. (2009). Novel methods for measuring cardiac autophagy in vivo. *Methods in Enzymology*, *453*, 325–342.

Punnonen, E. L., Pihakaski, K., Mattila, K., Lounatmaa, K., & Hirsimaki, P. (1989). Intramembrane particles and filipin labelling on the membranes of autophagic vacuoles and lysosomes in mouse liver. *Cell and Tissue Research*, *258*, 269–276.

Razi, M., & Tooze, S. A. (2009). Correlative light and electron microscopy. *Methods in Enzymology*, *452*, 261–275.

Rosenfeldt, M., Nixon, C., Liu, E., Mah, L. Y., & Ryan, K. M. (2012). Analysis of macroautophagy by immunohistochemistry. *Autophagy*, *8*, 963–969.

Rubinsztein, D. C., Codogno, P., & Levine, B. (2012). Autophagy modulation as a potential therapeutic target for diverse diseases. *Nature Reviews. Drug Discovery*, *11*, 709–730.

Shen, Y., Li, D. D., Wang, L. L., Deng, R., & Zhu, X. F. (2008). Decreased expression of autophagy-related proteins in malignant epithelial ovarian cancer. *Autophagy*, *4*, 1067–1068.

Shibata, M., Yoshimura, K., Furuya, N., Koike, M., Ueno, T., Komatsu, M., et al. (2009). The MAP1-LC3 conjugation system is involved in lipid droplet formation. *Biochemical and Biophysical Research Communications*, *382*, 419–423.

Shintaku, M. (2011). Immunohistochemical localization of autophagosomal membrane-associated protein LC3 in granular cell tumor and schwannoma. *Virchows Archiv*, *459*, 315–319.

Sivridis, E., Koukourakis, M. I., Zois, C. E., Ledaki, I., Ferguson, D. J., Harris, A. L., et al. (2010). LC3A-positive light microscopy detected patterns of autophagy and prognosis in operable breast carcinomas. *American Journal of Pathology*, *176*, 2477–2489.

Swanlund, J. M., Kregel, K. C., & Oberley, T. D. (2010). Investigating autophagy: Quantitative morphometric analysis using electron microscopy. *Autophagy*, *6*, 270–277.

White, E. (2012). Deconvoluting the context-dependent role for autophagy in cancer. *Nature Reviews. Cancer*, *12*, 401–410.

Yla-Anttila, P., Vihinen, H., Jokitalo, E., & Eskelinen, E. L. (2009). Monitoring autophagy by electron microscopy in mammalian cells. *Methods in Enzymology*, *452*, 143–164.

Yoshioka, A., Miyata, H., Doki, Y., Yamasaki, M., Sohma, I., Gotoh, K., et al. (2008). LC3, an autophagosome marker, is highly expressed in gastrointestinal cancers. *International Journal of Oncology*, *33*, 461–468.

Zois, C. E., Giatromanolaki, A., Sivridis, E., Papaiakovou, M., Kainulainen, H., & Koukourakis, M. I. (2011). "Autophagic flux" in normal mouse tissues: Focus on endogenous LC3A processing. *Autophagy*, *7*, 1371–1378.

CHAPTER SIX

Methods to Measure the Enzymatic Activity of PI3Ks

Elisa Ciraolo[1], Federico Gulluni[1], Emilio Hirsch[2]
Molecular Biotechnology Center, Department of Molecular Biotechnology and Health Sciences, University of Torino, Torino, Italy
[1]Equal contributors.
[2]Corresponding author: e-mail address: emilio.hirsch@unito.it

Contents

Abstract

Phosphoinositide-3-kinase (PI3K) signaling has been implicated in a panoply of cellular responses including survival, proliferation, protein synthesis, migration, and vesicular trafficking. In addition, alterations in the enzymatic activity of PI3Ks have been involved in the pathogenesis of multiple diseases, ranging from cancer to chronic inflammation. The emerging interest in PI3K as a pharmacological target has prompted the development of several molecules with inhibitory activity. In this context, the quantification of the second messenger generated by PI3Ks, phosphoinositide-3-phosphate, offers an opportunity to directly test variations in the lipid kinase activity of PI3K in physiological as well as pathological conditions. Here, we will describe common methods to measure the lipid kinase activity of PI3K *in vitro* and new techniques to follow the production of phosphoinositide-3-phosphate *in vivo*. These methods are relevant to study the alterations of the PI3K systems at the interface between signaling and oncometabolism.

Methods in Enzymology, Volume 543
ISSN 0076-6879
http://dx.doi.org/10.1016/B978-0-12-801329-8.00006-4

1. INTRODUCTION

Class I, class II, and class III phosphoinositide-3-kinases (PI3Ks) belong to a large family of enzymes able to phosphorylate the D-3 position of different phosphoinositides (PtdIns). These families comprise enzymes highly related to each other, but they can be distinguished by their preferential *in vivo* and *in vitro* substrate, by their structure and mechanism of activation/regulation (Vanhaesebroeck, Guillermet-Guibert, Graupera, & Bilanges, 2010).

Class I PI3Ks are heterodimers composed of a catalytic and a regulatory subunit. While PI3Kα and β are ubiquitous and abundantly expressed, the expression of PI3Kδ and γ is mainly restricted to leukocytes. Recently, PI3Kγ expression has been found in several other tissues such as heart and brain. Class I PI3Ks are divided into two subfamilies depending on the mechanism of activation. Class IA PI3Ks are activated by growth factor receptor tyrosine kinases (RTK), whereas class IB PI3Ks are predominantly triggered by G-protein-coupled receptors (GPCR). In the past years, several studies positioned PI3Kβ in between class IA and IB given that this enzyme can be recruited by both RTK and GPCR. In addition, a cooperative activation of PI3K by Ras and Rho family small GTPases has been recently described (Yang et al., 2012). Upon receptor activation, class I PI3Ks are capable of phosphorylating *in vivo* the D-3 position of PtdIns(4,5)P2 thus producing the PtdIns(3,4,5)P3 second messenger (Fig. 6.1) which is involved in the regulation of a plethora of different cellular responses (Tolias & Cantley, 1999) including proliferation, survival, and motility. PI3Kα and PI3Kβ have been mainly associated to growth factor signaling (Hirsch, Ciraolo, Ghigo, & Costa, 2008), while PI3Kδ and PI3Kγ play an essential role in leukocytes function by regulating a number of processes required for maturation, activation, and migration of both myeloid and lymphoid cells. Alterations in class I PI3K signaling pathway favors tumorigenesis and aberrant activation of PI3Ks often correlates with resistance to chemo- and radiotherapy (Ciraolo, Morello, & Hirsch, 2011; Hirsch, Ciraolo, Franco, Ghigo, & Martini, 2013).

Class II PI3Ks are monomers of high molecular mass that differ from class I and class III for their more extended N- and C-terminal domains (Falasca & Maffucci, 2012). In mammals, three different class II members have been identified, the ubiquitously expressed PI3K-C2α and PI3K-C2β and the liver-specific PI3K-C2γ (Falasca & Maffucci, 2012). Contrary to class I, class

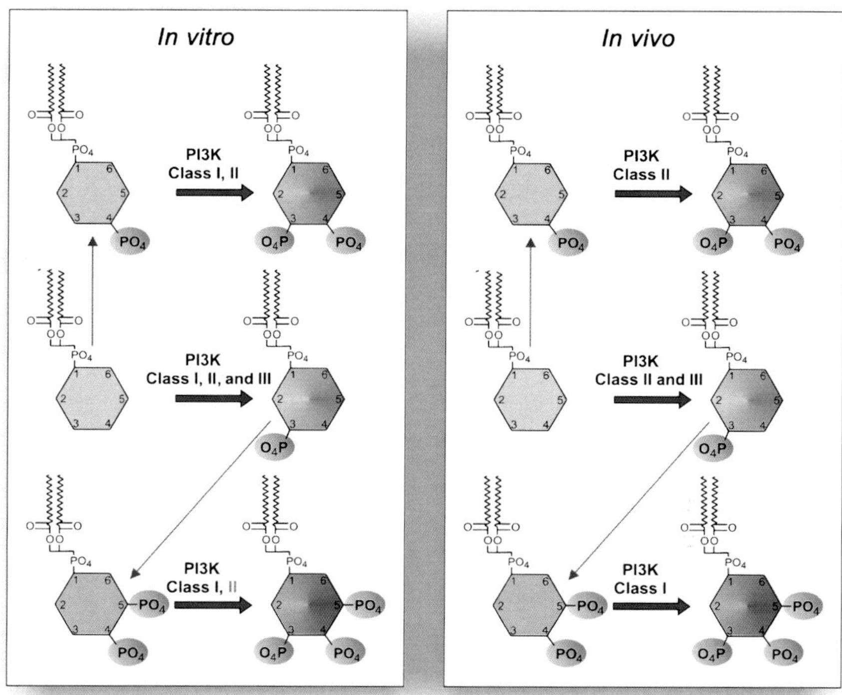

Figure 6.1 *PI3K-mediated production of phosphoinositides.* Position 3 on the inositol ring can be phosphorylated by PI3K family enzyme to produce PtdIns(3)P, PtdInd(3,4)P2, and PtdIns(3,4,5)P3. The three classes of PI3K can be distinguished by their specific product. *In vitro* class I PI3Ks are able to produce PtdIns(3), PtdIns(3,4)P2 and PtdIns(3,4,5)P3, while the PtdIns(3,4,5)P3 is their unique *in vivo* lipid product. Class II PI3Ks are able to produce, *in vitro* and *in vivo*, primarily PtdIns(3)P, PtdIns(3,4)P2 and PtdIns(3,4,5)P3 only *in vitro* in presence of specific substrates. The unique member of class III can produce uniquely PtdIns(3)P, both *in vitro* and *in vivo*. (See the color plate.)

II PI3Ks play a fundamental role in the regulation of vesicular trafficking. Several stimuli such as hormones, growth factor, chemokines, and cytokines may activate class II PI3Ks through different membrane receptor including RTKs (epidermal growth factor receptor and platelet-derived growth factor receptor) (Arcaro et al., 2000) and GPCR (Falasca & Maffucci, 2012; Maffucci et al., 2005). At cellular level, PI3K-C2α has been demonstrated to regulate full translocation of the glucose transporter GLUT4 to the plasma membrane of muscle cells, a key event for the regulation of glucose homeostasis (Falasca et al., 2007). In addition, a recent report established PI3K-C2α as an essential spatiotemporal regulator of clathrin-mediated endocytosis through the localized production of PtdIns(3,4)P2 at the plasma membrane (Posor et al., 2013). On the other hand, PI3K-C2β activity has been associated to cell migration, and its downregulation inhibits the LPA-dependent migration of ovarian and cervical cancer cell lines (Maffucci et al., 2005). The third member of class II PI3K, PI3K-C2γ represents the less characterized enzyme, and at the present, no evidence regarding its specific intracellular function are available.

Vacuolar protein sorting 34 represents the unique member of class III, and it is mainly involved in generation of PtdIns(3)P, a central phospholipid for membrane trafficking processes. The localized production of PtdIns(3)P recruits effector proteins containing FYVE or PX (Phox homology) domains that control membrane docking and fusion during the formation of internal vesicles required for recycling and autophagy pathways (Backer, 2008; Birkeland & Stenmark, 2004).

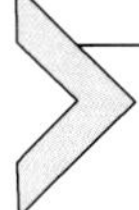

2. ANALYSIS OF PI3K LIPID KINASE ACTIVITY IN IMMUNOPRECIPITATE

As introduced above, *in vitro* all PI3Ks are capable of phosphorylating the D-3 position of a preferential phosphatidylinositol and their structure and their mechanism of activation/regulation determine the *in vivo* and *in vitro* production of a specific lipid product. In this section, we describe different methods to measure the PI3K lipid kinase activity associated to immunoprecipitates.

2.1. Measurement of class I PI3K lipid kinase activity

Class I PI3Ks are heterodimeric enzymes constituted by a regulatory and a catalytic subunit. Class IA is represented by three different catalytic subunits, p110α, p110β, and p110δ and five regulatory subunits, p85α, p55α, p50α, p85β, and p55γ. On the other hand, PI3Kγ is the unique member of class IB, and it can be formed by the association of the p110γ catalytic subunit with either p101 or p84/87 regulatory subunit.

All class IA PI3Ks are obligated heterodimers, and the p85 and p110 subunits are present in equimolar amounts in mammalian cell lines and tissue (Geering, Cutillas, Nock, Gharbi, & Vanhaesebroeck, 2007). Conversely, p110γ lipid kinase activity can be stimulated even in the absence of the regulatory subunit (Leopoldt et al., 1998). Accordingly, to measure the lipid kinase activity of class IA PI3Ks, it is necessary to coexpress the two subunits, while for class IB, it is possible to measure the kinase activity by using the p110γ monomer.

Here, we describe a method to measure the lipid kinase activity associated to the p85 regulatory subunits. However, this method can be applied to others regulatory subunits by using specific antibodies.

2.1.1 Immunoprecipitation

1. Use NIH-3T3 cells to immunoprecipitate p85 regulatory subunits.
2. Grow NIH-3T3 in 10 cm or larger cell culture dishes in presence of the proper medium, Dulbecco's modified Eagle's medium (DMEM) supplemented with 10% calf serum, 2 m*M* glutamine, 100 μg/ml streptomycin, and 100 units/ml penicillin at 37 °C and in a humid atmosphere of 5% CO_2. To maintain the culture, once cells are confluent, split them into either a new 10-cm dish by using 1 × trypsin/EDTA.
3. When a 10-cm dish is confluent, transfer the dish on ice and wash cells twice with ice-cold phosphate buffered saline (PBS). Discard the PBS carefully and add 500 μl of ice-cold lysis buffer and incubate for 10–15 min at 4 °C with moderate shaking. Scrape cells from the dish, collect the lysate, and transfer it in a microcentrifuge tube. Centrifuge in a refrigerated centrifuge at 10,000 × *g* for 10 min.
4. Transfer supernatant in a new tube and measure protein concentration by using the Bradford protein assay. Read the relative absorbance in a visible light spectrophotometer at 595 nm and compare the absorbance value with a standard curve.
5. Preclear the lysate to reduce unspecific binding of protein to sepharose beads by incubating protein extracts with sepharose beads for 1 h at 4 °C under rotation.
6. Based on the host specie of the anti-p85 antibody, choose the appropriate protein coupled sepharose beads as indicated in Table 6.1. Incubate 1–2 mg of cell lysate with 50 μl of bead slurry in the presence of either anti-p85 antibody or an irrelevant antibody as negative control. Incubate at 4 °C for 30–60 min on a rotator.
7. Spin at 2000 × *g* for 2 min in a refrigerated centrifuge and transfer the supernatant to a new tube. Add 50 μl of new beads slurry and 5 μg of the anti-p85 antibody to the precleared lysate. Incubate the sample for 2 h at 4 °C under rotation.

Table 6.1 Beads-species specificity (Bonifacino, Dell'Angelica, & Springer, 2001)

Species	Immunoglobulin isotype	Protein A	Protein G
Human	IgG1	+++	+++
	IgG2	+++	+++
	IgG3	–	+++
	IgG4	+++	+++
	IgE	–	+
	IgA	–	+
	IgM	Use antihuman IgM	Use antihuman IgM
Mouse	IgG1	+	+++
	IgG2a	+++	+++
	IgG2b	++	++
	IgG3	+	+
	IgM	Use antimouse IgM	Use antimouse IgM
Rat	IgG1	–	+
	IgG2a	–	+++
	IgG2b	–	++
	IgG2c	+	++
Hamster	All isotypes	+	++
Rabbit	All isotypes	+++	++
Goat	All isotypes	–	++
Sheep	All isotypes	–	++
Cow	All isotypes	++	+++
Guinea pig	All isotypes	+++	++
Pig	All isotypes	+	++
Horse	All isotypes	++	+++
Chicken	All isotypes	–	++

8. After incubation, centrifuge immune complexes bound to the beads for 4 min at 2000 × *g* at 4 °C. Discard the supernatant and keep the beads.
9. Wash the beads twice with 1 ml of ice-cold lysis buffer and keep the beads on ice.
10. Prepare immunoprecipitates for lipid kinase assay by washing beads twice with 1 ml of ice-cold washing buffer and twice with 1 ml of kinase buffer.

11. Carefully remove the supernatant and dry the beads with a flat-end syringe, without by disrupting the protein G sepharose beads. Resuspend the sepharose beads with 40 μl of kinase buffer.

2.1.2 Lipid kinase assay

1. Before starting, determine the specific activity (S_A) of the radioactive ^{32}P-γATP by considering the radioactive decay. Usually, for commercial ^{32}P-γATP, the S_A is specified as the calibration date (provided by supplier). This has to be considered in mass-dependent application. The S_A can be calculated considering formula I (Fig. 6.2A), for any day prior the calibration date, or formula II (Fig. 6.2B) and Table 6.2 for any day after the calibration date.
2. Prepare substrate solution I as described in Section 5 and keep it on ice.

A

$$S_A = \frac{S_{Acal}}{D_F + \frac{S_{Acal}\,(1-D_F)}{S_{ATheo}}}$$

B

$$S_A = \frac{D_F}{\frac{1}{S_{acal}} - \frac{(1-D_F)}{S_{ATheo}}}$$

Figure 6.2 *Formula for ^{32}P-γATP-specific activity calculation.* (A) Formula to calculate the 32P-γATP-specific activity for any day prior the calibration date. (B) Formula to calculate the ^{32}P-γATP-specific activity for any day after the calibration date. S_{Acal}, specific activity on the calibration date; D_F, fraction of current radioactivity remaining on the calibration date (Table 6.2). S_{ATheo}, 9120 Ci/mmol value for the theoretical-specific activity of carrier-free ^{32}P.

Table 6.2 Decay chart

Day	0	1	2	3	4	5	6	7	8	9
10	1.000	0.953	0.906	0.865	0.824	0.785	0.784	0.712	0.678	0.646
20	0.616	0.587	0.559	0.532	0.507	0.483	0.460	0.436	0.418	0.396
30	0.379	0.361	0.344	0.328	0.312	0.297	0.283	0.270	0.257	0.245
40	0.233	0.222	0.212	0.202	0.192	0.183	0.174	0.166	0.158	0.151
50	0.144	0.137	0.130	0.124	0.118	0.113	0.107	0.102	0.096	0.093
60	0.054	0.052	0.049	0.047	0.045	0.043	0.041	0.039	0.037	0.035

This table is used to calculate the residual-specific activity of 32P-γATP on any day after the calibration date.

3. Prepare substrate solution II as described in Section 5 and keep it on ice.
4. Add to the immunoprecipitates 10 μl of substrate solution I and 10 μl of substrate solution II.
5. Allow the kinase reaction to occur by incubating the samples at 30 °C for 10 min under constant mixing.
6. Stop enzymatic reaction by adding 100 μl of 1 N HCl and vortexing for few seconds.
7. Extract lipids with 200 μl of $CHCl_3$/MetOH (1:1 v/v), vortex, and centrifuge for 4 min at 2000 × *g*. After centrifugation, two different liquid phases can be detected. Collect the organic phase, which contains the phosphorylated lipids, on the bottom of the tube and transfer it in a new 1.5 ml tube.
8. Dry lipids in a vacuum concentration for 30 min at room temperature.
9. Resuspend dried pellet with 30–40 μl of a $CHCl_3$/MetOH (2:1 v/v) by mixing gently.

2.1.3 Thin layer chromatography

1. Use a silica gel thin layer chromatography (TLC) plate to monitor the amount of phosphorylated phosphoinositides produced in the Kinase reaction. Handle the TLC plate carefully with gloves to avoid contamination of the silica gel layer.
2. Preheat the TLC in an oven at 57 °C for 15–30 min to dry the TLC and eliminate possible humidity that can interfere with the chromatography.
3. Spot each sample drop by drop with a glass syringe, at least 1.5 cm of distance from one side of the plate. Be careful not to overlap one sample to another.
4. Dry TCL plate at room temperature to allow the solvent to completely evaporate.
5. Prepare the developing tank by adding the developing solvent. Once the tank is covered, wait for 10 min to allow the atmosphere in the tank to be saturated with the vapor of the developing solvent.
6. Insert the TLC plate into the prepared developing tank with the spots toward the bottom of the tank and in contact with the developing solvent.
7. Cover the developing tank with a lid and seal with a silicone vacuum grease to avoid solvent dispersion in the environment. The developing solvent will be absorbed by the silica gene, and it will ascend by capillary action. Because different analytes will be dragged by the solvent at different rates, separation of lipids is achieved. In this case, since only PtdIns(3)P is produced by the kinase reaction, the TLC chromatography allows the separation of lipids from the unbound ^{32}P-γATP fraction, which will remain at the base of the plate.

8. After about 1.5–2 h (once the solvent front has reached 3 cm from the top), remove the developed TLC plates and leave it to dry in an oven for 10–15 min.
9. Expose the TLC plate to a radiographic film for a period depending on the intensity of the radioactive signal.

2.2. Measurement of class II PI3K lipid kinase activity

Class II PI3K comprises three different isoforms, PI3K-C2α, PI3K-C2β, and PI3K-C2γ that, differently from class I enzymes, are monomers with high molecular mass. Structurally, these enzymes lack the binding domain to the regulatory subunits, while they maintain a Ras-binding domain, a class I PI3K-like C2 domain and a catalytic domain (Cantley, 2002). Another distinct characteristic is the extension of the C-terminus that contains a PX domain involved in phophatidylinositol (PI) binding and an additional C2 domain able to bind divalent cations like calcium (Stahelin et al., 2006). While all class II members share the majority of functional domains, they differ for the N-terminus extension as well as the specific role of this region. For example, PI3K-C2α specifically possesses a clathrin-binding domain that, once bound to clathrin, causes PI3K-C2α lipid kinase activation (Gaidarov, Smith, Domin, & Keen, 2001). Conversely, PI3K-C2β possesses a proline-rich region which mediates the association of this enzyme with activated EGF receptor (Wheeler & Domin, 2001). As previously described, Class II PI3K is able to phosphorylate *in vitro* both PtdIns and PtdIns(4)P (Fig. 6.1). Nonetheless, their ability to phosphorylate the PtdIns(4)P has been demonstrated to be strictly dependent on the presence of a specific divalent cation. Another report showed that human PI3K-C2α and PI3K-C2β can also use PtdIns(4,5)P2 as substrate but only after the addition of respectively phosphatidylserine (PS) and phosphatidylcholine, PS or phosphatidylethanolamine to the reaction (Arcaro et al., 1998; Domin et al., 1997; Fig. 6.1). Likewise, class I PI3K, PI3K-C2β, and PI3K-C2α are able to phosphorylate PtdIns and PtdIns(4)P in the presence of Mg^{2+}. Furthermore, both enzymes use the Ca^{2+} as a cofactor for phosphate transfer on PtdIns, while the addition of Ca^{2+} severely abolishes their ability to produce PtdIns(3,4)P2 (Arcaro et al., 2000). No evidence about the cation specificity of the third class II PI3K isoform, PI3K-C2γ, are available.

The ability of class II PI3Ks to use different cations to phosphorylate the PtdIns and PtdIns(4)P should be considered in the analysis of their kinase

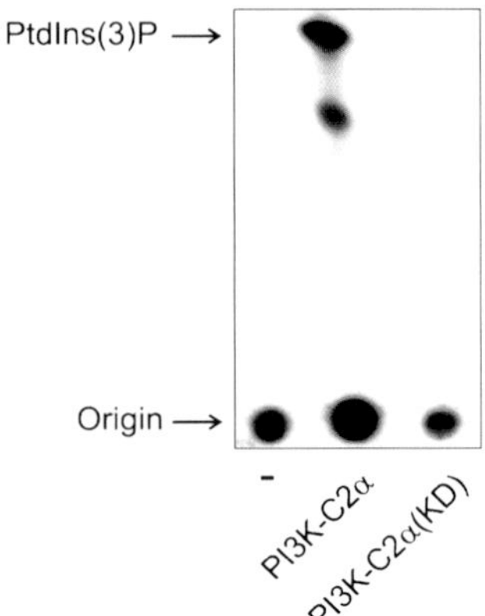

Figure 6.3 *Lipid kinase activity of immunoprecipitated PI3K-C2α.* Hek293 cells were transfected with GFP-fused *PIK3C2α* cDNA. The PI3K-C2α was immunoprecipitated with an antibody specific against the GFP, and the production of the PtdIns(3)P was measured (second line). The background (first line) was represented by the incubation of lipid substrate with the substrate solution II while the negative control was represented by the immunoprecipitation of a kinase inactive form of PI3K-C2α (PI3K-C2α(KD); third line).

activity. Here, we describe a method to estimate the PtdIns(3)P production by PI3K-C2α after its immunoprecipitation (Fig. 6.3).

2.2.1 Immunoprecipitation

Use NIH-3T3 cells to immunoprecipitate PI3K-C2α by using specific antibody. For the immunoprecipitation of PI3K-C2α, follow the methods described in Section 2.1.1.

2.2.2 Lipid kinase assay

For the lipid kinase assay of PI3K-C2α, follow the methods described in Section 2.1.2. Since PI3K-C2α is able to phosphorylate PtdIns both in presence of Mg^{2+} and Ca^{2+}, different kinase buffer can be used (see Section 5). For the reaction, use substrate solution III instead of substrate solution II because it is preferable to use 100 μ*M* ATP instead of 10 μ*M* ATP during the lipid kinase reaction.

2.2.3 Thin layer chromatography

For silica gel TLC, follow Section 2.1.3.

2.3. Measurement of class I PI3K lipid kinase activity associated to IRS protein

Class IA PI3K can be activated by RTK though the interaction of their phosphorylated YXXM motives or their substrates (i.e., IRS proteins for the insulin receptor, IR) with the Src Homology 2 (SH2) domain of the PI3K regulatory subunit (Burke & Williams, 2013). The binding of the regulatory subunit to phosphotyrosine induces a conformational change that releases its inhibitory action on the catalytic subunit, thus favoring the activation. Indeed, the incubation of the peptide-containing phosphotyrosine with the p85/p110 complex increases the lipid kinase activity *in vitro* (Pons et al., 1995; Wymann & Pirola, 1998). In this context, it is possible to evaluate the activation status of a specific class I PI3K isoform once recruited and activated by an RTK. The IR as well as the insulin-like growth factor-1 receptor trigger PI3K activation through an adaptor, the IR substrate (IRS). In mammals, six IRS isoforms have been described. Upon IR activation, they are recruited to the receptor and subsequently phosphorylated (Metz & Houghton, 2011). Hence, the phosphomotifs on IRS allow the recruitment and activation of PI3K (Backer et al., 1992; Myers et al., 1992). In this section, we describe the method to measure the PI3K activity associated to the IR and in particular to IRS-1 protein.

2.3.1 Immunoprecipitation

1. Grow NIH-3T3 cell as described in Section 2.1.1.
2. Prepare two different 10 cm or larger cell culture dishes and ensure that cells are 70% confluent or less.
3. The day after, wash cells twice with PBS and then maintain cell in serum-starved condition, overnight, by adding DMEM supplemented with 2 m*M* glutamine, 100 μg/ml streptomycin, and 100 units/ml penicillin.
4. After starvation, discard starving medium and stimulate one plate for 5 min with DMEM supplemented with 1 μ*M* of insulin. Use the second plate as negative control by adding free DMEM. After stimulation, transfer both plates (stimulated and control) on ice, quickly discard the stimulus, and wash them with ice-cold PBS.
5. Remove PBS, add 500 μl of ice-cold lysis buffer each plate, and incubate for 10–15 min a 4 °C with moderate shaking. Scrape cells from the dish, collect the lysate, and transfer it in a microcentrifuge tube. Centrifuge samples in a refrigerated centrifuge at 10,000 × *g* for 10 min.

6. Transfer supernatant in a new tube and measure protein concentration by using the Bradford protein assay as described in Section 2.1.1, step 4.
7. For immunoprecipitation, use an antibody specific for IRS-1 protein and follow the procedure described in Section 2.1.1, steps 5–9.

2.3.2 Lipid kinase assay

For the lipid kinase assay of IRS immunoprecipitates, the protocol is comparable to that described in Section 2.1.2.

2.3.3 Thin layer chromatography

For silica gel TLC, follow the method described in Section 2.1.3.

3. ANALYSIS OF THE ACTIVITY OF SMALL MOLECULE PI3K INHIBITORS

Recent studies have demonstrated a fundamental role for the PI3K signaling pathway in tumor onset and maintenance (Ciraolo et al., 2011). In addition, because of the frequent activation in cancer, in recent years, the PI3K signaling pathway has been extensively investigated to identify the targets of choice for pharmacological treatment (Rodon, Dienstmann, Serra, & Tabernero, 2013). In particular, PI3K enzymes are considered the new targets for cancer treatment, and different pharmaceutical companies have now a research and development program on PI3K inhibitors.

Depending on the pharmacokinetic properties and isoform selectivity, the available PI3K inhibitors can be divided into three different classes: the unselective pan-PI3K, the isoform-selective, and the dual PI3K/mTOR inhibitors. Pan-PI3K inhibitors, Wortmannin and LY29402, represent the first generation of molecules that potently inhibit PI3Ks (Ciraolo et al., 2011). Wortmannin is a fungal furanosteroid metabolite able to interact with the ATP-binding pocket of PI3K by forming a covalent interaction between the C-20 position of the furan ring and the primary amine group of the conserved Lys833 residue of p110γ and the Lys802 of p110α (Wipf & Halter, 2005; Wymann et al., 1996). Although for several years Wortmannin represented an extreme useful molecule for the study of PI3K function, its high reactivity as well as the poor selectivity can lead to deleterious off-target effects. This compound displays inhibitory activity also against the

PI3K-related enzymes such as the protein kinase ataxia telangiectasia mutated, the ataxia telangiectasia and Rad3-related protein, the myosin light chain kinase, and DNA-dependent protein kinase (Marone, Cmiljanovic, Giese, & Wymann, 2008). In contrast to Wortmannin, LY294002 is a synthetic compound and an ATP competitive PI3K inhibitor with an inhibitory concentration (IC50) value against PI3Kβ, PI3Kα, PI3Kδ, and PI3Kγ of 0.31, 0.73, 1.06, and 6.60 μM respectively (Ciraolo et al., 2011; Vlahos, Matter, Hui, & Brown, 1994). Like Wortmannin, LY294002 does not inhibit PI3Ks exclusively, but have several off-target effects on other enzymes, such as Casein Kinase-2 (Gharbi et al., 2007).

In the past years, the advances in the understanding of the biological function of PI3K and the publication of the crystal structure of different class I isoforms have prompted chemists to generate new PI3K inhibitors with higher potency, lower toxicity to be used as therapeutic agents (Knight, 2010; Maira, Finan, & Garcia-Echeverria, 2010; Vadas, Burke, Zhang, Berndt, & Williams, 2011). In addition, genetic approaches have partially validated the inhibition of single PI3K isoform in complex disease conditions. PI3Kγ and PI3Kδ are mainly restricted to the hematopoietic system and they are often implicated in several pathologies, such as chronic inflammation, allergy, and autoimmune diseases. On the other hand, since PI3Kα and PI3Kβ are ubiquitously expressed and their function is implicated in the regulation of cell growth and glucose homeostasis, their inhibition can raise a potential toxicity (Blajecka, Borgstrom, & Arcaro, 2011; Vanhaesebroeck et al., 2010). Therefore, the current challenge is represented by the generation of isoform-selective PI3K inhibitors for the treatment of specific diseases, such as the use of the PI3Kγ-selective inhibitor, AS65240, for the treatment of the rheumatoid arthritis (Camps et al., 2005).

Therefore, during the developing process of new molecular entities, the analysis of the PI3K lipid product has become fundamental for the isolation and identification of new pan or isoform-selective inhibitory molecules. Therefore, in recent years, several lipid kinase assay formats adapted to automated and high-throughput screening platform have been developed. Although the radioactive lipid kinase assay (described below) may be suitable for testing new molecular entities, the use of the radioactive ^{32}P isotope has several disadvantages in terms of licensing restriction, potential health risks associated with the exposure and lack of feasibility for high-throughput screening. Indeed, different companies offer new nonradioactive screening strategies with similar sensitivity and mostly based on fluorescence-based and luminescent kinase assay.

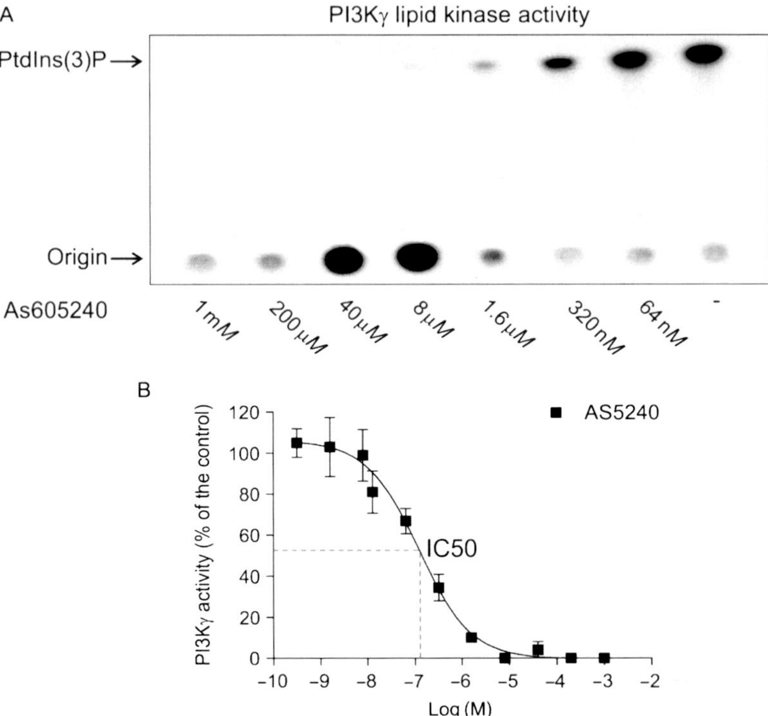

Figure 6.4 *Effect of AS605240 on PI3Kγ lipid kinase activity.* (A) Lipid kinase assay of PI3Kγ recombinant protein in presence or absence of different concentrations of the PI3Kγ-selective inhibitor, AS605240. (B) The resulting dose–response curve where the percentage of the residual PI3Kγ lipid kinase activity was related to the logarithmic concentration (log[M]). From the curve, an IC50 can be derived.

Here, we describe how the standard radioactive lipid kinase assay can be applied to the analysis of the inhibitory potential of new molecular entities; in particular, we show the inhibitory effect of AS605240 on PI3Kγ activity (Fig. 6.4A).

3.1. Dose–response effects of the PI3Kγ inhibitor AS605240

Although immunoprecipitated PI3Ks can be used for the screening of inhibitory molecules, we suggest to use recombinant PI3K enzymes produced through a baculovirus expression systems. This method ensures high functionality of the recombinant PI3K, high percent of purity, higher-specific activity, high reproducibility between different screenings, continuous supply, and high level of consistency in batch-to-batch production.

1. Prepare the PI3Kγ recombinant protein. Dilute the recombinant PI3Kγ in kinase buffer up to a final concentration of 6 μg/ml.

2. Dissolve the AS605240 (Camps et al., 2005) powder in DMSO to have a concentration of 50 m*M*. Dilute the compound in DMSO to have the following stock solutions: 10 m*M*, 2 m*M*, 400 μ*M*, 80 μ*M*, 16 μ*M*, 3.2 μ*M*.
3. Dilute the stock solutions in kinase buffer in order to have 2 m*M*, 400 μ*M*, 80 μ*M*, 16 μ*M*, 3.2 n*M*, 640 n*M*, and 128 n*M* working solutions (see Table 6.3). In this way, the amount of DMSO is maintained constant for each concentration.
4. Prepared substrate solution I and substrate solution II as described in Section 2.1.2.
5. In a microcentrifuge tube, mix 10 μl of recombinant protein and 30 μl of kinase buffer containing the inhibitor at the different concentrations. Leave the PI3Kγ inhibitor mixture for 15 min at room temperature.
6. Add to the PI3Kγ inhibitor mixture 10 μl of substrate solution I and 10 μl of substrate solution II. Start the enzymatic reaction at 30 °C for 10 min under constant mixing.
7. Proceed as described in Section 2.1.2, steps 6–9.

3.1.1 Thin layer chromatography

The procedure for the TLC is comparable to what described in Section 2.1.3.

3.1.2 Dose–response curve

Once the TCL plate has been exposed to the radiographic film, the intensity of the radioactive signal can be calculated and converted into an absolute

Table 6.3 Dilution scheme of AS605240

Samples	Stock solution (DMSO)		Working solution (kinase buffer)		Final concentration (reaction mix)
1	50 m*M*	→	2 m*M*	→	1 m*M*
2	10 m*M*		400 μ*M*		200 μ*M*
3	2 m*M*		80 μ*M*		40 μ*M*
4	400 μ*M*		16 μ*M*		8 μ*M*
5	80 μ*M*		3.2 μ*M*		1.6 μ*M*
6	16 μ*M*		640 n*M*		320 n*M*
7	3.2 μ*M*		128 n*M*		64 n*M*
8	100% DMSO		4% DMSO		2% DMSO

number. The residual PI3Kγ lipid kinase activity in presence of the AS605240 at different concentrations can be related to the control intended as the recombinant protein in the presence of the vehicle alone. To describe the relationship between response to drug treatment and drug dose or concentration, it is possible to generate a sigmoidal dose–response curve and calculate the half-maximal IC50 for the inhibitor. In the sigmoidal dose–response, the *X*-axis will contain the concentration (Molar—Fig. 6.4A) of the drug expressed on a logarithmic scale, and the *Y*-axis will plot the response intended as percentage of the residual PI3Kγ lipid kinase activity (Fig. 6.4B). The determined IC50 value will allow to define the drug/inhibitory potency of the compound.

3.2. Class II PI3K inhibitors

Pharmacological inhibitors selectively targeting class II PI3Ks have not been described yet. In particular, although both LY294002 and Wortmannin can inhibit class II PI3Ks activity *in vitro*, the sensitivity of class II PI3Ks to these two drugs is significantly lower than that of class I PI3Ks (Domin, Gaidarov, Smith, Keen, & Waterfield, 2000; Maffucci et al., 2005). Among class II PI3Ks, PI3K-C2α is the most refractory to inhibition by Wortmannin *in vitro*. Little inhibition of enzyme activity was achieved using 10 n*M* Wortmannin, but PI3K-C2α activity was still evident using 100 n*M* inhibitor and abolished only using 1 μ*M* Wortmannin, (Domin et al., 2000) a dose that is about 200 × higher than the IC50 for class I PI3Ks (4 n*M*) (Walker et al., 2000). Moreover, sensitivity to LY294002 is a key feature of PI3K-C2β that shows a different sensitivity to this inhibitor compared to class I PI3Ks. Of note, 1 μ*M* LY294002 has no effect on the activity of PI3K-C2β, while it significantly inhibits class I PI3Ks. In particular, a concentration of 10 μ*M* LY294002 completely blocks class I PI3Ks, whereas it has only a partial inhibitory effect on PI3K-C2β (Maffucci et al., 2005). These data clearly suggest that specific and effective inhibition of class II PI3Ks could be achieved with novel inhibitors specifically designed on the catalytic pocket of these enzymes. To screen new potential class II PI3Ks inhibitors, you can refer to Section 2.2 for measurement of class II PI3Ks lipid kinase activity.

4. ANALYSIS OF THE LOCALIZED SYNTHESIS OF PI3K LIPID PRODUCTS

The PI3K lipid products are produced at cellular membranes where they act as second messengers that regulate several cellular processes such as vesicular trafficking, cytoskeletal remodeling, and signal transduction.

In recent year, several methods to study the generation of the PI3K lipid product directly in living cells have been developed thus allowing the analysis of their localized synthesis. Taking advantage of the conserved protein module selected during the evolution for the high affinity binding to specific phosphoinositides, it is possible to use these interaction domains to examine within the cells the spatial and temporal distribution/synthesis of phosphoinositides in real time. Up to few years ago, the only protein domains considered suitable for the generation of intracellular inositide-binding probes were the pleckstrin homology (PH) domains. Nonetheless, the number of protein domains known to bind phosphoinositides is constantly increasing.

The most exploited inositide-binding probes for the analysis of PI3K family activation are the Akt-PH-GFP, TAPP1-PH-GFP, and EEA1-2x-FYVE-GFP (Table 6.4 and Fig. 6.5) that are able to recognize the PI3K lipid product with high affinity and specificity. In particular, the Akt-PH probe

Table 6.4 A panel of inositide-binding probes for the analysis of PI3K family activation

Phosphoinositide-binding probes	Recognized phosphoinositide	Localization	Derived from	For the activity of
Akt-PH-GFP	PtdIns(3,4,5)P3 (Watton & Downward, 1999) PtdIns(3,4)P2 (Gray, Van Der Kaay, & Downes, 1999)	Plasma membrane (Gray et al., 1999) Endosomes (Corvera & Czech, 1998)	Akt (Gray et al., 1999)	Class I PI3K
TAPP1-PH-GFP	PtdIns(3,4)P2 (Manna, Albanese, Park, & Cho, 2007)	Clathrin-coated vesicles (Posor et al., 2013) Endosomes (Watt et al., 2004)	Tandem PH domain-containing Protein 1 (TAPP1) (Watt et al., 2004)	Class II PI3K
EEA1-2x-FYVE-GFP	PtdIns(3)P (Corvera, D'Arrigo, & Stenmark, 1999; Gillooly et al., 2000)	Early/recycling endosomes (Corvera et al., 1999; Gillooly et al., 2000)	Early endosomal antigen 1 (EEA1) (Itoh & Takenawa, 2002)	Class III PI3K and Class II PI3K

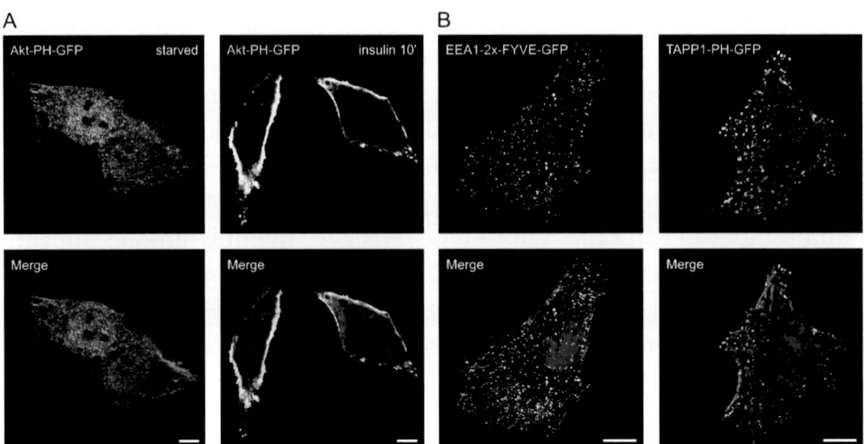

Figure 6.5 *Phosphoinositide-binding probes and their intercellular localization.* (A) Analysis of the intracellular localization of Akt-PH-GFP probe in HeLa cells during serum deprivation and after 10 min of insulin stimulation. After the addition of insulin, it is possible to appreciate the relocalization of the probe from cytosol (left panel) to plasma membrane (right panel) as a consequence of class I PI3K activation. (B) Analysis of the intracellular localization of EEA1-2x-FYVE-GFP (left panel) and TAPP1-PH-GFP (right panel) in HeLa cells cultured in Dulbecco's modified Eagle's medium (DMEM) supplemented with 10% fetal bovine serum. (See the color plate.)

was derived from the PH domain of Akt (Watton & Downward, 1999), one of the main effector proteins of class I PI3Ks, that is, able to recognize both the PtdIns(3,4,5)P3 and PtdIns(3,4)P2. After insulin stimulation, it is possible to appreciate the enrichment of this probe at plasma membrane as a result of class I PI3K activation (Gray et al., 1999; Fig. 6.5A). On the other hand, with TAPP1-PH, it is possible to appreciate the enrichment of the PtdIns(3,4)P2 at plasma membrane as a result of PI3K class II alpha activity at clathrin-coated vesicles during endocytosis (Fig. 6.5B; Posor et al., 2013). Finally, the EEA1-2x-FYVE is one of the first PtdIns(3)P-binding domain identified and largely used to analyze lipid kinase activity of class III PI3K in vesicular trafficking and phagocytosis (Fig. 6.5B; Corvera et al., 1999; Gillooly et al., 2000).

In this section, we will describe a method to evaluate the localized synthesis of a specific pool of Class I PI3K lipid products, both *in vitro* and *in vivo*, using these three GFP-fused inositide-binding probes.

4.1. Evaluation of PtdIns(3,4,5)P3 in fixed cells

After insulin stimulation, class I PI3Ks, and in particular PI3Kα and PI3Kβ, are recruited to the plasma membrane and activated to produce PtdIns(3,4,5)

P3. Through the use of the Akt-PH-GFP-binding probe, it is possible to visualize the enrichment at the plasma membrane of such lipid second messenger (Gray et al., 1999). Here, we describe the method to visualize PtdIns(3,4,5)P3 after 10 min of insulin stimulation and cell fixation (Fig. 6.5A). Nonetheless, it is possible to follow the production and the localization of the PtdIns(3,4,5)P3 in a time course by observing living cell transfected with the probe under a time-lapse microscope.

4.1.1 Preparation of poly-lysine-coated cover slips and cell transfection

1. Rinse 25-mm coverslips with 98% ethanol in the cell culture hood and air dry. Prepare a 6-well microtiter for cell culture and transfection by adding the 25-mm coverslip and coat them with 1 ml of poly-lysine solution for 1 h, in order to avoid the autofluorescence given by several plasticware for cell.
2. Gently aspirate the solution and air dry before plating cells.
3. Plate ~50,000 HeLa cells directly onto the poly-lysine-coated coverslips. Grow HeLa cells to 70% confluence in 2 ml of the proper medium, DMEM supplemented with 10% fetal bovine serum, 2 m*M* glutamine, 100 μg/ml streptomycin, and 100 units/ml penicillin at 37 °C and in a humid atmosphere of 5% CO_2.
4. Transfect the HeLa cells with an expression DNA plasmid containing the Akt-PH-GFP probe sequence using the Effectene Tranfection Reagent (Qiagen). For cell transfection, different protocols are available, and for each cell type, the reagent and the procedure that gives the best result can differ considerably. Thus, the user should refer to the manufacturers's instructions for different cell transfection methods. The optimal level of transfection has to be determined for each cell type and expression construct. It is preferable to not exceed with protein expression because high expression levels of lipid-binding probes are often toxic for the cells due to interference with endogenous proteins involved in lipid binding.
5. Let the cells grow for 24 h to allow the expression of the transfected protein.
6. Starve cells for 6–8 h by adding serum-free DMEM medium to render them quiescent before microscopy.
7. Stimulate cells by adding 1 μ*M* of insulin for 10–15 min and after that immediately fix the cells
8. The addition of insulin will lead to a redistribution of the Akt-PH-GFP-binding probe from the cytoplasm to the plasma membrane as a result of PtdIns(3,4,5)P3 production by class I PI3Ks (Fig. 6.5A).

4.1.2 *Observing the GFP signal by microscopy*

Once fixed, transfected and stimulated cells can be directly observed under the microscope without any other additional treatment. EGFP fluorescence can persists in fixed cells for several days, and fixed cells can be processed for an additional immunostaining which is often helpful to determine colocalization with specific compartment markers. A problem that should be considered is that often the GFP-fused inositole-binding probes can result in an insufficient GFP brightness thus impeding a proper visualization of the localized PtdIns(3,4,5)P3. The use of specific antibodies against the GFP portion of the probe can obviate the problem and fix cells can be subjected to immunostaining as described below.

1. Rinse fixed cells twice with 2 ml of PBS.
2. Add 2 ml of 0.1% saponin solution and incubate for 10 min.
3. Wash the cells 2 × with 2 ml of PBS for 5 min each wash.
4. Add 2 ml of blocking solution to block unspecific antibody binding. Incubate for 30 min at room temperature.
5. Remove the blocking solution. Do not wash cells.
6. Add the primary antibodies diluted appropriately in blocking solution. Incubate for 1 h at room temperature.
7. Wash the cells 2 × with 2 ml of blocking solution for 5 min each wash.
8. Add the fluorescent secondary antibodies diluted in blocking solution. Incubate for 1 h at room temperature protected from light.
9. Wash the cells 3 × with 2 ml of PBS for 5 min each wash.
10. Air dry until the coverslips are only damp.
11. Mount the coverslips with the cells down on a glass slide using Prolong Golden Antifade Reagent.

4.2. Time-lapse imaging of live cells with a confocal microscope

The use of fixed cell does not allow to estimate small variations of phosphoinositides production during a time course, and in some cases, fixation and permeabilization procedures may distort cellular morphology. Therefore, the use of living cells is the most reliable way of assessing proper morphology and to evaluate the response to stimulus in real time and the following protocol can be used.

1. For cell culture and transfection procedure, refer to Section 4.1.1.
2. Starve cells for 6–8 h by adding serum free.
3. Prepare time-lapse microscope by turning on the air stream incubator in order to allow thermal equilibration of the stage and set the right

microscope objective and filter wheel. Let microscope to equilibrate for at least 1 h before starting the experiment.

4. Turn on the lasers, computer, camera, filter wheel, and shutter controllers and wait 20 min before starting the experiment. Apply a drop of immersion oil to the dove prism and place the confocal Petri dish on the stage over the drop of oil. Make sure that there are no air bubbles in the immersion oil because they may disperse the laser beam and cause background epifluorescence illumination. Observe the cells for green fluorescence with epi-illumination by using 488 nm light for excitation and 509 nm emission filters. Visualize the cells with the CCD camera and adjust the laser power and the exposure time considering that the cells will become severalfold brighter after stimulation because the transition of the probe from a dispersed cytoplasmic localization to a more localized positioning at the plasma membrane. Indeed, an optimal setting should allow utilization of the full dynamic range of the camera without significant photobleaching over the time course of the experiment.
5. Set the software to record time-lapsed images every 5–20 s and at a scanning speed of 1.5–2.5 s to capture the cellular response with proper image resolution.
6. Start acquisition of the steady state of the cells for 2–5 min.
7. Add the insulin stimulus (the final concentration should be 1 μ*M*) without interrupting the acquisition. The addition of insulin should activate class I PI3K and at the same time cause a redistribution of the Akt-PH-GFP probes. Proceed with the recording for not <15 min and up to 30 min.
8. After completing the acquisition, quantify the results off line with either a line intensity histogram through a selected line spanning the image or with a pixel intensity histogram over the entire area, or only a selected area, of the cells. To quantify the translocation response of the Akt-PH-GFP probe after stimulation, quantify translocation in terms of fluorescence changes relative to a baseline fluorescence ($\Delta F/F_0$) given that often the fluorescence intensity significantly varies between the cells due to differences in the expression of the fluorescent proteins.

5. MATERIALS

1. *PBS*: 137 m*M* NaCl, 237 m*M* KCl, 10 m*M* NA_2HPO_4, and 1.76 m*M* KH_2PO_4, pH 7.4, in bidistilled water. Sterilize by autoclaving.

2. *Trypsin/EDTA solution* (10 ×): 5.0 g/l porcine trypsin and 2 g/l EDTA in 0.9% sodium chloride. Sterilize through filtering prior to use.
3. *Lysis buffer for protein extraction*: 20 m*M* Tris–HCl pH 8.0, 138 m*M* NaCl, 5 m*M* EDTA, 2.7 m*M* KCl, 1 m*M* $MgCl_2$, 1 m*M* CaCl2, 5% glycerol, and 1% Nonidet p-40. The choice ofthe detergent is extremely important to maintain the kinase activity of the kinase of interest. Prefer nonionic detergent (e.g., Triton X-100, NP-40) to the anionic detergent (e.g., sodium dodecyl sulphate (SDS), and sodium deoxycholate) that denatures the protein, often irreversibly. Just prior to use add phosphatase and protease inhibitor: 1 m*M* sodium vanadate, 20 m*M* NaF, 10 m*M* sodium pyrophosphate, 1 μg/ml aprotinin, 1 μg/ml leupeptin, and 1 μg/ml pepstatin.
4. *Washing buffer*: 01 *M* Tris–HCl, pH 7.4 and 0.5 *M* LiCl.
5. *Kinase buffer*: 20 m*M* Hepes–HCl, pH 7.4 and 5 m*M* $MgCl_2$.
6. *Kinase buffer with* Ca^{2+}: 20 m*M* Hepes–HCl, pH 7.4 and 5 m*M* $CaCl_2$.
7. *PI stock solution*: dissolve 1 mg/ml of L-α-PI in CHCl3–MetOH (2:1 v/v) and store in aliquots at −80 °C.
8. *PS stock solution*: dissolve 1 mg/ml of L-α-PS in CHCl3–MetOH (9:1 v/v) and store in aliquots at −80 °C.
9. *Substrate solution I*: to be prepared just prior to use. Mix one volume of PI stock solution and one volume of PS stock solution in a 1.5 microcentrifuge tube. Dry the lipids very carefully under a stream of nitrogen gas. Resuspend dried lipids with kinase buffer to a final concentration of 1 mg/ml and sonicate for 15 s at moderate amplitude.
10. *Substrate solution II*: to be prepared just prior to use. Add to a proper volume (depending on the number of samples) of kinase buffer the cold ATP up to a final concentration of 0.06 μ*M* and the radioactive ^{32}P-γ ATP (^{32}P-γATP-specific activity of 6000 Ci/mmol) up to a final concentration of 500 μCi/ml.
11. *Substrate solution III*: to be prepared just prior to use. Add to a proper volume (depending on the number of samples) of kinase buffer the cold ATP up to a final concentration of 0.60 μ*M* and the radioactive ^{32}P-γ ATP (^{32}P-γATP-specific activity of 6000 Ci/mmol) up to a final concentration of 500 μCi/ml.
12. *Developing solvent*: prepare a solution of $CHCl_3$–MeOH–H_2O–NH_4OH(25%), 45:35:8.5:1.5 v/v.
13. *Poly-lysine solution*: prepare a 1:100 dilution of poly-lysine in sterile deionized water.

14. *4% Paraformaldehyde*: dissolve paraformaldehyde powder in heated PBS solution (60 °C) at a concentration of 4% (w\v). Heat while stirring at 60 °C until the powder will be completely dissolved. Take care that the solution does not boil. Once the paraformaldehyde is dissolved, the solution should be cooled, filtered, and adjusted the pH to 7.4 with NaOH.
15. *Blocking solution*: freshly prepare 2% solution of BSA in PBS.
16. *Saponin solution*: prepare an 1% solution of saponin by dissolving 0.1 g of saponin powder in 10 ml of PBS. Aliquote and store at −20 °C. To prepare 10 ml of the 0.1% saponin solution, dilute 1 ml of the 1% solution in 9 ml of PBS.

ACKNOWLEDGMENTS

This work was supported by a grant from Fondazione San Paolo and Associazione Italiana Ricerca Cancro (AIRC) and Bando Futuro in ricerca 2010.

REFERENCES

Arcaro, A., Volinia, S., Zvelebil, M. J., Stein, R., Watton, S. J., Layton, M. J., et al. (1998). Human phosphoinositide 3-kinase C2beta, the role of calcium and the C2 domain in enzyme activity. *The Journal of Biological Chemistry*, *273*(49), 33082–33090.

Arcaro, A., Zvelebil, M. J., Wallasch, C., Ullrich, A., Waterfield, M. D., & Domin, J. (2000). Class II phosphoinositide 3-kinases are downstream targets of activated polypeptide growth factor receptors. *Molecular and Cellular Biology*, *20*(11), 3817–3830.

Backer, J. M. (2008). The regulation and function of Class III PI3Ks: Novel roles for Vps34. *The Biochemical Journal*, *410*(1), 1–17. http://dx.doi.org/10.1042/BJ20071427, PII: BJ20071427.

Backer, J. M., Myers, M. G., Jr., Shoelson, S. E., Chin, D. J., Sun, X. J., Miralpeix, M., et al. (1992). Phosphatidylinositol 3'-kinase is activated by association with IRS-1 during insulin stimulation. *The EMBO Journal*, *11*(9), 3469–3479.

Birkeland, H. C., & Stenmark, H. (2004). Protein targeting to endosomes and phagosomes via FYVE and PX domains. *Current Topics in Microbiology and Immunology*, *282*, 89–115.

Blajecka, K., Borgstrom, A., & Arcaro, A. (2011). Phosphatidylinositol 3-kinase isoforms as novel drug targets. *Current Drug Targets*, *12*(7), 1056–1081.

Bonifacino, J. S., Dell'Angelica, E. C., & Springer, T. A. (2001). Immunoprecipitation. *Current Protocols in Immunology*. http://dx.doi.org/10.1002/0471142735.im0803s41, Chapter 8, Unit 8. 3.

Burke, J. E., & Williams, R. L. (2013). Dynamic steps in receptor tyrosine kinase mediated activation of class IA phosphoinositide 3-kinases (PI3K) captured by H/D exchange (HDX-MS). *Advances in Biological Regulation*, *53*(1), 97–110. http://dx.doi.org/10.1016/j.jbior.2012.09.005.

Camps, M., Ruckle, T., Ji, H., Ardissone, V., Rintelen, F., Shaw, J., et al. (2005). Blockade of PI3Kgamma suppresses joint inflammation and damage in mouse models of rheumatoid arthritis. *Nature Medicine*, *11*(9), 936–943. http://dx.doi.org/10.1038/nm1284.

Cantley, L. C. (2002). The phosphoinositide 3-kinase pathway. *Science*, *296*(5573), 1655–1657. http://dx.doi.org/10.1126/science.296.5573.1655.

Ciraolo, E., Morello, F., & Hirsch, E. (2011). Present and future of PI3K pathway inhibition in cancer: Perspectives and limitations. *Current Medicinal Chemistry*, *18*(18), 2674–2685.

Corvera, S., & Czech, M. P. (1998). Direct targets of phosphoinositide 3-kinase products in membrane traffic and signal transduction. *Trends in Cell Biology*, *8*(11), 442–446. http://dx.doi.org/10.1016/S0962-8924(98)01366-X.

Corvera, S., D'Arrigo, A., & Stenmark, H. (1999). Phosphoinositides in membrane traffic. *Current Opinion in Cell Biology*, *11*(4), 460–465. http://dx.doi.org/10.1016/S0955-0674(99)80066-0.

Domin, J., Gaidarov, I., Smith, M. E., Keen, J. H., & Waterfield, M. D. (2000). The class II phosphoinositide 3-kinase PI3K-C2alpha is concentrated in the trans-Golgi network and present in clathrin-coated vesicles. *The Journal of Biological Chemistry*, *275*(16), 11943–11950.

Domin, J., Pages, F., Volinia, S., Rittenhouse, S. E., Zvelebil, M. J., Stein, R. C., et al. (1997). Cloning of a human phosphoinositide 3-kinase with a C2 domain that displays reduced sensitivity to the inhibitor wortmannin. *The Biochemical Journal*, *326*(Pt 1), 139–147.

Falasca, M., Hughes, W. E., Dominguez, V., Sala, G., Fostira, F., Fang, M. Q., et al. (2007). The role of phosphoinositide 3-kinase C2alpha in insulin signaling. *The Journal of Biological Chemistry*, *282*(38), 28226–28236. http://dx.doi.org/10.1074/jbc.M704357200.

Falasca, M., & Maffucci, T. (2012). Regulation and cellular functions of class II phosphoinositide 3-kinases. *The Biochemical Journal*, *443*(3), 587–601. http://dx.doi.org/10.1042/BJ20120008.

Gaidarov, I., Smith, M. E., Domin, J., & Keen, J. H. (2001). The class II phosphoinositide 3-kinase C2alpha is activated by clathrin and regulates clathrin-mediated membrane trafficking. *Molecular Cell*, 7(2), 443–449. http://dx.doi.org/10.1016/S1097-2765(01)00191-5.

Geering, B., Cutillas, P. R., Nock, G., Gharbi, S. I., & Vanhaesebroeck, B. (2007). Class IA phosphoinositide 3-kinases are obligate p85-p110 heterodimers. *Proceedings of the National Academy of Sciences of the United States of America*, *104*(19), 7809–7814. http://dx.doi.org/10.1073/pnas.0700373104.

Gharbi, S. I., Zvelebil, M. J., Shuttleworth, S. J., Hancox, T., Saghir, N., Timms, J. F., et al. (2007). Exploring the specificity of the PI3K family inhibitor LY294002. *The Biochemical Journal*, *404*(1), 15–21. http://dx.doi.org/10.1042/BJ20061489.

Gillooly, D. J., Morrow, I. C., Lindsay, M., Gould, R., Bryant, N. J., Gaullier, J. M., et al. (2000). Localization of phosphatidylinositol 3-phosphate in yeast and mammalian cells. *The EMBO Journal*, *19*(17), 4577–4588. http://dx.doi.org/10.1093/emboj/19.17.4577.

Gray, A., Van Der Kaay, J., & Downes, C. P. (1999). The pleckstrin homology domains of protein kinase B and GRP1 (general receptor for phosphoinositides-1) are sensitive and selective probes for the cellular detection of phosphatidylinositol 3,4-bisphosphate and/or phosphatidylinositol 3,4,5-trisphosphate in vivo. *The Biochemical Journal*, *344*(Pt 3), 929–936.

Hirsch, E., Ciraolo, E., Franco, I., Ghigo, A., & Martini, M. (2013). PI3K in cancer-stroma interactions: Bad in seed and ugly in soil. *Oncogene*. http://dx.doi.org/10.1038/onc.2013.265.

Hirsch, E., Ciraolo, E., Ghigo, A., & Costa, C. (2008). Taming the PI3K team to hold inflammation and cancer at bay. *Pharmacology & Therapeutics*, *118*(2), 192–205. http://dx.doi.org/10.1016/j.pharmthera.2008.02.004.

Itoh, T., & Takenawa, T. (2002). Phosphoinositide-binding domains: Functional units for temporal and spatial regulation of intracellular signalling. *Cellular Signalling*, *14*(9), 733–743, PII: S0898656802000281.

Knight, Z. A. (2010). Small molecule inhibitors of the PI3-kinase family. *Current Topics in Microbiology and Immunology*, *347*, 263–278. http://dx.doi.org/10.1007/82_2010_44.

Leopoldt, D., Hanck, T., Exner, T., Maier, U., Wetzker, R., & Nurnberg, B. (1998). Gbetagamma stimulates phosphoinositide 3-kinase-gamma by direct interaction with two domains of the catalytic p110 subunit. *The Journal of Biological Chemistry*, *273*(12), 7024–7029.

Maffucci, T., Cooke, F. T., Foster, F. M., Traer, C. J., Fry, M. J., & Falasca, M. (2005). Class II phosphoinositide 3-kinase defines a novel signaling pathway in cell migration. *The Journal of Cell Biology*, *169*(5), 789–799. http://dx.doi.org/10.1083/jcb.200408005.

Maira, S. M., Finan, P., & Garcia-Echeverria, C. (2010). From the bench to the bed side: PI3K pathway inhibitors in clinical development. *Current Topics in Microbiology and Immunology*, *347*, 209–239. http://dx.doi.org/10.1007/82_2010_60.

Manna, D., Albanese, A., Park, W. S., & Cho, W. (2007). Mechanistic basis of differential cellular responses of phosphatidylinositol 3,4-bisphosphate- and phosphatidylinositol 3,4,5-trisphosphate-binding pleckstrin homology domains. *The Journal of Biological Chemistry*, *282*(44), 32093–32105. http://dx.doi.org/10.1074/jbc.M703517200, PII: M703517200.

Marone, R., Cmiljanovic, V., Giese, B., & Wymann, M. P. (2008). Targeting phosphoinositide 3-kinase: Moving towards therapy. *Biochimica et Biophysica Acta*, *1784*(1), 159–185. http://dx.doi.org/10.1016/j.bbapap.2007.10.003.

Metz, H. E., & Houghton, A. M. (2011). Insulin receptor substrate regulation of phosphoinositide 3-kinase. *Clinical Cancer Research*, *17*(2), 206–211. http://dx.doi.org/10.1158/1078-0432.CCR-10-0434.

Myers, M. G., Jr., Backer, J. M., Sun, X. J., Shoelson, S., Hu, P., Schlessinger, J., et al. (1992). IRS-1 activates phosphatidylinositol 3'-kinase by associating with src homology 2 domains of p85. *Proceedings of the National Academy of Sciences of the United States of America*, *89*(21), 10350–10354.

Pons, S., Asano, T., Glasheen, E., Miralpeix, M., Zhang, Y., Fisher, T. L., et al. (1995). The structure and function of p55PIK reveal a new regulatory subunit for phosphatidylinositol 3-kinase. *Molecular and Cellular Biology*, *15*(8), 4453–4465.

Posor, Y., Eichhorn-Gruenig, M., Puchkov, D., Schoneberg, J., Ullrich, A., Lampe, A., et al. (2013). Spatiotemporal control of endocytosis by phosphatidylinositol-3,4-bisphosphate. *Nature*, *499*(7457), 233–237.

Rodon, J., Dienstmann, R., Serra, V., & Tabernero, J. (2013). Development of PI3K inhibitors: Lessons learned from early clinical trials. *Nature Reviews. Clinical Oncology*, *10*(3), 143–153. http://dx.doi.org/10.1038/nrclinonc.2013.10.

Stahelin, R. V., Karathanassis, D., Bruzik, K. S., Waterfield, M. D., Bravo, J., Williams, R. L., et al. (2006). Structural and membrane binding analysis of the Phox homology domain of phosphoinositide 3-kinase-C2alpha. *The Journal of Biological Chemistry*, *281*(51), 39396–39406. http://dx.doi.org/10.1074/jbc.M607079200.

Tolias, K. F., & Cantley, L. C. (1999). Pathways for phosphoinositide synthesis. *Chemistry and Physics of Lipids*, *98*(1–2), 69–77.

Vadas, O., Burke, J. E., Zhang, X., Berndt, A., & Williams, R. L. (2011). Structural basis for activation and inhibition of class I phosphoinositide 3-kinases. *Science Signaling*, *4*(195), re2. http://dx.doi.org/10.1126/scisignal.2002165.

Vanhaesebroeck, B., Guillermet-Guibert, J., Graupera, M., & Bilanges, B. (2010). The emerging mechanisms of isoform-specific PI3K signalling. *Nature Reviews. Molecular Cell Biology*, *11*(5), 329–341. http://dx.doi.org/10.1038/nrm2882.

Vlahos, C. J., Matter, W. F., Hui, K. Y., & Brown, R. F. (1994). A specific inhibitor of phosphatidylinositol 3-kinase, 2-(4-morpholinyl)-8-phenyl-4H-1-benzopyran-4-one (LY294002). *The Journal of Biological Chemistry*, *269*(7), 5241–5248.

Walker, E. H., Pacold, M. E., Perisic, O., Stephens, L., Hawkins, P. T., Wymann, M. P., et al. (2000). Structural determinants of phosphoinositide 3-kinase inhibition by wortmannin, LY294002, quercetin, myricetin, and staurosporine. *Molecular Cell*, *6*(4), 909–919.

Watt, S. A., Kimber, W. A., Fleming, I. N., Leslie, N. R., Downes, C. P., & Lucocq, J. M. (2004). Detection of novel intracellular agonist responsive pools of phosphatidylinositol 3,4-bisphosphate using the TAPP1 pleckstrin homology domain in immunoelectron microscopy. *The Biochemical Journal*, *377*(Pt 3), 653–663. http://dx.doi.org/10.1042/BJ20031397, PII: BJ20031397.

Watton, S. J., & Downward, J. (1999). Akt/PKB localisation and 3' phosphoinositide generation at sites of epithelial cell-matrix and cell-cell interaction. *Current Biology*, *9*(8), 433–436, PII: S0960982299801924.

Wheeler, M., & Domin, J. (2001). Recruitment of the class II phosphoinositide 3-kinase C2beta to the epidermal growth factor receptor: Role of Grb2. *Molecular and Cellular Biology*, *21*(19), 6660–6667.

Wipf, P., & Halter, R. J. (2005). Chemistry and biology of wortmannin. *Organic & Biomolecular Chemistry*, *3*(11), 2053–2061. http://dx.doi.org/10.1039/b504418a.

Wymann, M. P., Bulgarelli-Leva, G., Zvelebil, M. J., Pirola, L., Vanhaesebroeck, B., Waterfield, M. D., et al. (1996). Wortmannin inactivates phosphoinositide 3-kinase by covalent modification of Lys-802, a residue involved in the phosphate transfer reaction. *Molecular and Cellular Biology*, *16*(4), 1722–1733.

Wymann, M. P., & Pirola, L. (1998). Structure and function of phosphoinositide 3-kinases. *Biochimica et Biophysica Acta*, *1436*(1–2), 127–150.

Yang, H. W., Shin, M. G., Lee, S., Kim, J. R., Park, W. S., Cho, K. H., et al. (2012). Cooperative activation of PI3K by Ras and Rho family small GTPases. *Molecular Cell*, *47*(2), 281–290. http://dx.doi.org/10.1016/j.molcel.2012.05.007, PII: S1097-2765(12)00390-5.

CHAPTER SEVEN

Luciferase-Based Reporter to Monitor the Transcriptional Activity of the *SIRT3* Promoter

F. Kyle Satterstrom[*,†], **Marcia C. Haigis**[†,1]
[*]Harvard School of Engineering and Applied Sciences, Cambridge, Massachusetts, USA
[†]Department of Cell Biology, Harvard Medical School, Boston, Massachusetts, USA
[1]Corresponding author: e-mail address: marcia_haigis@hms.harvard.edu

Contents

Abstract

Sirtuin 3 (SIRT3) is a major regulator of oncometabolism. Indeed, the activity of SIRT3 significantly affects the response to oxidative stress, glycolytic proficiency, and tumorigenic potential of malignant cells. Thus, a system to accurately measure the transcriptional activity of the *SIRT3* promoter could facilitate the identification of novel antineoplastic agents or have diagnostic applications. Here, we describe all the steps involved in the development of a luciferase-based reporter system to measure the activation of the human *SIRT3* promoter, encompassing the design of appropriate primers, the cloning of the promoter fragment, and its site-directed mutagenesis. We validated

Methods in Enzymology, Volume 543
ISSN 0076-6879
http://dx.doi.org/10.1016/B978-0-12-801329-8.00007-6

this system in human embryonic kidney 293T cells, taking advantage of the renowned ability of the transcription factor estrogen-related receptor α to transactivate SIRT3. Moreover, here we demonstrate that SIRT3 expression is responsive to rapamycin, a small inhibitor of mammalian target of rapamycin that has been extensively employed as a caloric restriction mimetic. Finally, we provide an overview of the complementary molecular biology techniques that might be employed to further verify the reliability of this system.

1. INTRODUCTION

Sirtuins are mammalian homologs of the yeast enzyme silent information regulator 2 (SIR2), a regulator of chromatin packing and transcription which may also be involved in the regulation of replicative lifespan (Lin, Defossez, & Guarente, 2000). The seven mammalian sirtuins possess NAD^+-dependent activities including deacetylation, deacylation, and ADP-ribosylation (as reviewed in Houtkooper, Pirinen, & Auwerx, 2012). Their best-known biological functions involve helping the cell adapt to nutrient challenges and stress. For example, SIRT1 and SIRT3 levels are upregulated during calorie restriction to promote fatty acid oxidation (Chakrabarti et al., 2011; Cohen et al., 2004; Hallows et al., 2011; Hirschey et al., 2010). Recently, the sirtuins have come under close study by cancer biologists because several of them—at least SIRT3, SIRT4, and SIRT6—function as tumor suppressors through their regulation of cellular metabolism (e.g., Finley et al., 2011; Jeong et al., 2013; Kim et al., 2010; Sebastián et al., 2012).

SIRT3 in particular seems to lie at the nexus of metabolism and tumor suppression. In addition to promoting fatty acid metabolism, SIRT3 activates superoxide dismutase 2, one of the cell's main defenses against reactive oxygen species (ROS; Tao et al., 2010). By fighting ROS, SIRT3 plays dual roles: first, it helps to promote genomic integrity (Kim et al., 2010), and second, it maintains hypoxia-inducible factor 1α (HIF1α) in a destabilized state, preventing it from inducing the glycolytic gene expression and Warburg effect which are typical of cancer cells (Bell, Emerling, Ricoult, & Guarente, 2011; Finley et al., 2011). When SIRT3 activity is low, tumor formation is more likely—in fact, one copy of *SIRT3* is deleted in approximately 20% of all human tumors (Finley et al., 2011). Because SIRT3 plays such an important role in metabolism and tumor suppression, it is important to develop quantitative and sensitive methods to monitor its expression.

This chapter details the construction of a luciferase-based reporter system for assaying the activity of the *SIRT3* promoter. The system was inspired by previous work (Bellizzi et al., 2007) but greatly expands and improves upon it. The first part of the protocol describes cloning the *SIRT3* promoter into an otherwise promoterless luciferase reporter plasmid. The construct can then be mutagenized site-specifically to test the effects of ablating certain sequence elements. Assaying *SIRT3* promoter activation with the construct involves transfecting it into cells under different conditions and measuring the activity of the luciferase reporter gene; we present two example applications to illustrate validation and use of the system, including the novel upregulation of SIRT3 expression by rapamycin, a small molecule inhibitor of the nutrient-responsive kinase mammalian target of rapamycin (mTOR; Zoncu, Efeyan, & Sabatini, 2011). The protocol concludes by describing verification of reporter results by quantitative PCR and Western blot. Though tailored specifically for *SIRT3* in this chapter, these same methods could be applied to any target gene of interest.

2. PROTOCOL FOR CONSTRUCTION OF REPORTER PLASMID

2.1. Primer design

Perhaps the most crucial step in creating a reporter construct is to decide which DNA fragment to clone. Because this fragment will be bounded by the primers used to amplify it, the first step is therefore a decision about where to place the primers.

2.1.1 Place the reverse (downstream) primer

The reverse primer is more straightforward to place than the forward (upstream) primer. In most cases, place the reverse primer in the region after the transcriptional start site for the gene of interest but before the translation start codon. This will allow the construct to respond to the transcriptional machinery that normally controls the gene but instead produce the reporter enzyme after translation. The location of a gene's transcriptional and translational start sites can be read from DNA sequence information annotated by the National Center for Biotechnology Information (NCBI), displayed via the NCBI's own Web site (searchable by gene at http://www.ncbi.nlm.nih.gov/gene) or the University of California, Santa Cruz Genome Browser (http://genome.ucsc.edu/index.html).

Table 7.1 Location of salient features within the *SIRT3–PSMD13* promoter (using the NM_012239.5 transcript annotation for *SIRT3* and NM_002817.3 transcript annotation for *PSMD13* from the National Center for Biotechnology Information)

Feature	Location relative to *SIRT3* transcription start site
PSMD13 start codon	−687 to −685
First base of forward primer	−682
Last base of forward primer	−663
PSMD13 transcription start site	−446
Last base of reverse primer	+8
First base of reverse primer	+29
SIRT3 start codon	+35 to +37
First base of Bellizzi reverse primer	+150
Last homologous base of Bellizzi reverse primer	+169

We followed these guidelines and placed our downstream primer between the *SIRT3* transcription start site and the translation start codon (Table 7.1).

2.1.2 Place the forward (upstream) primer

Placement of the forward (upstream) primer depends on the size of the promoter and the locations of known upstream regulatory sequences. Use DNA sequence analysis algorithms such as MAPPER (http://genome.ufl.edu/mapper/) or TFSearch (http://www.cbrc.jp/research/db/TFSEARCH.html) to identify upstream regions containing potential transcription factor binding sites, and combine this information with any regulatory sequences already known to exist in the promoter. When possible, place the primer to include these regions in the amplified fragment. For genes with lengthy promoters or vague promoter boundaries, the number of kilobases between the forward and reverse primers may depend on how large a region you are able to amplify via PCR. Approximately 1 in 10 human genes has a bidirectional promoter—that is, a short (generally less than 1 kb) stretch of noncoding DNA between the 5′ ends of two genes on opposite strands that may be functionally related and coregulated (Lin et al., 2007). This situation establishes more definite bounds for the promoter region.

SIRT3 shares a bidirectional promoter with the proteasome subunit gene *PSMD13*. The intergenic sequence is not prohibitively long (only 721 bp between start codons for *PSMD13* and the "canonical" *SIRT3* isoform, using the NCBI NC_000011.9 human chromosome 11 sequence and UniProt.org for isoform information). We therefore placed the forward primer between the *PSMD13* translation start codon and the transcription start site (Fig. 7.1, Table 7.1). This symmetric design allowed for the amplified fragment to be used in assaying either *SIRT3* or *PSMD13* promoter activity, depending on its orientation in the reporter vector.

2.1.3 Design the exact primer sequences

Having selected the rough locations, design the exact primer sequences using the NCBI Primer Blast tool (http://www.ncbi.nlm.nih.gov/tools/primer-blast/). This tool designs primer pairs that have minimal sequence similarity to other regions in the genome while maintaining desirable estimated hybridization properties to a given input sequence. To enable subsequent cloning, add the recognition site for a restriction enzyme to the 5′ ends of the primers, as well as an overhang of three bases to allow the restriction enzyme to bind. The forward and reverse primers may optionally be designed to have recognition sites for different restriction enzymes, which will direct the insert to adopt a definite orientation during its later ligation into the vector (ensure that the order of recognition sites in the vector's multiple cloning site matches the planned directionality of the insert). Check that the recognition sequences do not occur in the region you plan to amplify.

For *SIRT3*, the NCBI Primer Blast algorithm identified a primer pair which fit the desired regions. We added a recognition site (ggtacc) for the restriction enzyme *Kpn*I to the 5′ end of each primer, as well as a three-base (ctc) overhang. The specific primers we used for the reversible insert were thus:

Forward: 5′-ctcggtaccCAGGAAGACCCCCGGCACAG-3′

Reverse: 5′-ctcggtaccGCGCAGTCCAAGGAGTCCTCCG-3′

Their positions within the *SIRT3–PSMD13* promoter are given in Table 7.1. (It should be noted that our construct does not contain bases -97 to -20, as we sequenced human DNA from multiple sources and none contained that fragment.)

We additionally made a *SIRT3* reporter using the reverse primer from Bellizzi et al. (2007), which is 5′-ctcggtaccATCGTCCCTGCCGCCAAGCA-3′. This primer falls after the *SIRT3* translation start codon. Transcription factor binding sites or other regulatory elements may in some instances overlap the start codon (Xi et al., 2007), necessitating its inclusion in the

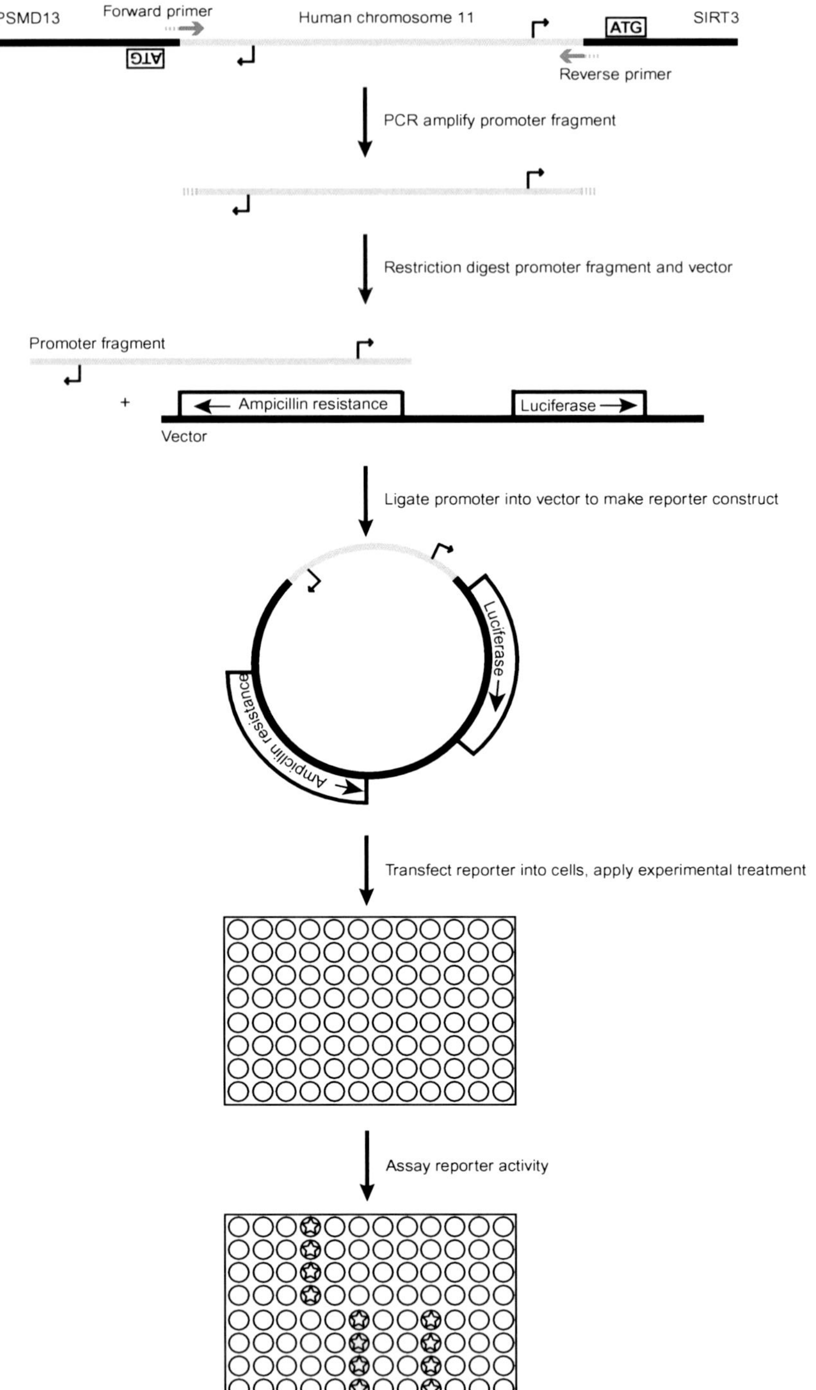

Figure 7.1 Overview of promoter fragment amplification, cloning into reporter plasmid, and experimental use.

amplified promoter fragment. In this scenario, it is important to design the reverse primer so that the reporter gene encoded by the construct remains in frame.

2.2. Promoter fragment amplification

Following primer design, the next step is to amplify the promoter fragment. This process involves generating a starting material to use as template, amplifying the promoter fragment via PCR, and verifying correct amplification.

2.2.1 Extract template genomic DNA

A single 10-cm dish or T75 flask of cultured cells that is near confluence will provide ample DNA for use as template. Collect the cells and proceed to DNA extraction. We extracted DNA from a pellet of human embryonic kidney 293T cells using a DNEasy kit (Qiagen, cat. # 69504), following the manufacturer's instructions.

2.2.2 Amplify promoter fragment via PCR

Amplify the promoter fragment using the primers designed above and the template DNA extracted in the previous step. We used a FastStart PCR Master kit (Roche, cat. # 04710444001) with 50 ng template DNA in 20 μl total volume, following the manufacturer's instructions. The recommended length of the elongation step is 1 min for every 1 kb of DNA to be amplified (up to 3 min); because the *SIRT3–PSMD13* intergenic promoter sequence is slightly less than 1 kb, we used a 1-min elongation step.

2.2.3 Verify amplification

Check for successful amplification by running a small amount of PCR product (2 μl of the reaction, plus 2 μl of DNA loading dye; New England Biolabs, cat. # B7021S) on a 2% agarose gel with SYBR Safe DNA Gel Stain (Life Technologies, cat. # S33102) incorporated into the gel at a 1:10,000 dilution. Include control lanes (e.g., 5 μl of 100 bp DNA ladder alongside 5 μl of 1 kb DNA ladder; New England Biolabs, cat. #s N3231S and N3232S) for size comparison. Bands may be visualized by exciting the gel with blue or ultraviolet light (if using UV light, use proper shielding to limit personal UV exposure). If the PCR bands are clean and of the expected size, proceed to cloning the promoter fragment into the vector.

2.2.4 Troubleshoot

If the PCR results are not clean, try one or more troubleshooting strategies. To optimize the reaction, use annealing temperatures 5 °C warmer or cooler than recommended, or design entirely new primers. To control for template quality, include reactions with a pair of primers known to amplify well. If these steps do not lead to improved PCR results, then it may help to use as template a bacterial artificial chromosome (BAC) with the segment of the genome that includes the promoter of interest. BACs generally contain only a few hundred kilobases, greatly increasing the chances that primers will encounter the desired region for amplification rather than a different genomic location of similar sequence.

Though our PCR results with genomic template were usable, we found that our reactions were cleaner when using BACs as template. We tested two BACs which contain *SIRT3*: CTD-2344F1 (a clone with human DNA from sperm, developed by the California Institute of Technology, available from Life Technologies) and RP11-652O18 (a clone with human male DNA from white blood cells, developed by the Roswell Park Cancer Institute, available from Life Technologies).

2.3. Inserting the promoter into the vector

After verifying that a band of the desired length is the major PCR product, insert it into the reporter vector. This process involves digesting both the PCR products and the vector plasmid with a restriction enzyme. The vector is additionally treated with alkaline phosphatase to prevent recircularization prior to the ligation reaction, which inserts the amplified promoter fragment into the vector.

2.3.1 Purify PCR product

Remove excess primer from the PCR product by prepping the reactions with a PCR purification kit (Qiagen, cat. # 28104), eluting into 40 μl of water in the final step. This typically gives PCR products with concentrations of 100–200 ng/μl, of which only ~1% is residual template.

2.3.2 Digest PCR product with restriction enzyme

Cleave the PCR products with the restriction enzyme corresponding to the recognition sites that were added to the primers designed above. To digest the PCR products and create the insert for our *SIRT3* reporter construct, we used 35.5 μl of each product and 0.5 μl of the restriction enzyme *Kpn*I-HF (New England Biolabs, cat. # R3142S) with 4 μl of NEBuffer 4.

We then incubated in a 37 °C water bath for 1.5 h. The 0.5 μl of restriction enzyme was 10 units, which was sufficient to cut 10 μg of DNA in 1 h at that temperature.

2.3.3 Similarly cleave the vector

As with the PCR products, cleave the vector. We used the vector pGL3 basic (Promega, cat. # E1751). A similar vector may also be used, provided it is a promoterless vector with a luciferase reporter gene immediately following the recognition site(s) for the restriction enzyme(s) used in amplifying the promoter fragment. We cleaved with *Kpn*I-HF, again using 10 units of enzyme with approximately 10 μg of DNA in a 37 °C water bath for 1.5 h.

2.3.4 Treat the vector with alkaline phosphatase

Treat the vector with alkaline phosphatase (from calf intestine; New England Biolabs, cat. # M0290S) to ensure that it does not self-ligate and recircularize after digestion. An efficient way to do this is to prep the post-digest vector DNA using the same purification kit as above (Qiagen, cat. # 28104) and elute with 40 μl of water directly into a microtube containing 5 μl of the 10 × buffer, 2 μl of water, and 3 μl of alkaline phosphatase (or approximately 0.5 units/μg of DNA). Then mix and incubate in a 37 °C water bath for 1 h.

2.3.5 Gel extract the PCR products and vector

To prepare pure DNA samples for ligation, run the cleaved fragments (PCR product and vector alike) on a 2% agarose gel with 1:10,000 SYBR Safe. Under blue or UV light, carefully excise the desired DNA bands (if using UV light, work quickly to avoid excessive DNA cross-linking). Extract the DNA from the bands with a gel extraction kit (Qiagen, cat. # 28704), eluting into 40 μl of water. Yields are typically ~20 ng/μl for the PCR products.

2.3.6 Perform ligation

Ligate the fragment into the vector using T4 DNA ligase (New England Biolabs, cat. # M0202S). Each reaction is 20 μl total volume and should contain 1 μl of a 1:10 dilution of the ligase and 2 μl of the 10 × buffer. Use 50 ng of vector DNA with a 5:1 molar ratio of insert to vector and incubate at 16 °C in a thermal cycler for 30 min. Include two control ligation reactions: one with no DNA, and one with vector DNA but no ligase. These will be important for troubleshooting the electroporation step. (Note that the molarity of DNA need not be calculated explicitly. To calculate the nanograms of insert to use per nanogram of vector, multiply the desired 5:1 molar ratio by the insert/vector length ratio. In our case, the promoter fragment was only 0.13 × the length of the vector, so we used approximately 0.65 ng insert/ng vector DNA.)

2.4. Transforming the bacteria with the reporter construct

At this point, the reporter construct may be finished, but usable quantities of it must be produced in order to verify proper incorporation of the insert. This is accomplished by electroporating and growing bacteria, followed by extracting DNA from individual colonies. This section follows standard biological procedures with bacteria.

2.4.1 Electroporate

Transform electrocompetent bacteria with the reporter construct by electroporation. To do this, add 1 μl of ligation product to 60 μl of DH5α *Escherichia coli* bacteria, place into an electroporation cuvette (Bio-Rad, cat. # 165-2089), and electroporate with a Gene Pulser (Bio-Rad) or a similar instrument. Resuspend the bacteria in 600 μl LB medium (Sigma, cat. # L7275) and incubate for 1 h at 37 °C. Next, spread 200 μl of each onto a 10-cm dish containing LB agar (Sigma, cat. # L2897) and 100 μg/ml ampicillin (EMD Millipore, cat. # 171254) and grow overnight at 37 °C. Include control plates which contain bacteria treated with the no-DNA and the no-ligase control reactions from the previous step.

2.4.2 Pick and grow colonies

The next day, pick colonies from the plates to seed mini (~7 ml) LB–ampicillin cultures (final concentration of ampicillin 100 μg/ml, as in the agar plates). The control plate with no DNA should not grow colonies unless the ampicillin selection is faulty, and the control with no ligase should not grow more than a few colonies unless the vector was able to recircularize itself efficiently, which would indicate a problem with alkaline phosphatase treatment. After seeding the cultures, grow them overnight (with shaking) at 37 °C.

2.4.3 Extract DNA

After growing overnight, set aside ~500 μl of each culture at 4 °C for later use. Prep the rest using a Wizard Plus SV Miniprep DNA Purification kit

(Promega, cat. # A1460). Eluting each sample into 50 μl of water typically gives a yield of about 5 μg DNA, which is sufficient to test for incorporation of the insert into the clone.

2.4.4 Verify incorporation of insert by restriction digest

First, test for incorporation of the insert by cleaving 1 μg of each clone with restriction enzyme, following the manufacturer's instructions as above. Run on a 2% agarose gel (with SYBR Safe, adding loading dye to the DNA and running ladder lanes as above) and visualize. If the insert has been incorporated, the lane will have one band for the insert and one band for the vector. If the insert is not present in the clone, the vector will be the only visible band.

2.4.5 Verify proper insert sequence and orientation by sequencing

For the clones which contain an insert, verify proper sequence and proper insert orientation by sequencing. The forward and reverse cloning primers may be used as sequencing primers; additional internal primers may also be designed to obtain better read coverage of the entire insert. Because the construct will be used as template, primers can be designed with tools such as Primer3 (http://bioinfo.ut.ee/primer3/) without worrying about potential homology to other locations in the genome.

We designed and used multiple internal primers to verify the sequence of our *SIRT3* constructs:

5′-GTGGGCGCCTGTGGTCGAAC-3′ (starts at position -441)

5′-TCACCGCCATCCGGGTTGAA-3′ (starts at position -277)

5′-AGGTTTGACCTCCGGGGCGA-3′ (starts at position -221)

Our sequencing runs were performed by the Dana-Farber/Harvard Cancer Center DNA Sequencing Facility and required 1 μg of template per reaction.

2.4.6 Grow larger amounts of correct construct(s)

After identifying which constructs contain the correct insert in the desired orientation, take 50 μl of the saved aliquot from these samples and seed larger (~50 ml) LB–ampicillin cultures. Grow them overnight (with shaking) at 37 °C, then prep the DNA from each. We used Plasmid Plus Midi Kits (Qiagen, cat. # 12945) because the vacuum manifold allowed for quick sample preparation (although it was necessary to clear the bacterial lysate by centrifugation prior to the syringe filtering step in order to prevent the DNA-binding column from becoming overloaded with debris). We then eluted into 200 μl of elution buffer. Yields were on the order of 100 μg DNA.

2.5. Mutagenesis

To disrupt specific sequence elements by introducing site-specific mutations into the promoter construct, use the QuikChange II Site-Directed Mutagenesis Kit (Agilent, cat. # 200523). This kit uses mutagenized primers to amplify a new version of the construct in a PCR reaction followed by cleaving methylated template DNA with the restriction enzyme *Dpn*I, leaving behind only the newly synthesized mutagenized strands which can then be used to transform bacteria.

2.5.1 Design primers

As with the original construct design, primer design is important. This depends on identifying the specific sequence element to alter and deciding upon the specific mutation(s) to introduce. Once these decisions are made, use the Agilent mutagenesis primer design tool (http://www.genomics.agilent.com/primerDesignProgram.jsp) to design primer pairs which introduce the desired mutations while adhering to the parameters of the kit. The *SIRT3* promoter contains a binding site for the transcription factor estrogen-related receptor α (ERRα), and this site plays a role in mediating induction of SIRT3 expression by the transcription factor peroxisome proliferator-activated receptor gamma coactivator 1-α (PGC-1α) (Kong et al., 2010). We created a version of our plasmid with an abrogated ERRα element by mutating the native binding site (TGCCATTG) to TGTAATTG (as Kong et al. did with murine *SIRT3*). We used the following mutagenic primers (with the binding site underlined):

Forward: ERRα mutagenic primer: 5′-ccgcgcacttggctgtgtaattgaggcgtt aaagag-3′

Reverse: ERRα mutagenic primer: 5′-ctctttaacgcctcaattacacagccaagtgcg cgg-3′

2.5.2 Amplify mutagenized construct

Create a mutagenized construct by using the mutagenic primers in a PCR reaction. The manufacturer's instructions can largely be followed, though we slightly modified the PCR protocol to better handle a large amplification product. In our mutagenesis reactions, we used a 7-min elongation step and repeated the denaturation–annealing–elongation cycle 18 ×. We then added 1 μl of *Dpn*I restriction enzyme directly to each amplification reaction and incubated at 37 °C for 1 h to degrade the original template, and we subsequently checked for the presence of amplified plasmid by running 5 μl of each reaction (mixed with 5 μl of DNA sample buffer) on a 2% agarose gel (with SYBR Safe). We ran an equal volume of a 1:50 template dilution to use as comparison.

2.5.3 Transform bacteria

To introduce the mutagenized vector into bacteria, we found that electroporation of DH5α electrocompetent cells worked more efficiently than the heat shock system described in the manufacturer's protocol. We electroporated and then grew these bacteria in the same manner as described above for cloning the construct. We also verified the introduction of the correct mutation by sequencing. Note that the internal sequencing primers designed above are especially important when high-quality sequences are needed to examine a few specific base pairs.

3. ASSAYING *SIRT3* PROMOTER ACTIVITY WITH THE CONSTRUCT

3.1. Assay protocol

The completed constructs may be used to assay promoter activity when transfected into cells growing in an opaque tissue culture plate. The assay involves transfecting the cells with the construct along with a second, constitutively expressed plasmid (to control for transfection efficiency), applying one or more treatments to the cells, waiting 24–48 h, and measuring luciferase activity after the addition of a substrate for the luciferase reporter. The resulting data indicate which treatments stimulate activity from the promoter fragment cloned into the reporter construct.

3.1.1 Seed cells

Grow cells on an opaque 96-well plate (Corning, cat. # 3917). Seeding densities will vary by cell line; for 293T cells, use 10,000 cells per well. It may be helpful to simultaneously grow cells on a clear plate (Corning, cat. # 3595) at an identical density to allow for visual inspection of population confluence and health through a normal light microscope.

3.1.2 Transfect cells

Prepare the transfection mixture by adding 0.2–0.4 μl per well of FuGene 6 transfection reagent (Promega, cat. # E2691) to 4 μl per well of serum-free medium, then adding 0.03–0.06 μg per well of construct and 0.01–0.03 μg per well of a pRL renilla control vector (used to control for cell number and transfection efficiency, Promega, cat. # E2261). The exact amounts of these two plasmids should be empirically determined for each assay in order to obtain optimal luciferase and renilla signals. (Note that the constitutive promoter of the renilla vector is quite strong and, in some cases, a larger construct/renilla ratio may be necessary than the typical values given here.) Mix according to the manufacturer's instructions, then add to cells.

3.1.3 Add chemicals or apply other treatment

Depending on the hypothesis being tested, add chemicals, transfect additional plasmids, or apply another treatment to the cells.

3.1.4 Measure luciferase activity

Twenty-four to forty-eight hours after the initial transfection, assay luciferase activity using a Dual-Luciferase Reporter Assay kit (Promega, cat. # E1910). Begin by aspirating the medium from the cells, then add 25 μl per well of the included passive lysis buffer (dilute the 5 × stock to a 1 × working solution with distilled water). Place the cells on a shaker with gentle movement for at least 20 min to ensure lysis.

Following lysis, add 100 μl of the provided luciferase assay reagent to each well. Use a multichannel pipettor to do this quickly, as the luminescence signal degrades with time. Measure the luminescence of each well in a luminescence-equipped plate reader such as a Cary Varian Eclipse fluorescence spectrophotometer with a gate time of 200 ms and an emission slit of 2.5 nm around a target emission wavelength of 560 nm (the first two of these parameters may be adjusted to optimize signal capture). Next, add 100 μl per well of the provided "Stop & Glo" reagent to quench the luciferase reaction and begin the renilla luminescence. Measure the luminescence of each well again, changing the target emission wavelength from 560 to 500 nm.

For data analysis, divide the first (luciferase) luminescence reading by the second (renilla) luminescence reading. This provides a signal value for each well that is normalized by transfection efficiency. Average by condition to obtain a readout of the relative activity of the luciferase promoter construct across different treatments.

3.2. Example application: Importance of ERRα for *SIRT3* promoter activity in 293T cells

We used the reversible *SIRT3–PSMD13* promoter construct to verify the effect of ERRα binding on *SIRT3* promoter activity in 293T cells. To do this, we generated both forward and reverse *SIRT3* promoter reporter constructs, making versions for each with either wild-type sequence or a mutated ERRα binding site, as described above. We also used Gateway cloning techniques to generate an HA-tagged ERRα overexpression plasmid (starting from the HsCD00079872 ORF in a pDONR221 entry clone from the PlasmID database of the Dana-Farber/Harvard Cancer Center DNA Resource Core).

We then seeded an opaque 96-well plate with 10,000 293T cells per well. Cells were grown in Dulbecco's Modified Eagle Medium (DMEM; Life Technologies, cat. # 11995) with 10% fetal bovine serum (FBS; HyClone, cat. # SH30910.03), and 1% penicillin–streptomycin supplement (Life Technologies, cat. # 15140). The next day, we mixed and applied to each well 4 μl DMEM, 0.2 μl FuGene6 transfection reagent, 0.03 μg of renilla plasmid, and 0.06 μg of a reporter plasmid construct. We additionally combined and added 4 μl DMEM, 0.2 μl FuGene6, and 0.02 μg ERRα-HA expression plasmid or vector control (as well as 0.02 μg PCDNA3.1 empty vector for comparison to wells overexpressing PGC1α - data not shown). Forty-eight hours later, we assayed luciferase activity as described above and saw that ERRα overexpression increased promoter activity from the wild-type sequence promoter construct in the *SIRT3* direction (Fig. 7.2A), but not from the mutated promoter in the *SIRT3* direction (Fig. 7.2A) or from either sequence variant in the *PSMD13* direction (Fig. 7.2B). This verified that the previously reported activation of the *SIRT3* promoter by ERRα was exhibited by our system.

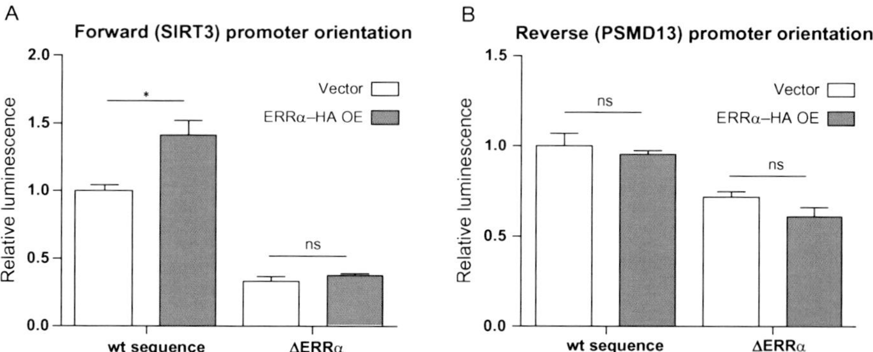

Figure 7.2 Luciferase reporter results after transfection of wild-type sequence or mutated estrogen-related receptor α (ERRα) reporter constructs plus vector or ERRα-HA overexpression constructs in human 293T cells. Promoters were oriented in the (A) forward or (B) reverse orientation relative to *SIRT3*. $n = 4$ wells per condition. *: $p < 0.05$.

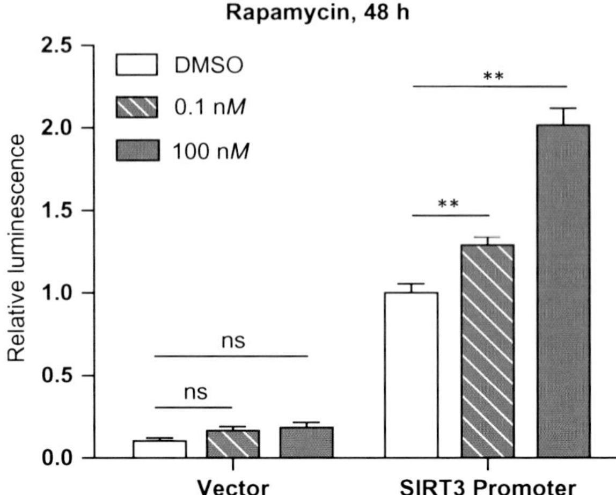

Figure 7.3 Luciferase reporter results after transfection of the longer-form *SIRT3* reporter construct or vector control plus DMSO or the indicated concentration of rapamycin. $n = 4$ wells per condition. **: $p < 0.01$.

3.3. Example application: Effect of rapamycin on *SIRT3* promoter activity in 293T cells

Because SIRT3 expression is known to be upregulated by calorie restriction, we also tested the responsiveness of our longer *SIRT3* promoter construct to rapamycin (Sigma, cat. # R0395), a small molecule that inhibits the nutrient-sensing TOR pathway and has been used as a calorie restriction mimetic (Harrison et al., 2009; Zoncu et al., 2011). To do this, we seeded 293T cells into 24 wells of a 96-well plate at 10,000 cells per well. The next day, we mixed and applied to each well 4 μl DMEM, 0.4 μl FuGene6 transfection reagent, 0.03 μg renilla plasmid, and 0.06 μg of a reporter plasmid construct or promoterless vector control. The following day, we treated with DMSO or one of two concentrations of rapamycin, to final concentrations of 0, 0.1, or 100 nm of rapamycin in 0.01% DMSO. Forty-eight hours later, we assayed luciferase activity as described above and observed that increasing rapamycin dosage led to increased activity from the *SIRT3* promoter construct (Fig. 7.3).

4. VERIFYING REPORTER RESULTS WITH COMPLEMENTARY METHODS

Multiple molecular biology methods may be used to verify the results of the luciferase reporter system. Quantitative polymerase chain (qPCR) reaction is conceptually similar, as it assays promoter activity by measuring the amount of RNA transcript detected by a pair of primers specific to the gene of interest. A standard Western blot can also be used to estimate sirtuin expression by measuring protein level.

4.1. Quantitative PCR

4.1.1 Quantitative PCR protocol

Begin by extracting RNA with an RNEasy Mini kit (Qiagen, cat. # 74104). One well of a 6-well plate (or 10 mg of tissue) is sufficient for each biological replicate. Reverse transcribe RNA into cDNA with an iScript cDNA synthesis kit (Bio-Rad, cat. # 170-8891), using 1 μg of RNA as a starting material.

Quantitative PCR may be performed using 2 × LightCycler 480 SYBR Green I Master mix (Roche, cat. # 04707516001) on a LightCycler 480 thermal cycler (Roche). Halve the 20 μl reaction volume called for by the manufacturer's protocol and use 5 μl of master mix, 0.5 μl each of the forward and reverse primers (at 10 μ*M* stock concentration), and 4 μl of a 1:50 working cDNA dilution for each qPCR reaction. Run two technical replicates for each biological replicate. In addition, prepare a set of six standards, beginning with a 1:4 dilution of pooled cDNA and serially diluting 1:4 five times to create a basis for software estimation of the relative concentrations of target RNA in each sample. Run each plate through 45 amplification cycles, following the manufacturer's instructions, and end with a melting curve to verify homogeneity of the amplified DNA.

Before concentration estimates for the gene of interest in each sample can be compared to one another, they must be normalized by the levels of a reference gene within the same sample. We found that β2 microglobulin (B2M), ribosomal protein S16 (RPS16), and peptidylprolyl isomerase A (PPIA) were generally stable in expression across different treatments in human 293T cells and made good reference genes. In cases where detection of small perturbations in target gene level was important, we ran all three and used the geometric mean of their expression as reference.

4.1.2 Verification of ERRα results by qPCR

To assay SIRT3 expression in 293T cells following overexpression of the ERRα-HA construct described above, we seeded 100,000 cells per well in a six-well plate. The next day, we treated each well with a mix of 8 μl FuGene6 and 0.5 μg ERRα-HA or vector (and an equal amount of PCDNA3.1 empty vector). Two days later, we extracted RNA, synthesized cDNA, and analyzed SIRT3 expression by qPCR. Consistent with our luciferase data, we saw that SIRT3 mRNA was upregulated with ERRα overexpression (Fig. 7.4A and B). For these data, $n=3$ and B2M was used as the reference gene.

We also created an ERRα knockdown 293T cell line using the shRNA construct TRCN0000330191 from the RNAi consortium (via the Dana-Farber/Harvard Cancer Center RNAi Core facility). We conducted qPCR on these cells and saw the complementary result, that ERRα knockdown led to lower SIRT3 transcript levels (Fig. 7.4C and D). For this data, $n=3$ and B2M was used as the reference gene. Primer sequences used are given in Table 7.2; the ERRα pair is from Schreiber, Knutti, Brogli, Uhlmann, and Kralli (2003).

4.1.3 Verification of rapamycin results by qPCR

To verify the rapamycin luciferase results, we inhibited the TOR pathway in three different ways. We inhibited by treatment with rapamycin for 8 h at 100 n*M* (Fig. 7.4E); by treatment with the small molecule Torin1, which inhibits the catalytic site of mTOR (Thoreen et al., 2009), for 8 h at 250 n*M* (Fig. 7.4F); and by 24-h amino acid deprivation (Fig. 7.4G). In all three cases, inhibition of the TOR pathway led to greater expression of SIRT3, validating our luciferase results. In this case, $n=4$, and we used the geometric mean of B2M, RPS16, and PPIA expression as a reference. Primer sequences used are given in Table 7.2.

4.2. Western blot

Additional verification may be performed by Western blot, which looks at protein rather than mRNA levels. We used this to further verify the upregulation of SIRT3 expression by rapamycin treatment because it represented a novel finding.

4.2.1 Western blot protocol

Briefly, obtain whole cell lysate by applying 1% NP40 buffer (1% NP40 detergent with 150 m*M* NaCl, 50 m*M* Tris (pH 8), 1 m*M* dithiothreitol, plus 1 × Roche cOmplete Mini protease inhibitor tablets (cat. # 11836170001) and 1% each of Sigma phosphatase inhibitor cocktail #s 2 (cat. # P5726) and 3 (cat. # P0044)) to the pellet of collected cells. Use 150 μl of lysis buffer for mostly confluent cells in a 35-mm well, or 300 μl for cells in a 10-cm dish (providing a more concentrated lysate). Separate the proteins by running the lysate on a polyacrylamide gel (Bio-Rad, cat. # 345-0043), transfer to nitrocellulose membrane (Bio-Rad, cat. # 162-0112), and blot for the protein of interest using standard molecular biology techniques.

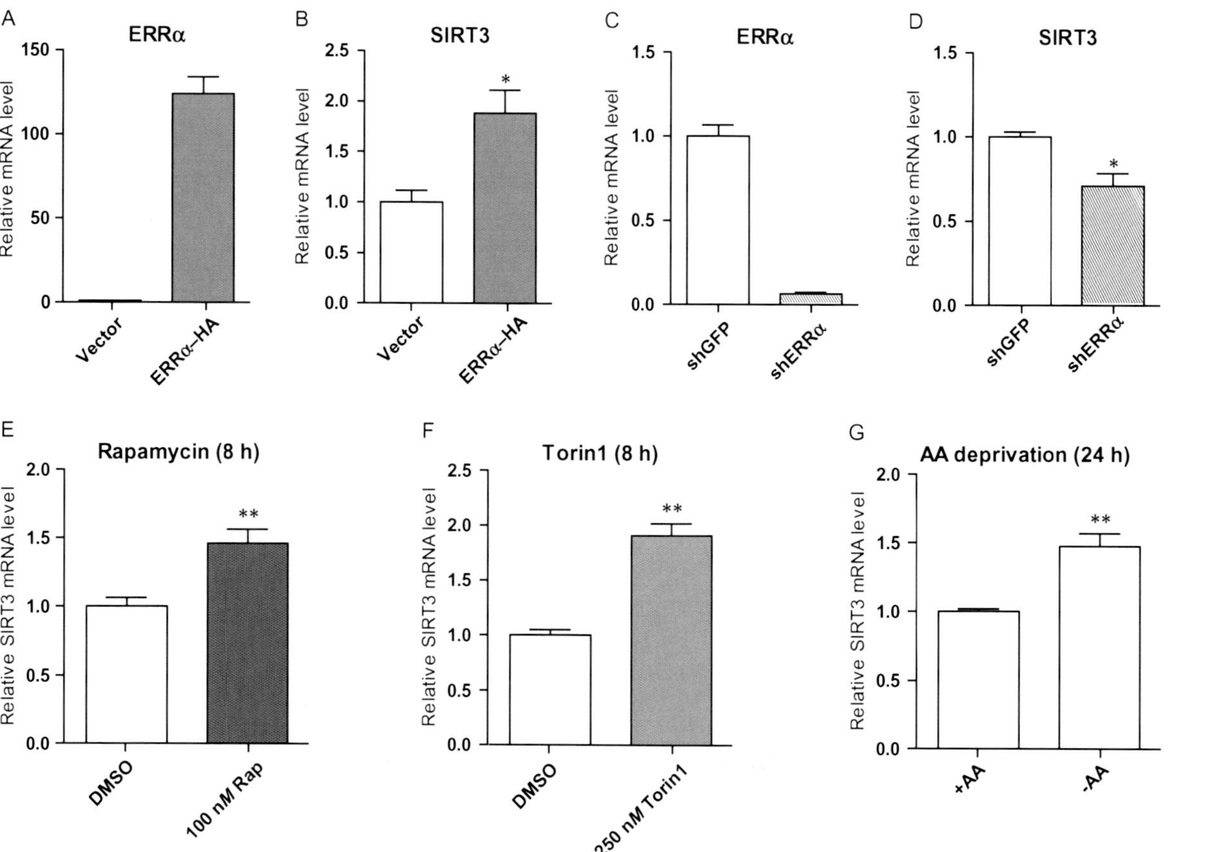

Figure 7.4 (A) Verification of ERRα overexpression in 293T cells and (B) effect on SIRT3 expression; (C) verification of ERRα knockdown in 293T cells, and (D) effect on SIRT3 expression; induction of SIRT3 expression by (E) 8-h 100 n*M* rapamycin treatment, (F) 8-h 250 n*M* Torin1 treatment, and (G) 24-h amino acid deprivation. *: $p < 0.05$, **: $p < 0.01$. (See the color plate.)

Table 7.2 Quantitative PCR primer sequences used in this study

qPCR primer	**Sequence**
SIRT3 forward	5′-AGCCCTCTTCATGTTCCGAAGTGT-3′
SIRT3 reverse	5′-TCATGTCAACACCTGCAGTCCCTT-3′
ERRα forward	5′-AAGACAGCAGCCCCAGTGAA-3′
ERRα reverse	5′-ACACCCAGCACCAGCACCT-3′
B2M forward	5′-AGATGAGTATGCCTGCCGTGTGAA-3′
B2M reverse	5′-TGCTGCTTACATGTCTCGATCCCA-3′
RPS16 forward	5′-AGATCAAAGACATCCTCATCCAG-3′
RPS16 reverse	5′-TGAGTTTTGAGTCACGATGGG-3′
PPIA forward	5′-AGCATACAGGTCCTGGCATCTTGT-3′
PPIA reverse	5′-CAAAGACCACATGCTTGCCATCCA-3′

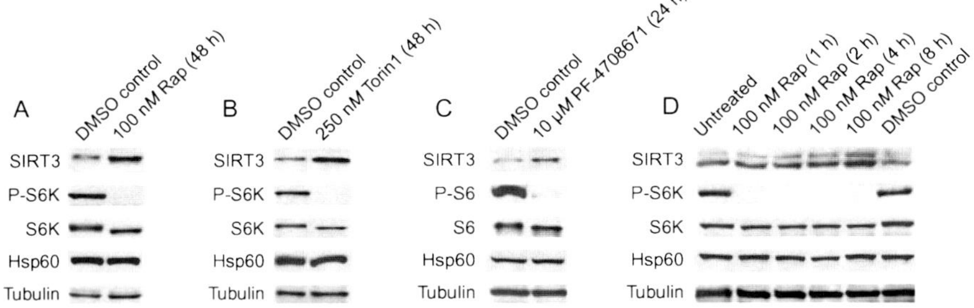

Figure 7.5 SIRT3 protein level after treatment with (A) rapamycin, (B) Torin1, or (C) PF-4708671, an inhibitor of p70-S6 kinase (S6K); (D) rapamycin-induced increase in SIRT3 protein level over 8 h.

4.2.2 Verification of rapamycin results by western blot

For detection of human SIRT3, we used Cell Signaling C73E3 (cat. # 2627) as the primary antibody at a dilution of 1:1000 in 3% BSA with 0.02% sodium azide (note that this antibody detects human but not mouse SIRT3, while Cell Signaling D22A3 (cat. # 5490) detects both). We observed that our Western blot reproduced at the protein level the induction of SIRT3 expression that we observed at the RNA level. In 293T cells grown in DMEM with 10% FBS and 1% penicillin–streptomycin, inhibition of the TOR pathway by 100 n*M* rapamycin for 48 h (Fig. 7.5A) or 250 n*M* Torin1 for 48 h (Fig. 7.5B) led to an increase in SIRT3 expression, as did inhibition of the mTOR target p70-S6 kinase (S6K) via the S6K inhibitor PF-4708671 (Pearce et al., 2010) at 10 μ*M* for 24 h (Fig. 7.5C). An 8-h time course of 100 n*M* rapamycin treatment displayed gradual SIRT3 induction (Fig. 7.5D). In these experiments, we used phosphorylation of S6K (Cell Signaling, cat. # 9234, 1:1000) as a marker of mTOR pathway activation (except in the case of direct S6K inhibition, where we used phosphorylation of ribosomal protein S6 (Cell Signaling, cat. # 4856, 1:1000)). HSP60 (Abcam, cat. # ab3080, 1:5000) was used as a control for total amount of mitochondria, and the loading controls were total S6K (Cell Signaling, cat. # 9202, 1:1000) or total S6 (Cell Signaling, cat. # 2317, 1:1000) and α-tubulin (Santa Cruz Biotechnology, cat. # sc-8035, 1:5000).

5. SUMMARY

In this chapter, we have detailed the development of a luciferase-based plasmid reporter system to measure activation of the human *SIRT3* promoter, we have validated its activity, and we have shown that SIRT3 expression is upregulated by inhibition of the nutrient-sensing TOR pathway in 293T cells. Not only can this system be used with cell lines in culture, as described above, but it could also see future utilization as a diagnostic tool for use with primary tumor cell lines to assay the level of *SIRT3* activation. Furthermore, future studies using such a system may identify small molecule compounds and proteins that activate or inhibit SIRT3 expression. Because SIRT3 exerts a range of metabolic effects on tumors, including fighting against the Warburg effect, the knowledge gained by implementations of this system could have important diagnostic, prognostic, and therapeutic consequences.

ACKNOWLEDGMENTS

F. K. S. was supported by NIH Training Grant No. T32 DK007260. M. C. H. was supported by an American Cancer Society New Scholar Award and the Glenn Foundation for Medical Research.

REFERENCES

Bell, E. L., Emerling, B. M., Ricoult, S. J., & Guarente, L. (2011). SIRT3 suppresses hypoxia inducible factor 1α and tumor growth by inhibiting mitochondrial ROS production. *Oncogene*, *30*(26), 2986–2996.

Bellizzi, D., Dato, S., Cavalcante, P., Covello, G., Di Cianni, F., Passarino, G., et al. (2007). Characterization of a bidirectional promoter shared between two human genes related to aging: SIRT3 and PSMD13. *Genomics*, *89*(1), 143–150.

Chakrabarti, P., English, T., Karki, S., Qiang, L., Tao, R., Kim, J., et al. (2011). SIRT1 controls lipolysis in adipocytes via FOXO1-mediated expression of ATGL. *Journal of Lipid Research*, *52*(9), 1693–1701.

Cohen, H. Y., Miller, C., Bitterman, K. J., Wall, N. R., Hekking, B., Kessler, B., et al. (2004). Calorie restriction promotes mammalian cell survival by inducing the SIRT1 deacetylase. *Science*, *305*(5682), 390–392.

Finley, L. W., Carracedo, A., Lee, J., Souza, A., Egia, A., Zhang, J., et al. (2011). SIRT3 opposes reprogramming of cancer cell metabolism through HIF1α destabilization. *Cancer Cell*, *19*(3), 416–428.

Hallows, W. C., Yu, W., Smith, B. C., Devries, M. K., Ellinger, J. J., Someya, S., et al. (2011). SIRT3 promotes the urea cycle and fatty acid oxidation during dietary restriction. *Molecular Cell*, *41*(2), 139–149.

Harrison, D. E., Strong, R., Sharp, Z. D., Nelson, J. F., Astle, C. M., Flurkey, K., et al. (2009). Rapamycin fed late in life extends lifespan in genetically heterogeneous mice. *Nature*, *460*(7253), 392–395.

Hirschey, M. D., Shimazu, T., Goetzman, E., Jing, E., Schwer, B., Lombard, D. B., et al. (2010). SIRT3 regulates mitochondrial fatty-acid oxidation by reversible enzyme deacetylation. *Nature*, *464*(7285), 121–125.

Houtkooper, R. H., Pirinen, E., & Auwerx, J. (2012). Sirtuins as regulators of metabolism and healthspan. *Nature Reviews. Molecular Cell Biology*, *13*(4), 225–238.

Jeong, S. M., Xiao, C., Finley, L. W., Lahusen, T., Souza, A. L., Pierce, K., et al. (2013). SIRT4 has tumor-suppressive activity and regulates the cellular metabolic response to DNA damage by inhibiting mitochondrial glutamine metabolism. *Cancer Cell*, *23*(4), 450–463.

Kim, H. S., Patel, K., Muldoon-Jacobs, K., Bisht, K. S., Aykin-Burns, N., Pennington, J. D., et al. (2010). SIRT3 is a mitochondria-localized tumor suppressor required for maintenance of mitochondrial integrity and metabolism during stress. *Cancer Cell*, *17*(1), 41–52.

Kong, X., Wang, R., Xue, Y., Liu, X., Zhang, H., Chen, Y., et al. (2010). Sirtuin 3, a new target of PGC-1alpha, plays an important role in the suppression of ROS and mitochondrial biogenesis. *PLoS One*, *5*(7), e11707.

Lin, J. M., Collins, P. J., Trinklein, N. D., Fu, Y., Xi, H., Myers, R. M., et al. (2007). Transcription factor binding and modified histones in human bidirectional promoters. *Genome Research*, *17*(6), 818–827.

Lin, S. J., Defossez, P. A., & Guarente, L. (2000). Requirement of NAD and SIR2 for life-span extension by calorie restriction in *Saccharomyces cerevisiae*. *Science*, *289*(5487), 2126–2128.

Pearce, L. R., Alton, G. R., Richter, D. T., Kath, J. C., Lingardo, L., Chapman, J., et al. (2010). Characterization of PF-4708671, a novel and highly specific inhibitor of p70 ribosomal S6 kinase (S6K1). *Biochemical Journal*, *431*(2), 245–255.

Schreiber, S. N., Knutti, D., Brogli, K., Uhlmann, T., & Kralli, A. (2003). The transcriptional coactivator PGC-1 regulates the expression and activity of the orphan nuclear receptor estrogen-related receptor alpha (ERRalpha). *Journal of Biological Chemistry*, *278*(11), 9013–9018.

Sebastián, C., Zwaans, B. M., Silberman, D. M., Gymrek, M., Goren, A., Zhong, L., et al. (2012). The histone deacetylase SIRT6 is a tumor suppressor that controls cancer metabolism. *Cell*, *151*(6), 1185–1199.

Tao, R., Coleman, M. C., Pennington, J. D., Ozden, O., Park, S. H., Jiang, H., et al. (2010). SIRT3-mediated deacetylation of evolutionarily conserved lysine 122 regulates MnSOD activity in response to stress. *Molecular Cell*, *40*(6), 893–904.

Thoreen, C. C., Kang, S. A., Chang, J. W., Liu, Q., Zhang, J., Gao, Y., et al. (2009). An ATP-competitive mammalian target of rapamycin inhibitor reveals rapamycin-resistant functions of mTORC1. *Journal of Biological Chemistry*, *284*(12), 8023–8032.

Xi, H., Yu, Y., Fu, Y., Foley, J., Halees, A., & Weng, Z. (2007). Analysis of overrepresented motifs in human core promoters reveals dual regulatory roles of YY1. *Genome Research*, *17*(6), 798–806.

Zoncu, R., Efeyan, A., & Sabatini, D. M. (2011). mTOR: From growth signal integration to cancer, diabetes and ageing. *Nature Reviews. Molecular Cell Biology*, *12*(1), 21–35.

CHAPTER EIGHT

Metabolomic Profiling of Cultured Cancer Cells

Marie Scoazec[*,†], **Sylvere Durand**[*,†], **Alexis Chery**[*,†], **Lorenzo Galluzzi**[‡,§,¶,1,2], **Guido Kroemer**[*,†,‡,§,||,1,2]

[*]Metabolomics and Cell Biology Platform, Gustave Roussy; Villejuif, France
[†]INSERM, U848; Villejuif, France
[‡]Université Paris Descartes/Paris V, Sorbonne Paris Cité; Paris, France
[§]Equipe 11 labellisée par la Ligue Nationale contre le Cancer, Centre de Recherche des Cordeliers; Paris, France
[¶]Gustave Roussy; Villejuif, France
[||]Pôle de Biologie, Hôpital Européen Georges Pompidou, AP-HP, Paris, France
[1]Share senior coauthorship.
[2]Corresponding authors: e-mail address: deadoc@vodafone.it; kroemer@orange.fr

Contents

Abstract

Quantitative proteomics approaches have been developed—and now begin to be implemented on a high-throughput basis—to fill-in the large gap between the genomic/transcriptomic setup of (cancer) cells and their phenotypic/behavioral traits, reflecting a significant degree of posttranscriptional regulation in gene expression as well as a robust posttranslational regulation of protein function. However, proteomic profiling assays not only fail to detect labile posttranslational modifications as well as unstable protein-to-protein interactions but also are intrinsically incapable of assessing the enzymatic activity, as opposed to the mere abundance, of a given protein. Thus, determining the abundance of theoretically all the metabolites contained in a cell/tissue/organ/organism may significantly improve the informational value of proteomic approaches. Several techniques have been developed to this aim, including high-performance liquid chromatography (HPLC) coupled to quadrupole time-of-flight (Q-TOF) high-resolution mass spectrometry (HRMS). This approach is particularly advantageous for metabolomic profiling as it offers elevated accuracy and improved sensitivity. Here, we describe a simple procedure to determine the complete complement of intracellular metabolites in cultured malignant cells by HPLC coupled to Q-TOF HRMS. According to this method, (1) cells are collected and processed to minimize

Methods in Enzymology, Volume 543
ISSN 0076-6879
http://dx.doi.org/10.1016/B978-0-12-801329-8.00008-8

contaminations as well as fluctuations in their metabolic profile; (2) samples are separated by HPLC and analyzed on a Q-TOF spectrometer; and (3) data are extracted, normalized, and deconvoluted according to refined mathematical methods. This protocol constitutes a simple approach to determine the intracellular metabolomic profile of cultured cancer cells. With minimal variations (mostly related to sample collection and processing), this method is expected to provide reliable metabolomic data on a variety of cellular samples.

ABBREVIATIONS

CGH comparative genomic hybridization
EIS electrospray ion source
HGP Human Genome Project
HP-921 hexakis(1H,1H,3H-tetrafluoropropoxy)phosphazine
HPLC high-performance liquid chromatography
HRMS high-resolution mass spectrometry
NSCLC non-small cell lung carcinoma
Q-TOF quadrupole time-of-flight

1. INTRODUCTION

The so-called Human Genome Project (HGP) was launched in the late 1980s (on funds from the US Department of Energy's Office of Health and Environmental Research) with the aims of determining the sequence of base pairs making up human DNA and characterizing human genes from both a physical and functional perspective (International Human Genome Sequencing, 2004). The project, which remains the largest collaborative project ever launched for biological purposes, was declared completed on April 2003, although significant amounts of data had already been made available to the scientific community 2 years earlier (International Human Genome Sequencing, 2004; Lander et al., 2001; Venter et al., 2001). When the HGP was just beginning to run full speed, which means in the early 1990s, interest grew around the possibility of comparing the copy number of specific genetic loci in two related samples, for instance malignant tissues and their nontransformed counterparts (Kallioniemi et al., 1992). Upon the completion of the HGP, this approach, which is now known as comparative genomic hybridization (CGH), could be implemented on a genome-wide, high-throughput and high-resolution scale (microarray-based CGH), quickly becoming part of oncological diagnostic procedures (Pinkel & Albertson, 2005). Along similar lines, the sequence data generated by the

HGP allowed for the implementation of high-throughput platforms for global gene expression profiling, that is, for the assessment of changes in the abundance of messenger RNAs (mRNAs) on a genome-wide basis, although similar approaches had been undertaken earlier (based on sequence data from cDNA microarray screenings) (Moch et al., 1999).

Both CGH arrays and gene expression arrays raised great expectations, in particular among experimental oncologists and clinicians, hoping that specific genomic/transcriptomic alterations would be of diagnostic/prognostic relevance and/or guide therapeutic choices. This was indeed the case in a few instances, including the amplification of *v-erb-b2* avian erythroblastic leukemia viral oncogene homolog 2 (*ERBB2*), which is nowadays assessed in breast carcinoma patients as an indicator of dismal prognosis and possible susceptibility to ERBB2-targeting therapeutics (Coussens et al., 1985; Galluzzi et al., 2012; Vacchelli et al., 2013). Nonetheless, most often variations in gene copy number and/or alterations in the abundance of specific gene transcripts do not correlated with the phenotypic/behavioral features of cancer cells. At least in part, this reflects not only the significant degree of posttranscriptional regulation to which the expression of the majority of genes is subjected (Craig, Bathurst, & Herries, 1980; Klausner & Harford, 1989; Rosen et al., 1986), but also the fact that, perhaps with a few exceptions, the activity of all proteins is under the control of multiple posttranslational modifications, including phosphorylation, acetylation, ubiquitination, and sumoylation events (just to mention a few examples), as well as of protein-to-protein interactions (Ahearn, Haigis, Bar-Sagi, & Philips, 2012; Sims & Reinberg, 2008; Vucic, Dixit, & Wertz, 2011). For this latter reason, recently developed proteomic approaches fill-in the gap between functional genomics and phenotypic/functional parameters only to a limited extent. Indeed, while these strategies allow for the comparative assessment of the abundance of virtually any protein in two samples, they are intrinsically incapable of providing insights into its activity (Seow et al., 2000). Thus, although proteomic profiles have been attributed a prognostic/predictive significance in some clinical setting (Voss, Ahorn, Haberl, Dohner, & Wilgenbus, 2001), this may not always be the case. Moreover, technological platforms for the implementation of high-throughput proteomics are still in their infancy (Nilsson et al., 2010).

In this context, metabolomic profiling, that is, the quantification of the complete complement of small metabolites contained in a given cell/tissue/organ/organisms, may significantly integrate and expand the informational value of proteomics (Griffin & Shockcor, 2004). By directly measuring the relative abundance of the substrates and products of specific reactions, indeed, this approach allows us to obtain indirect (but reliable) insights into the activity of a given enzyme or metabolic circuitry. Thus, the combination of proteomics and metabolomics is expected to generate databases with an improved prognostic/predictive/therapeutic potential (Griffin & Shockcor, 2004). As it stands, several methods are available for the high-throughput metabolomic profiling of material of various origin (e.g., cultured cells, tissues, and biological fluids), each of which is characterized by specific advantages and drawbacks. A detailed discussion of the pros and cons associated with the use of these techniques can be found in Dunn, Broadhurst, Atherton, Goodacre, and Griffin (2011), Dunn and Ellis (2005), Goodacre, Vaidyanathan, Dunn, Harrigan, and Kell (2004), and Xiao, Zhou, and Ressom (2012).

Here, we describe a rapid and simple method for the determining the complete metabolic complement of cultured malignant cells based on high-performance liquid chromatography (HPLC) coupled to a quadrupole time-of-flight (Q-TOF) high-resolution mass spectrometry (HRMS) (Fig. 8.1). This approach is particularly suitable for metabolomic profiling in that it offers a high accuracy together with an increased sensitivity as compared, for instance, to nuclear magnetic resonance-based methods (Pan & Raftery, 2007). For exemplificative purposes, here we will refer to a widely employed model of human non-small cell lung carcinoma (NSCLC), namely, A549 cells (Criollo et al., 2007; Vitale et al., 2010). With adequate variations (especially concerning sample collection and processing), this method can provide reliable metabolomic data on a variety of cellular samples, including malignant cells of alternative origin as well as nontransformed cells, be them cultured or isolated from tissues/organs/organisms (Lainey et al., 2012, 2013; Michaud et al., 2011).

2. CELL CULTURE, TREATMENTS, AND SAMPLE COLLECTION

1. Human NSCLC A549 cells are routinely maintained in Glutamax®-containing DMEM/F12 medium (Gibco®-Life Technologies™) supplemented with 10% fetal bovine serum (FBS), 10 m*M* HEPES buffer, 100 units/mL penicillin G sodium, and 100 μg/mL streptomycin sulfate, at 37 °C in a humidified 5% CO_2-enriched atmosphere using 75 cm^2 supports for cell culture (see Notes 1–3).

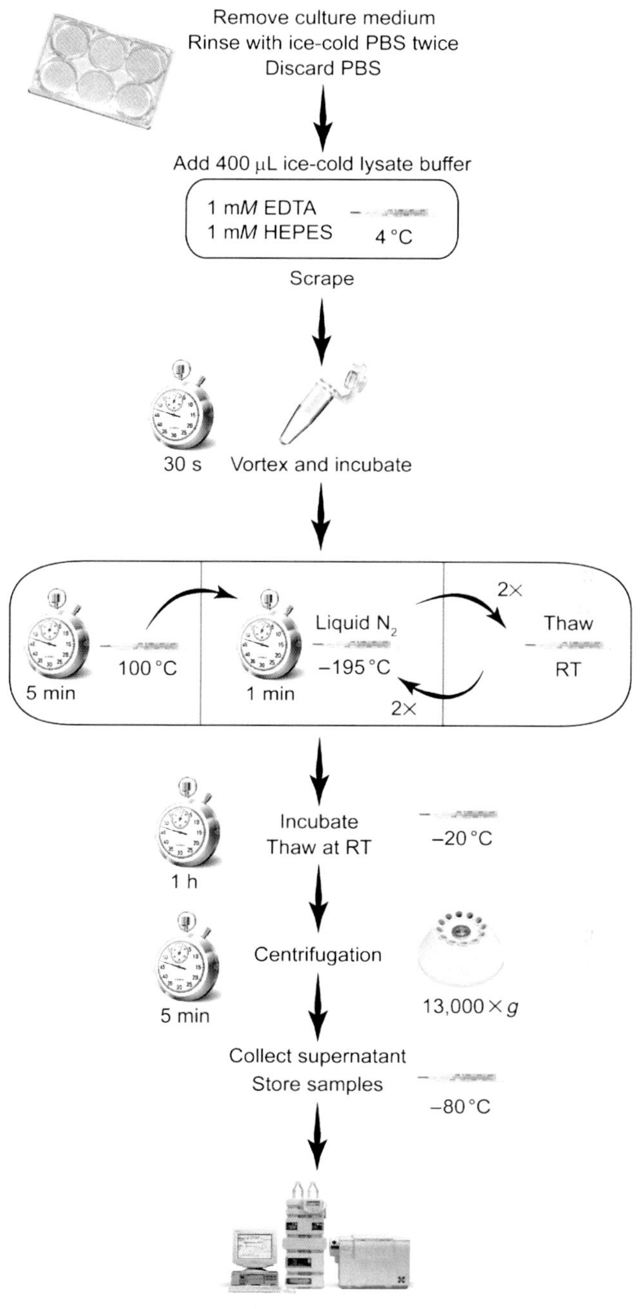

Figure 8.1 *Collection and processing of samples for the metabolomic profiling of cultured cancer cells.* Workflow for the preparation of lysates for high-performance liquid chromatography (HPLC) coupled to quadrupole time-of-flight (Q-TOF) high-resolution mass spectrometry (HRMS), starting from human non-small cell lung carcinoma A549 cell cultures. With minimal variations, this method is compatible with the preparation of samples for HPLC/Q-TOF HRMS from a variety of cell types. RT, room temperature.

2. Before cultures attain complete confluence, cells are regularly passaged upon trypsin/EDTA detachment, dilution, and reseeding (see Notes 4 and 5). This not only ensures the persistence of maintenance cultures (see Notes 6) but also provides cells for experimental determinations, which are generally carried out in six-well plates (see Notes 7–9).
3. When cells have adhered to the support and have resumed proliferation, which normally occurs 16–24 h upon seeding (see Note 10), supernatants are replaced with a complete culture medium containing the agents under investigation at the desired final concentration (see Notes 11 and 12) or with an alternative medium compatible with the desired experimental interventions (e.g., glucose-free medium).
4. At the end of the stimulation period, culture supernatants are discarded and adherent cells are thoroughly washed with 500 μL/well ice-cold PBS (see Notes 13 and 14), followed by the addition of 400 μL/well lysis buffer (1 m*M* EDTA, 1 m*M* HEPES in H_2O) (see Notes 15 and 16).
5. Upon the addition of lysis buffer, adherent cells are detached by scraping, and the content of four to six replicate wells is gathered within 1.5 mL microcentrifuge tubes and thoroughly vortexed for 30 s (see Notes 17 and 18).
6. Samples are boiled for 5 min and subjected twice to a freeze/thaw cycle involving incubation in liquid N_2 for 1 min and thawing at room temperature (see Note 19).
7. Samples are frozen at −20 °C for 1 h and then thawed at room temperature.
8. Finally, samples are centrifuged for 15 min at 13,000 × *g* and 4 °C and supernatants are collected and stored at −80 °C until analysis by HPLC/QTOF-HRMS.

3. CHROMATOGRAPHY, MASS SPECTROMETRY, AND DATA ANALYSIS

1. Samples for chromatography are dispensed in 100 μL aliquots in amber HPLC vials with 250 μL pulled-point glass inserts (Agilent Technologies Inc., Santa Clara, CA, USA) (see Note 20).
2. Chromatographic runs are performed on a 1260 Infinity HPLC device (Agilent Technologies Inc.) consisting of a G1379B degasser, a G1367E thermostated high-performance autosampler (HiP-ALS), a G1312B Binary Pump system, and a G1316A Thermostated Column Compartment (see Note 21).

3. Analysis and detection are performed by means of a 6520 Accurate-Mass Q-TOF spectrometer (Agilent Technologies Inc.). The system is equipped with an Agilent Multimode Dual Electrospray Ion Source (EIS) that can be operated in positive and negative ionization mode and allows automatic mass correction by the addition of lockmass standard (see Notes 21 and 22).
4. The HPLC/Q-TOF-HRMS apparatus is operated by MassHunter Acquisition B.02.01 software (Agilent Technologies Inc.)
5. The following protocols are complementary to each other for the metabolomic profiling of samples obtained as detailed in Section 2 (see Notes 23–25):
   a. An alkyl (C18) reversed-phase bonded Zorbax SB-AQ column (2.1 mm × 150 mm, 3.5 μm) is used with mobile phase A = 0.2% acetic acid in H_2O and mobile phase B = acetonitrile, with a flow rate of 0.3 mL/min. Gradient conditions:
      - Equilibration: 98% phase A and 2% phase B
      - Elution: gradient from 2% to 95% phase B in 10 min
      - Wash: 95% phase B for 1 min
      - Reequilibration: 98% phase A and 2% phase B for 8 min

      Samples are kept at 4 °C and the volume of injection is 10 μL. The column temperature is 40 °C. The EIS can be operated either in positive or in negative ion mode. Source parameters are gas temperature = 300 °C, gas flow = 10 L/min, nebulizer pressure = 40 psi, EIS capillary voltage = 3500 or 4000 V (for positive and negative ion mode, respectively), and fragmentor voltage = 180 V. Reference ions for mass calibration are purine (121.050873 $[M+H]^+$ and 119.036319 $[M-H]^-$) and hexakis(1H,1H,3H-tetrafluoropropoxy)phosphazine (HP-921; 922.009798 $[M+H]^+$).
   b. An alkyl (C18) reversed-phase bonded Zorbax SB-AQ column (2.1 mm × 150 mm, 3.5 μm) (Agilent Technologies Inc.) is used with mobile phase A = 10 m*M* ammonium acetate (NH_4Ac) plus 0.01% acetic acid in H_2O and mobile phase B = acetonitrile. Flow rate, gradient, injection volume, and EIS conditions are as in Step 3.5a.
6. Finally, MS data are processed using an in-house set of tools to convert raw MS data into a matrix that is compatible with the statistical analyses and interpretation, as previously detailed (Beckmann, Parker, Enot, Duval, & Draper, 2008; Enot, Haas, & Weinberger, 2011; Enot, Lin, et al., 2008; Overy et al., 2008) (see Note 26). In brief:

a. The MassHunter Qualitative software package (Agilent Technologies Inc.) is used to segment the signals obtained from each sample into a series of regions defined by mass to charge ratio (m/z) and retention time.

b. Extracted ion chromatograms are generated, weighted on quality, and subjected to time domain alignment.

c. Peaks are identified and integrated to obtain both qualitative and quantitative information.

d. Chemical species are selected for subsequent analysis based on the following selection criteria: missing value rate <25%, 95% signal-to-blank ratio >5, coefficient of variation from pooled samples <20%, and removal of correlated coeluting species (difference in retention time = 0.001 min and correlation >0.9) with the lowest overall intensities.

e. For the presentation of profiles, we generally adopt the following strategy: chemical species that are regularly observed in our datasets and validated by an analytical chemist are denoted with their name followed by retention time and measurement conditions, whereas the remaining signals are labeled with m/z ratio and retention time.

4. CONCLUDING REMARKS

The completion of the HGP and the development of technical platforms that would allow for the high-throughput assessment of gene copy number and global transcriptional profile in several samples, including clinical specimens from cancer patients, generated great expectations, which have been met only in part. In the post-genomic era, preferential attention is being given to the proteomic and metabolomic profiles of malignant cells, as these are expected to influence in a straightforward manner phenotypic as well as functional cellular features. Here, we reported a simple and rapid protocol to determine the complete metabolic complement of cultured neoplastic cells. We have successfully employed this approach in several preclinical studies (Galluzzi et al., 2012; Lainey et al., 2012, 2013; Michaud et al., 2011; Michels, Vitale, Galluzzi, et al., 2013; Michels, Vitale, Senovilla, et al., 2013), and we expect that—with minimal variations—it can also be employed to determine the metabolic complement of tissues, organs, and small organisms.

5. NOTES

1. Optimal growth conditions vary according with the cell line of choice. As the intracellular metabolome is extremely sensitive to environmental cues, optimal growth conditions, as recommended by the American Type Culture Collection (ATCC, Manassas, USA), are required, unless changes in the composition of the growth medium are part of experimental variables.
2. Depending on specific necessities (e.g., limited room within incubators, need for elevated amounts of cells), smaller or larger cell culture supports can be used in substitution of 75 cm^2 flasks.
3. While A549 cells grow rapidly in standard, untreated flasks, pretreated (e.g., collagen- or gelatin-coated), supports may be required for other cell types.
4. In slightly underconfluent conditions, a 75 cm^2 flask contains $8–10 \times 10^6$ A549 cells. This may significantly vary for other cell lines, mainly depending on cell size and on the ability of cells to form tridimensional clumps.
5. Passaging excessively diluted as well as excessively confluent cultures should be avoided as it may result in a genetic drift or in the overcrowding-dependent stress of the cell population, respectively, both of which negatively affect experimental determinations.
6. It is a good experimental practice to define a maximal number of passages and discard maintenance cultures once this threshold has been reached, to avoid genetic drifts of the cell population. This strategy calls for an adequate N_2 stock of cells, which should be constituted from an homogeneous and healthy population before the beginning of experimental assessments.
7. Freshly thawed cells should be allowed to grow in culture for two to three passages before being employed in experimental assessments.
8. Petri dishes with a diameter of 10 cm represent a suitable alternative in this setting, as they provide increased surface for cell culture while facilitating scraping (see also Note 17).
9. The amounts of cells to be seeded depend on multiple variables, including cell size, proliferation rate, growth surface, and total assay duration. Ideally, control cultures at the end of assays should exhibit confluence levels that never exceed 80–85%. This is particularly important for metabolomic determinations, given the intrinsic sensitivity of this

method. Indicatively, for assays lasting no more than 72 h, $200–400 \times 10^3$ A549 cells/well can be seeded (in 1 mL complete growth medium) in six-well plates.

10. Rapidly adhering cells (such as A549 cells and human cervical carcinoma HeLa cells) can be treated after 12–16 h, whereas at least 24–36 h is required for cells with prolonged adaptation times (e.g., human colorectal carcinoma HCT 116 cells). As the intracellular metabolome is extremely sensitive to microenvironmental perturbations, premature treatment should be avoided. Observing cells cultures under the light microscope generally suffices to assess their adherence/adaptation status.
11. This is particularly important when the experimental setup involves chemical stimuli that are stocked in highly lipophilic agents such as *N*,*N*-dimethylformamide (DMF) and dimethylsulfoxide (DMSO), as these organic solvents can be toxic if directly administered to the cell layer (see also Note 12).
12. Even in reduced amounts, DMF and DMSO can promote specific cellular responses (e.g., differentiation). It is therefore of the utmost importance to determine/take into account the impact of solvents on the intracellular metabolome, as these may act as confounding factors at data analysis.
13. Careful washing is required to minimize the contamination of samples with extracellular metabolites.
14. The use of ice-cold buffers is important as it blunts metabolic reactions until lysis. Along similar lines, it is important to work quickly from this step onward to minimize unwarranted, pre- and postmortem drifts in the intracellular metabolome.
15. According to our experience, the lysis buffer is stable for several weeks if properly stored at 4 °C.
16. Starting with lysis, all steps are performed with analytical grade reagents.
17. Although scraping may not result in the complete detachment of adherent cells, in our experience, it affects the intracellular metabolome to a lesser degree and in a more reproducible manner than trypsinization.
18. Generally, 1×10^6 cells are sufficient to perform reliable metabolomic assessments. As the amount of cells contained in each well is dictated by multiple variables (including cell number at seeding, proliferation rates, and the presence of cytostatic/cytotoxic stimuli), it is useful to allow for

five to six replicates of each experimental and control condition, one of which can be used to quantify cell number. This is important to ensure that lysates are obtained from a sufficient amount of starting material.

19. This step ensures the complete breakdown of virtually all subcellular structures.

20. Amber vials should be preferred over clear vials because they preserve light-sensitive metabolites, such as vitamin B12 and folates. Glassware is preferable over plasticware as it is associated with a limited loss of potentially contaminating species.

21. While the cell collection and sample preparation protocols provided here are expected to be compatible with a majority of HPLC/Q-TOF-HRMS systems, significant adjustments in the steps detailed from here onward may be required to obtain reliable results with technical platforms other than the one described here.

22. This instrument is characterized by a mass accuracy <5 ppm, a mass resolution of 5000–10,000 (100–922 *m/z*), a measuring frequency of 10,000 transients/s, and a detection frequency of 2 GHz (200,000 points/transient).

23. The complementary use of these two protocols allows for the detection of an increased amount of metabolites as compared to each method employed alone. For instance, the use of NH_4Ac facilitates the detection acetyl-CoA.

24. The EIS positive mode allows for the detection of species that acquire an H^+ ion more with a higher probability than losing one (containing a higher number of basic than acidic sites); the EIS negative mode allows for the detection of species that lose an H^+ ion more with a higher probability than acquiring one (containing a higher number of acidic than basic sites).

25. Column temperature must be higher than room temperature to ensure the stability of the chromatographic run.

26. This approach has been developed for the processing and analysis of data generated by a 6520 Accurate-Mass Q-TOF spectrometer (Agilent Technologies Inc.) equipped with an Agilent Multimode EIS that can be operated in positive and negative ionization mode, and hence may not necessarily be adequate for treating datasets generated by other MS platforms. Additional methods for the processing and analysis of metabolomics data can be found in Chapter 16.

ACKNOWLEDGMENTS

We thank José-Miguel Bravo-San Pedro for assistance with the preparation of figures. Authors are financed by the Ligue contre le Cancer (équipe labellisée), Agence National de la Recherche (ANR), Association pour la recherche sur le cancer (ARC), Cancéropôle Ile-de-France, AXA Chair for Longevity Research, Institut National du Cancer (INCa), Fondation Bettencourt-Schueller, Fondation de France, Fondation pour la Recherche Médicale (FRM), the European Commission (ArtForce), the European Research Council (ERC), the LabEx Immuno-Oncology, the SIRIC Stratified Oncology Cell DNA Repair and Tumor Immune Elimination (SOCRATE), the SIRIC Cancer Research and Personalized Medicine (CARPEM), and the Paris Alliance of Cancer Research Institutes (PACRI).

REFERENCES

Ahearn, I. M., Haigis, K., Bar-Sagi, D., & Philips, M. R. (2012). Regulating the regulator: Post-translational modification of RAS. *Nature Reviews. Molecular Cell Biology*, *13*, 39–51.

Beckmann, M., Parker, D., Enot, D. P., Duval, E., & Draper, J. (2008). High-throughput, nontargeted metabolite fingerprinting using nominal mass flow injection electrospray mass spectrometry. *Nature Protocols*, *3*, 486–504.

Coussens, L., Yang-Feng, T. L., Liao, Y. C., Chen, E., Gray, A., McGrath, J., et al. (1985). Tyrosine kinase receptor with extensive homology to EGF receptor shares chromosomal location with neu oncogene. *Science*, *230*, 1132–1139.

Craig, R. K., Bathurst, I. C., & Herries, D. G. (1980). Post-transcriptional regulation of gene expression in guinea pig tissues. *Nature*, *288*, 618–619.

Criollo, A., Galluzzi, L., Maiuri, M. C., Tasdemir, E., Lavandero, S., & Kroemer, G. (2007). Mitochondrial control of cell death induced by hyperosmotic stress. *Apoptosis*, *12*, 3–18.

Dunn, W. B., Broadhurst, D. I., Atherton, H. J., Goodacre, R., & Griffin, J. L. (2011). Systems level studies of mammalian metabolomes: The roles of mass spectrometry and nuclear magnetic resonance spectroscopy. *Chemical Society Reviews*, *40*, 387–426.

Dunn, W. B., & Ellis, D. I. (2005). Metabolomics: Current analytical platforms and methodologies. *Trends in Analytical Chemistry*, *24*, 285–294.

Enot, D. P., Haas, B., & Weinberger, K. M. (2011). Bioinformatics for mass spectrometry-based metabolomics. *Methods in Molecular Biology*, *719*, 351–375.

Enot, D. P., Lin, W., Beckmann, M., Parker, D., Overy, D. P., & Draper, J. (2008). Preprocessing, classification modeling and feature selection using flow injection electrospray mass spectrometry metabolite fingerprint data. *Nature Protocols*, *3*, 446–470.

Galluzzi, L., Vacchelli, E., Fridman, W. H., Galon, J., Sautes-Fridman, C., Tartour, E., et al. (2012a). Trial Watch: Monoclonal antibodies in cancer therapy. *Oncoimmunology*, *1*, 28–37.

Galluzzi, L., Vitale, I., Senovilla, L., Olaussen, K. A., Pinna, G., Eisenberg, T., et al. (2012b). Prognostic impact of vitamin B6 metabolism in lung cancer. *Cell Reports*, *2*, 257–269.

Goodacre, R., Vaidyanathan, S., Dunn, W. B., Harrigan, G. G., & Kell, D. B. (2004). Metabolomics by numbers: Acquiring and understanding global metabolite data. *Trends in Biotechnology*, *22*, 245–252.

Griffin, J. L., & Shockcor, J. P. (2004). Metabolic profiles of cancer cells. *Nature Reviews Cancer*, *4*, 551–561.

International Human Genome Sequencing, C. (2004). Finishing the euchromatic sequence of the human genome. *Nature*, *431*, 931–945.

Kallioniemi, A., Kallioniemi, O. P., Sudar, D., Rutovitz, D., Gray, J. W., Waldman, F., et al. (1992). Comparative genomic hybridization for molecular cytogenetic analysis of solid tumors. *Science*, *258*, 818–821.

Klausner, R. D., & Harford, J. B. (1989). cis-trans models for post-transcriptional gene regulation. *Science*, *246*, 870–872.

Lainey, E., Sebert, M., Thepot, S., Scoazec, M., Bouteloup, C., Leroy, C., et al. (2012). Erlotinib antagonizes ABC transporters in acute myeloid leukemia. *Cell Cycle*, *11*, 4079–4092.

Lainey, E., Wolfromm, A., Marie, N., Enot, D., Scoazec, M., Bouteloup, C., et al. (2013). Azacytidine and erlotinib exert synergistic effects against acute myeloid leukemia. *Oncogene*, *32*, 4331–4342.

Lander, E. S., Linton, L. M., Birren, B., Nusbaum, C., Zody, M. C., Baldwin, J., et al. (2001). Initial sequencing and analysis of the human genome. *Nature*, *409*, 860–921.

Michaud, M., Martins, I., Sukkurwala, A. Q., Adjemian, S., Ma, Y., Pellegatti, P., et al. (2011). Autophagy-dependent anticancer immune responses induced by chemotherapeutic agents in mice. *Science*, *334*, 1573–1577.

Michels, J., Vitale, I., Galluzzi, L., Adam, J., Olaussen, K. A., Kepp, O., et al. (2013a). Cisplatin resistance associated with PARP hyperactivation. *Cancer Research*, *73*, 2271–2280.

Michels, J., Vitale, I., Senovilla, L., Enot, D. P., Garcia, P., Lissa, D., et al. (2013b). Synergistic interaction between cisplatin and PARP inhibitors in non-small cell lung cancer. *Cell Cycle*, *12*, 877–883.

Moch, H., Schraml, P., Bubendorf, L., Mirlacher, M., Kononen, J., Gasser, T., et al. (1999). High-throughput tissue microarray analysis to evaluate genes uncovered by cDNA microarray screening in renal cell carcinoma. *The American Journal of Pathology*, *154*, 981–986.

Nilsson, T., Mann, M., Aebersold, R., Yates, J. R., 3rd., Bairoch, A., & Bergeron, J. J. (2010). Mass spectrometry in high-throughput proteomics: Ready for the big time. *Nature Methods*, *7*, 681–685.

Overy, D. P., Enot, D. P., Tailliart, K., Jenkins, H., Parker, D., Beckmann, M., et al. (2008). Explanatory signal interpretation and metabolite identification strategies for nominal mass FIE-MS metabolite fingerprints. *Nature Protocols*, *3*, 471–485.

Pan, Z., & Raftery, D. (2007). Comparing and combining NMR spectroscopy and mass spectrometry in metabolomics. *Analytical and Bioanalytical Chemistry*, *387*, 525–527.

Pinkel, D., & Albertson, D. G. (2005). Array comparative genomic hybridization and its applications in cancer. *Nature Genetics*, *37*(Suppl.), S11–S17.

Rosen, C. A., Sodroski, J. G., Goh, W. C., Dayton, A. I., Lippke, J., & Haseltine, W. A. (1986). Post-transcriptional regulation accounts for the trans-activation of the human T-lymphotropic virus type III. *Nature*, *319*, 555–559.

Seow, T. K., Ong, S. E., Liang, R. C., Ren, E. C., Chan, L., Ou, K., et al. (2000). Two-dimensional electrophoresis map of the human hepatocellular carcinoma cell line, HCC-M, and identification of the separated proteins by mass spectrometry. *Electrophoresis*, *21*, 1787–1813.

Sims, R. J., 3rd., & Reinberg, D. (2008). Is there a code embedded in proteins that is based on post-translational modifications? *Nature Reviews. Molecular. Cell Biology*, *9*, 815–820.

Vacchelli, E., Eggermont, A., Galon, J., Sautes-Fridman, C., Zitvogel, L., Kroemer, G., et al. (2013). Trial watch: Monoclonal antibodies in cancer therapy. *Oncoimmunology*, *2*, e22789.

Venter, J. C., Adams, M. D., Myers, E. W., Li, P. W., Mural, R. J., Sutton, G. G., et al. (2001). The sequence of the human genome. *Science*, *291*, 1304–1351.

Vitale, I., Senovilla, L., Jemaa, M., Michaud, M., Galluzzi, L., Kepp, O., et al. (2010). Multipolar mitosis of tetraploid cells: Inhibition by p53 and dependency on Mos. *The EMBO Journal*, *29*, 1272–1284.

Voss, T., Ahorn, H., Haberl, P., Dohner, H., & Wilgenbus, K. (2001). Correlation of clinical data with proteomics profiles in 24 patients with B-cell chronic lymphocytic leukemia. *International Journal of Cancer*, *91*, 180–186.

Vucic, D., Dixit, V. M., & Wertz, I. E. (2011). Ubiquitylation in apoptosis: A post-translational modification at the edge of life and death. *Nature Reviews. Molecular Cell Biology*, *12*, 439–452.

Xiao, J. F., Zhou, B., & Ressom, H. W. (2012). Metabolite identification and quantitation in LC-MS/MS-based metabolomics. *Trends in Analytical Chemistry*, *32*, 1–14.

CHAPTER NINE

Pulsed Stable Isotope-Resolved Metabolomic Studies of Cancer Cells

Matthias Pietzke, Stefan Kempa[1]

Berlin Institute for Medical Systems Biology at the MDC Berlin-Buch, Berlin, Germany

[1]Corresponding author: e-mail address: stefan.kempa@mdc-berlin.de

Contents

Abstract

Metabolic reprogramming is a key step in oncogenic transformation, and it involves alterations in both bioenergetic and anabolic metabolism. Sustained by these metabolic alterations, malignant cells acquire the ability to re-enter the cell cycle and proliferate. The so-called central carbon metabolism (CCM) is the ultimate source for energy and building blocks enabling cellular growth and proliferation. The time-resolved monitoring of the conversion of stable isotope-labeled metabolites provides profound insights into the metabolic dynamics of malignant cells and enables the tracking of individual carbon routes within the CCM. Specifically, the analysis of isotope incorporation rates within short time frames by means of pulsed stable isotope-resolved metabolomics (pSIRM) can be used to determine the dynamics of glycolysis and glutaminolysis—two metabolic circuitries that are often deregulated in malignant cells. Here, we detail a pSIRM-based method that can be applied to the study of metabolic alteration in cultured cancer cells.

Methods in Enzymology, Volume 543
ISSN 0076-6879
http://dx.doi.org/10.1016/B978-0-12-801329-8.00009-X

ABBREVIATIONS

13BPG 1,3-bisphosphoglyceric acid
2PGA 2-phosphoglyceric acid
3PGA 3-phosphoglyceric acid
6PglcLactone 6-phospho-gluconolactone
AcoA acetyl-CoA
Acon cis-aconitate
aKG alpha-ketoglutarate
beta-Ala beta-alanine
CCM central carbon metabolism
Cit citrate
DHAP dihydroxyacetonephosphate
E4P erythrose-4-phosphate
F16BP fructose-1,6-bisphosphate
F1P fructose-1-phosphate
F6P fructose-6-phosphate
Fum fumarate
G1P glucose-1-phosphate
G6P glucose-6-phosphate
GAP glyceraldehyde-3-phosphate
Glc glucose
Glyc glycerol
Glyc3P glycerol-3-phosphate
IsoCit isocitrate
Lac lactate
m/z mass-to-charge ratio
Mal malate
MeOX methoxamine (−group)
OAA oxaloacetic acid
OPP oxidative pentose pathway
PEP phosphoenolpyruvic acid
PG6 6-phosphogluconic acid
pSer phosphoserine
Pyr pyruvate
R5P ribose-5-phosphate
Ru5P ribulose-5-phosphate
S7P sedoheptulose-7-phosphate
Ser serine
Suc succinate
SucCoA succinyl-CoA
TMS trimethylsilyl
X5P xylulose-5-phosphate

1. INTRODUCTION

1.1. Usage of isotopes to monitor metabolic dynamics

The *de novo* synthesis or breakdown of metabolites needs to be monitored to follow the dynamics of metabolism. Therefore, it is important to separate "newly" synthesized from "old" existing metabolites. The application of isotope-labeled metabolites in combination with an appropriate detection method allows such analyses. In this regard, the use of isotopes to decode cellular metabolism has a long tradition. Initially, studies were performed by Rudolph Schonheimer, who introduced the use of deuterated substances to analyze the metabolic activity *in vivo* (Schoenheimer & Rittenberg, 1935, 1938). The discovery of methods to synthesize ^{11}C and ^{14}C opened further perspectives for isotope labeling experiments. Subsequently, the application of ^{14}C-labeled compounds led to the discovery of the major pathways within the central carbon metabolism (CCM) of plants and animals. For example, ^{14}C isotopes were used to unravel the structure of metabolic pathways within photosynthesis (Calvin & Benson, 1949) or glycolysis (Entner & Doudoroff, 1952).

Isotopes are indeed the tool of choice to monitor dynamic metabolic processes—isotopically labeled metabolites are taken up and are metabolized by the cell with the same specificity and speed as their "normal" counterparts. However, it is known since decades that carboxylation in plants discriminates ^{12}C over ^{13}C to some per mill, an effect that can be used to differentiate sugar from C3 or C4 plants (Farquhar, Ehleringer, & Hubick, 1989). Recently, it was reported that pyruvate carboxylase shows a similar effect (Wasylenko & Stephanopoulos, 2013). Despite these effects, isotopic labeling is still the best method to monitor the metabolic activity.

1.2. The biochemical network

The biochemical network is composed of a defined set of enzymes within cells and organisms. The biochemical properties, abundance, and compartmentation of these enzymes finally determine the structure and functionality of the network. The identity and quantitative expression of individual enzyme variants are highly controlled at all levels of gene regulation: (post-) transcriptional and (post-) translational. The underlying structure of the metabolic network, for example, in neuronal, muscle, or liver cells is comparable; however, the individual networks fulfill individual functions according to the needs of the organ context. These differences are due to the isoenzyme equipment of the distinct cell types. Moreover, not all regulatory properties of isoenzymes of CCM are known to date. Therefore, it is not trivial to predict the metabolic performance of a defined network—making the direct quantitative measurement of dynamic metabolic processes an ultimate goal.

1.3. Concept of isotope-resolved metabolomics

Within the last decades, the improvement of bioanalytical techniques enabled the use of stable isotope-labeled metabolites to track the individual metabolic routes *in vitro* and *in vivo*. The isotope incorporation is measured as mass shift by the mass spectrometer or by the spin induced by the odd carbon numbers by NMR (Fan et al., 2012). The introduction of heavy isotopes can add three main information layers that are not detectable by "static metabolomics" (Fig. 9.1). The term isotopologue is referred to compounds with different amounts of incorporated carbon atoms, whereas the term isotopomers (Fig. 9.1B) refers to different entities of one compound with the same amount of ^{13}C, but at different positions (Hellerstein & Neese, 1999; MacNaught & Wilkinson, 1997). The formation of isotopomers and isotopologues is dependent on the used substrates and the pathway usage: u-^{13}C glucose typically introduces three carbon atoms into pyruvate that will introduce two carbon atoms into the TCA cycle *via* acetyl-CoA (pyruvate dehydrogenase) or three carbon atoms by the action of pyruvate carboxylase, to replenish the TCA cycle at the level of oxaloacetate. Furthermore, in many cases, the information about the fraction of the ^{13}C-labeled compounds is the most important one

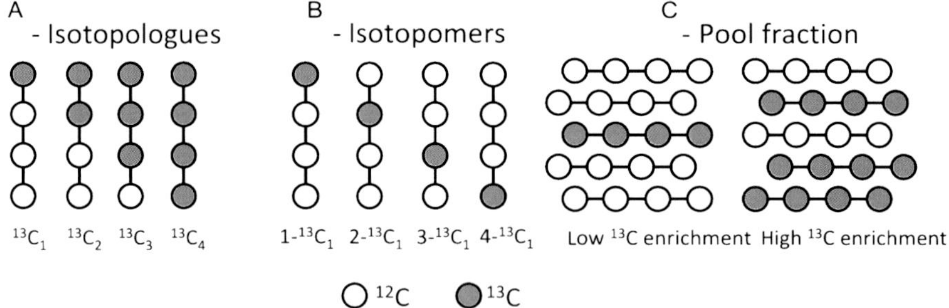

Figure 9.1 *The different layers of information that can be deduced from ^{13}C incorporation studies.* (A) Isotopologues differ in their amount of incorporated carbon atoms. (B) Isotopomers contain the same amount of carbon atoms on different positions. (C) ^{13}C pool fraction indicates the fraction of ^{13}C compounds in the total pool.

(Fig. 9.1C), as this represents the enzymatic activity within the network, in other words, the turnover of this compound.

Isotopologues can be differentiated with GC–MS by the intensity of mass shifts introduced by the heavy atoms. In addition, mass fragments originating from different parts of the molecules contain positional information and can be used to identify isotopomers. Finally, the amount of labeled compound in the pool size is encoded in the ratio of the heavy labeled to the unlabeled m/z, for example, 323–319 for glucose. Tools other than GC–MS can also be used to determine the heavy carbon incorporation. The mass shift found with LC–MS measurement identifies the isotopologues. If fragmentation is possible, the MS/MS fragments contain isotopomeric (positional) information as well. Carbon positions can also be elucidated from spin couplings within NMR measurements (Lane et al., 2009; Zwingmann, Richter-Landsberg, & Leibfritz, 2001).

1.4. Analysis of nonstationary isotope labeling

Fan and co-workers introduced the use of stable isotope-resolved metabolomics (SIRM) of cancer cell metabolism; ^{13}C carbon-labeled glucose was used to decode the metabolic properties of lung cancer cells *in vivo* (Fan et al., 2009). We have further developed this concept to dynamically analyze cellular metabolism by pulsed stable isotope-resolved metabolomics (pSIRM). The use of pulse labeling and the analysis of stable isotope incorporation in a nonstationary phase allow using fully labeled substrates, as they would give no information in stationary experiments (Noh & Wiechert, 2011). This simplifies the analysis of isotope incorporation specifically for glycolysis. For metabolic pathways, TCA cycle distinct positions of the heavy isotopes may encode the individual routes within the network.

Flux analysis with stable isotopes matured in the field of biotechnology to understand the effect of genetic engineering to bacteria and other microbes. This resulted in adaptations to the special characteristics of microbial metabolism: (i) many microbes have the capability to grow in minimal media with glucose as sole carbon source; (ii) due to their small size, they offer a high surface to volume ratio, which enables a rapid transport over the membranes; and (iii) they also possess short generation times. After some hours of ^{13}C incorporation and multiple cell doublings, a stable, "stationary" equilibrium of isotope enrichment is achieved. The method of choice for microbial flux measurements is to feed cells with a mixture of ^{13}C isotopes (e.g., 1-$^{13}C_1$-glucose, u-^{13}C-glucose, and ^{12}C glucose) and to elucidate the labeling pattern within protein bound amino acids. With the knowledge of the network structure, the label within the core metabolism (e.g., glycolysis, TCA cycle) is deduced from the resulting labeling pattern (Zamboni, Fendt, Ruhl, & Sauer, 2009). This approach is robust, the biomass delivers enough amino acids to measure label pattern with reliable intensity, and most descriptive fragment masses are identified. Further on, software tools were developed to calculate the fluxes (Zamboni, Fischer, & Sauer, 2005). In contrast to monitor soluble metabolites in their incomplete labeled state, quenching of cellular metabolism has to be performed within seconds after introducing ^{13}C into the system (Noh et al., 2007).

Mammalian cells possess different preconditions. The growth medium is more complex and contains many substrates. The generation time is significantly higher (~24 h), and further cellular processes can be energy demanding or require precursors of biomass production. Additionally, the compartmentation increases the complexity for mathematical modeling. A stationary labeling may be obtained after much longer incubation times and will be more difficult to interpret. In this case, the nonstationary or pulsed approach is the method of choice. Therefore, the incorporation of ^{13}C into metabolites is measured within short time scales (minutes) after introduction of ^{13}C substrates. This decreases the costs and the time necessary for the analysis and effectively shrinks the number of variables in the system. The number of compounds that are labeled within short time frames is smaller than the total numbers of measured metabolites, as only those metabolites will be found labeled that are closely connected to the highway of CCM.

1.5. The metabolic identity of cancer cells

Driven by the motivation to heal cancer, researchers started to investigate the metabolic difference of cancer cells, aiming to use them as therapeutic targets (Linehan & Rouault, 2013; Pelicano, Martin, Xu, & Huang, 2006). Especially, the glucose uptake is elevated in most cancer cells. This property is used for *in vivo* diagnosis *via* FDG-PET screening (Gambhir, 2002; Groheux, Espie, Giacchetti, & Hindie, 2013). Most cancers do not use their excess amounts of glucose effectively and secrete the majority of their glucose as lactate, even in the presence of sufficient oxygen, an effect known as the "Warburg effect" (Garber, 2004; Hsu & Sabatini, 2008; Pedersen, 2007; Warburg, 1956).

SIRM of cancer cells already revealed interesting differences in their pathway usage compared with other cells, for example, shuffling high amounts of carbon to glycine and serine by genetic amplification of phosphoglycerate dehydrogenase (Locasale et al., 2011). Recently, the presence of reductive carboxylation of ketoglutarate, described earlier as reaction of the reverse TCA cycle (Evans, Buchanan, & Arnon, 1966), was identified in hypoxic cancer cells (Wise et al., 2011). A systematic evaluation of cancer metabolism is still lacking, especially the *in vivo* usage of different substrates must be explored in much higher details by SIRM approaches.

2. EXPERIMENTAL PROCEDURES

Because of the dynamics of the metabolism of cancer cells, cell harvesting procedures have to be applied that preserve the metabolic homeostasis. A general protocol with regard to substrates, concentrations and incubation time cannot be given here as the experimental setup is highly dependent on the question. Nevertheless, we can mention some points that have to be addressed prior the experiment:

- Identification of the stable isotope-labeled substrate(s): For analyses of glycolysis and related reactions, ^{13}C-glucose may be the tracer of choice; to understand changes in the TCA cycle, ^{13}C-pyruvate or ^{13}C-glutamine might be useful; to monitor nitrogen flux into amino acid synthesis, ^{15}N-glutamine might be used (Fig. 9.2).
- Initial uptake experiments and characterization of the experimental system: Cells should be incubated with the substrates of choice to confirm the experimental strategy. Be aware that cells in cell culture may use every suitable compound if this represents the only carbon source; however, this must not reflect the natural metabolic activity.
- Identification of the experimental time frame: two things may be considered: (i) The metabolic intermediates should contain a reasonable label incorporation (minimum 5%), however, (ii) the stable isotope incorporation should not reach a stationary state.

For a typical pulse labeling experiment, we perform the stable isotope labeling ($^{13}C_6$ glucose) for 3–5 min. Within this time frame, the intermediates of upper glycolysis are labeled to very high extends and, for example, lactate contains 15–30% of label incorporation, for example, in HeLa cells. The appropriate incorporation time needed for ^{13}C stable isotope incorporation experiments (^{13}C glucose or ^{13}C glutamine) must be evaluated individually as the dynamics of these pathways can differ strongly and must not correlate with cell growth and proliferation. For example within two recent studies,

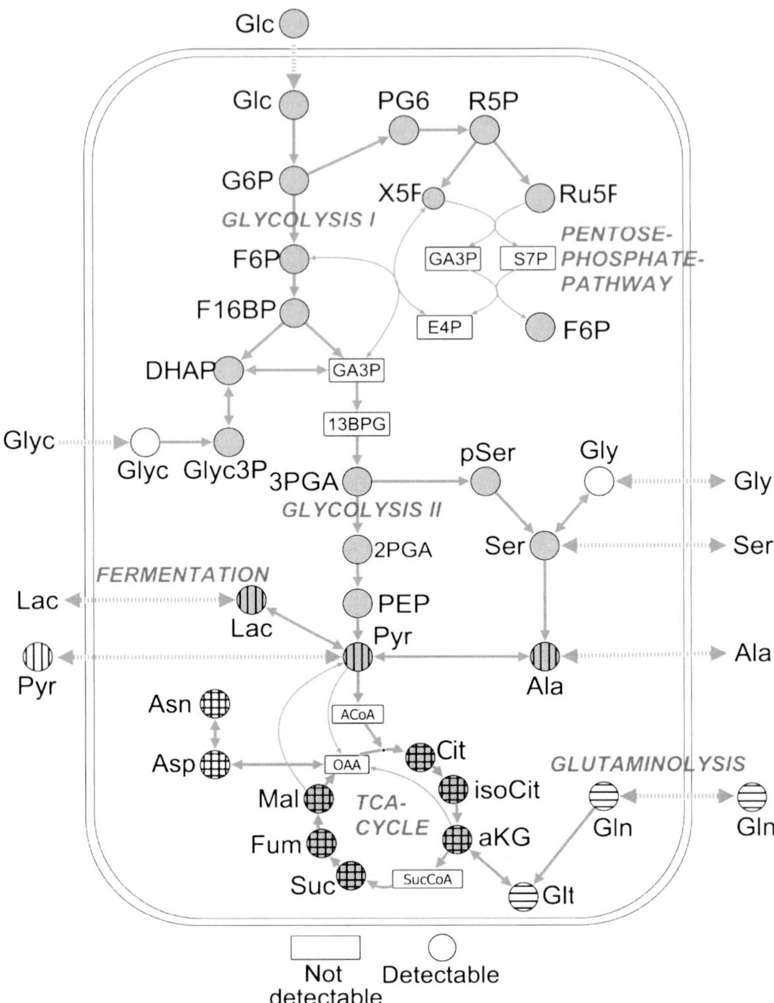

Figure 9.2 *Overview of the central carbon metabolism of mammalian cells.* The entry points of glucose, pyruvate, and glutamine and their mixing in products are indicated by gray color or line patterns.

we found the regulation of glycolysis not correlating with cell growth (Dorr et al., 2013; Liu et al., 2012).

2.1. Sampling and sample preparation

Growing cells in 10 cm cell culture dishes and harvesting at around 70% confluency (which will result in approximately two million cells for HeLa cells) delivers enough material for two to three measurements. Cellular metabolism is typically frozen with −20 °C cold methanol; cells are disrupted, extracted, and aliquoted, and extracts are dried under vacuum before derivatization and GC–MS measurements. Extraction with methanol and water or methanol + water + chloroform performs in general well (Dietmair, Timmins, Gray, Nielsen, & Kromer, 2010), and phase separation into polar and lipid phase effectively decreases the complexity of measurements and effectively pellets proteins and other insoluble cell material.

2.2. GC–MS-based analysis of central metabolites

A prerequisite for GC–MS-based metabolomics is the derivatization of polar compounds to increase their volatility and stability at high temperatures applied for the chromatographic separation. Silylation (adding trimethylsilyl (TMS) or tert-butyldimethylsilyl (TBDMS) groups) in combination with methoximation is the most versatile and broadly used method (Halket & Zaikin, 2003). Derivatization of dry samples with 20 μl of methoxamine hydrochloride (40 mg/ml in pyridine) for 90 min at 30 °C followed by silylation with 80 μl of *N*-methyl-*N*-trimethylsilyl-trifluoroacetamide (MSTFA) for 45 min at 37 °C is a common protocol used in many laboratories. As a consequence of derivatization, a high number of carbon, nitrogen, and silicon atoms are added increasing the mass and complexity of the molecules. The weight of pyruvate increases from 88 Da ($C_3H_4O_3$) to 189 Da ($C_7H_{15}O_3NSi$) resulting in an increase of the natural ^{13}C abundance of the $m+1$ fragment from 3.3% to 11.3% (Rosman & Taylor, 1998). Electron impact ionization is typically employed at the interface between gas-chromatographic separation and mass-spectrometric detection. This ionization is very effective, resulting in a fragmentation of the molecules. The typical fragmentation patterns are used for a library-assisted identification. As different mass fragments represent different parts of the molecule, they contain positional information as well.

In Fig. 9.2 and Table 9.1, GC–MS offers a superior coverage in the highly connected intermediates of CCM. It is a very efficient method to detect small-to-medium-sized molecules and even allows the separation of metabolites with the same mass (citrate/isocitrate, glucose-6-phosphate/fructose-6-phosphate). However, the size limitations do not allow monitoring molecules as acetyl-CoA or ATP.

Out of 38 CCM compounds typically detectable by GC–MS-based metabolome analyses, some can be found with stable isotope incorporation. The degree of isotope enrichment strongly depends on the used substrate,

Table 9.1 Overview of pathway coverage of the central carbon metabolism accessible by GC–MS measurements

Class	Max. Possible	Detectable by GC–MS	Found in Biological Samples
Glycolysis	12	10	10
	Glc, G1P, G6P, F6P, F16BP, DHAP, GAP, 3PGA, 13BPG, 2PGA, PEP, Pyr	Glc, G1P, G6P, F6P, F16BP, DHAP, 3PGA, 2PGA, PEP, Pyr	Glc, G1P, G6P, F6P, F16BP, DHAP, 3PGA, 2PGA, PEP, Pyr
TCA cycle	10	8	5
	ACoA, Cit, Acon, IsoCit, aKG, SucCoA Suc, Fum, Mal, OAA	Cit, Acon, IsoCit, aKG, Succ, Fum, Mal, OAA	Cit, aKG, Succ, Fum, Mal
PPP	7	4	2
	6PGlcLactone, PG6, Ru5P, R5P, S7P, E4P, X5P	PG6, Ru5P, R5P, E4P	PG6, R5P
Proteinogenic amino acids	20	17	16
	Ala, Arg, Asn, Asp, Cys, Gln, Glu, Gly, His, Ile, Leu, Lys, Met, Phe, Pro, Ser, Thr, Trp, Tyr, Val	Ala, Arg, Asn, Asp, Gln, Glu, Gly, Ile, Leu, Lys, Phe, Pro, Ser, Thr, Trp, Tyr, Val	Ala, Arg, Asn, Asp, Gln, Glu, Gly, Ile, Leu, Lys, Phe, Pro, Ser, Thr, Tyr, Val
C3 bodies	7	6	5
	Lac, glyceric acid, Glyc, Glyc3P, pSer beta-ala, dihydroxyacetone	Lac, glyceric acid, Glyc, Glyc3P, beta-ala, dihydroxyacetone	Lac, glyceric acid, Glyc, Glc3P, beta-ala
Total	56	45	38

incorporation time, and pathway usage of the cell. The metabolites with isotope incorporation after ^{13}C-glucose, ^{13}C-pyruvate, or ^{13}C-glutamine application are shown in Fig. 9.2. Due to the short incubation times (3–10 min), further metabolites displayed only traces of ^{13}C incorporation. The relative low number of metabolites with isotope incorporation simplifies the data interpretation and may allow estimating the real carbon uptake as the main metabolic routes can be quantified, although it is not a closed system. On the other hand, only central metabolic pathways are monitored, and the strategy is not appropriate for downstream metabolic pathways.

2.3. Incorporation of stable isotopes and its occurrence in MS fragment spectra

Introducing heavy isotopes as ^{13}C, ^{15}N, ^{18}O into metabolic intermediates produces a mass shift within the mass spectra. After EI ionization, different fragments of the molecule contain distinct numbers of carbon atoms. The base mass of the fragment (the most intense mass in the unlabeled compound) is denoted as m0; the mass with a mass shift of 1 Da is called $m+1$, and so on. A mass shift of +1 could be produced by a single ^{13}C carbon atom or by a single ^{15}N nitrogen atom.

The effect of fragmentation and the influence of introduced ^{13}C atoms for glucose are shown in Fig. 9.3. The mass spectrum is made of multiple fragment masses (top), and the natural isotope distribution is shown below the mass spectrum for three different mass ranges. The effect of a single ^{13}C atom or three ^{13}C atoms in the upper or lower part of the molecule depicts how the information is reflected in the mass spectra. For example, the fragment with a mass of 160 contains two carbon atoms from the top of the glucose molecule; its mass is shifted by one unit in 1-$^{13}C_1$-glucose, by two units in 123-$^{13}C_3$ and u-$^{13}C_6$, and it is neither affected by 6-$^{13}C_1$ or 456-$^{13}C_3$. In a similar way, the fragment with a mass of 217 contains three carbon atoms of the lower part of the molecule. Finally, the fragment with the mass 319 is made of the carbon atoms 3-4-5-6 of the glucose molecule.

3. DATA ANALYSIS

3.1. Calculation of isotope incorporation

After manual or automated identification (Hiller et al., 2013) of metabolites containing heavy atoms, the mass range(s) bearing information about isotopic enrichment can be extracted from the mass spectra (Kempa et al., 2009). The fractional abundance of each isotopologue is calculated by dividing the intensity of the individual isotopologue by the sum of the intensities of all isotopologues in the selected mass range (Hellerstein & Neese, 1999). This calculation eliminates differences in measured intensities among different samples, making the normalized values comparable. In the next step, the abundance of natural occurring isotopes is subtracted at every position to

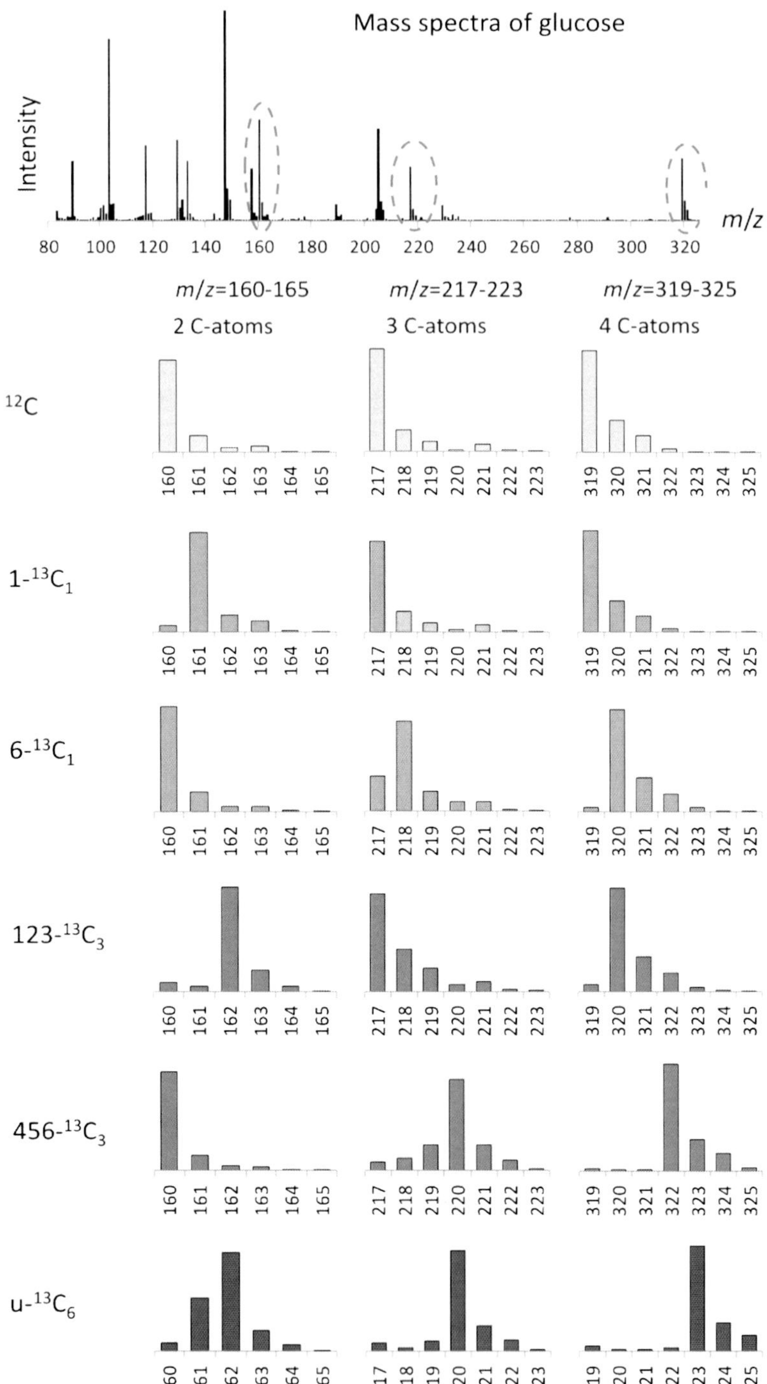

Figure 9.3 *Isotopomer patterns of different ^{13}C glucose species*. The mass spectrum for the 1MeOX, 5TMS derivative of pure glucose standards after GC–MS detection is shown at the top, and the impact of ^{13}C-induced mass shifts within some fragments is shown below the mass spectrum (electron impact ionization with −70 V).

calculate the correct mass isotopomer distribution (MID) by applying a series of linear equation systems or correction matrices for carbon only (Fernandez, Des Rosiers, Previs, David, & Brunengraber, 1996) or by including all possible isotopes (Nanchen, Fuhrer, & Sauer, 2007; Wittmann & Heinzle, 1999). These strategies need the knowledge of the chemical identity of the analyzed fragments; additionally, Jennings et al. reported a strategy to use measured spectra to calculate the enrichment over the natural abundance in even unknown compounds (Jennings & Matthews, 2005).

The beauty of using completely labeled substrates over of position-resolved isotopes is that multiple fragments of one metabolite contain similar information. The ^{13}C labeling information can be extracted from the best measured fragments, which means fragments with high intensity and specificity. A list of identified mass pairs after application of ^{13}C glucose or ^{13}C glutamine is given in Table 9.2.

3.2. Interpretation of carbon routing through CCM

Within the time frame of nonstationary ^{13}C carbon incorporation, the degree of label incorporation contains the information about the dynamics of the pathway. Together with the concentration of the metabolite, the carbon routing and pathway usage in the cell can be interpreted (Fig. 9.4).

The same is true for dividing the carbon flux at branching points of the metabolism. One must be aware that a complex metabolic network is monitored. Thus, the concentration changes of single compounds can have multiple reasons; however, if the same differences are also found in adjacent metabolites, the assumptions may be valid. For some compounds, the number of incorporated atoms allows to draw conclusions about the directionality of the flux. Citrate occupies a central position in the TCA cycle, which itself is an important hub between anabolic and catabolic processes. Citrate can be labeled at the $m+2$ position by carbons derived from glycolysis *via* acetyl-CoA (pyruvate dehydrogenase). Additionally by feeding pyruvate into the oxaloacetate pool (pyruvate carboxylase), a label at the $m+3$ position can be induced. By the action of the TCA cycle and anaplerotic replenishment of the TCA cycle by glutamine, a label at the $m+4$ position is introduced; additionally, the label at the $m+5$ position occurs by reductive carboxylation of ketoglutarate (Fig. 9.5).

For example, Bak et al. showed just from the label incorporation (in percent) after feeding u-^{13}C-glucose or u-^{13}C-lactate that glutamatergic

Table 9.2 Mass pairs extracted from in-depth analysis of mass isotopomer distributions after application of ^{13}C-glucose or ^{13}C-glutamine (taken from multiple cancer cell lines)

			Mass fragment (*m*/*z*)		
Compound	**Derivate**	**Abbr.**	**Unlabeled**	**Labeling with u-^{13}C-Glucose**	**Labeling with u-^{13}C Glutamine**
Alanine	3TMS	Ala	188	190	–
Aspartic acid	3TMS	Asp	232	235	–
Citric acid	4TMS	Cit	273	275	277
Dihydroxyacetone phosphate	1MeOX 3TMS	DHAP	400	403	–
Fructose	1MeOX 5TMS	Fru	217	220	–
Fructose-1,6-bisphosphate	1MeOX 7TMS	F16BP	217	220	–
Fructose-6-phosphate	1MeOX 6TMS	F6P	217	220	–
Fumaric acid	2TMS	Fum	245	247	249
Glucose	1MeOX 5TMS	Glc	319	323	–
Glucose-6-phosphate	1MeOX 6TMS	G6P	217	220	–
Gluconic acid-6-phosphate	7TMS	PG6	217	220	–
Glutamic acid	3TMS	Glu	246	–	250
Glutamine	3TMS	Gln	156	–	160
Glutaric acid	2TMS	Glut	261	–	266
Glutaric acid, 2-hydroxy	3TMS	Glut-OH	247	–	251
Glutaric acid, 2-oxo	1MeOX 2TMS	aKG	198	200	203
Glyceric acid-3-phosphate	4TMS	3PGA	357	359	–
Glycerol	3TMS	Glyc	218	221	–

Table 9.2 Mass pairs extracted from in-depth analysis of mass isotopomer distributions after application of ^{13}C-glucose or ^{13}C-glutamine (taken from multiple cancer cell lines)—cont'd

			Mass fragment (*m/z*)		
Compound	**Derivate**	**Abbr.**	**Unlabeled**	**Labeling with u-^{13}C-Glucose**	**Labeling with u-^{13}C Glutamine**
Glycerol-3-phosphate	4TMS	Glyc3P	357	359	–
Glycine	3TMS	Gly	276	277	–
Lactic acid	2TMS	Lac	117	119	–
Malic acid	3TMS	Mal	233	235	236
Phosphoenolpyruvic acid	3TMS	PEP	369	372	–
Pyruvic acid	1MeOX 1TMS	Pyr	174	177	–
Ribose-5-P	1MeOX 5TMS	R5P	217	220	–
Serine	3TMS	Ser	204	206	–
Succinic acid	2TMS	Suc	247	249	251

The list was generated after application of fully labeled substrates and derivatization using MSTFA as reagent.

neurons use a significant amount of lactate, which feeds into the TCA cycle (Bak, Schousboe, Sonnewald, & Waagepetersen, 2006). In resting neurons, the label incorporation from lactate was even higher than that from glucose.

Also the importance of the enzyme phosphoglycerate dehydrogenase for some cancer lines was analyzed in detail. By labeling cells with u-^{13}C-glucose, Locasale et al. (2011) could show that a high fraction of the labeling is ending up in serine or glycine. In a recent study, we could show that the central metabolism of oncogene driven cancer cells indeed possesses valuable targets (Liu et al., 2012). It could be demonstrated that MYC driven cancer cells depend on the signaling of AMPK related kinase 5 (ARK5). If the ARK5 is impaired, the energy metabolism of these cells was corrupted, and the cells died from energy depletion. Dynamic metabolome analyses revealed that the activity of ketoglutarate dehydrogenase was most probably impaired as seen by a dramatic decrease of the

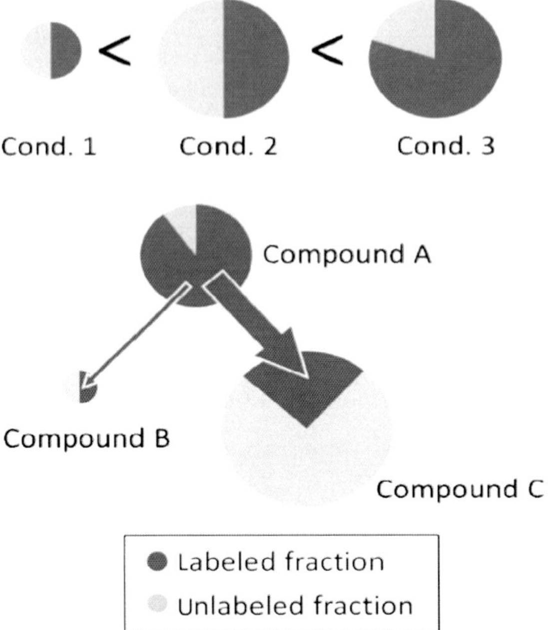

Figure 9.4 *Relation between stable isotope incorporation rates and pool size of labeled compounds.* Especially for nonstationary stable isotope incorporation experiments, already the percentage of label incorporation (dark color) in combination with the concentration of the metabolites (illustrated by size of the circles) allows the interpretation of the carbon routing through the pathways. A higher amount of compound or a higher label incorporation defines a higher labeled fraction (top) and can be used to derive directions of carbon flow through metabolism (bottom).

labeled pool size after glutamine labeling, which was not found in the "earlier" compounds in the TCA cycle. This was indeed confirmed by proteome data in which many subunits of complexes of the mitochondrial electron transport chain were down regulated. So, we speculated that specifically the down regulation of subunits of mitochondrial complex I (NADH dehydrogenase) resulted in an accumulation of NADH and subsequently in an allosteric inhibition of ketoglutarate dehydrogenase.

3.3. Using pSIRM data for instationary metabolic flux analysis

The applications of pSIRM analyses may be manifold. In the past years, the overwhelming flexibility of the central metabolism of cancer cells was reported. Especially, the variety of different substrates used and the flexible

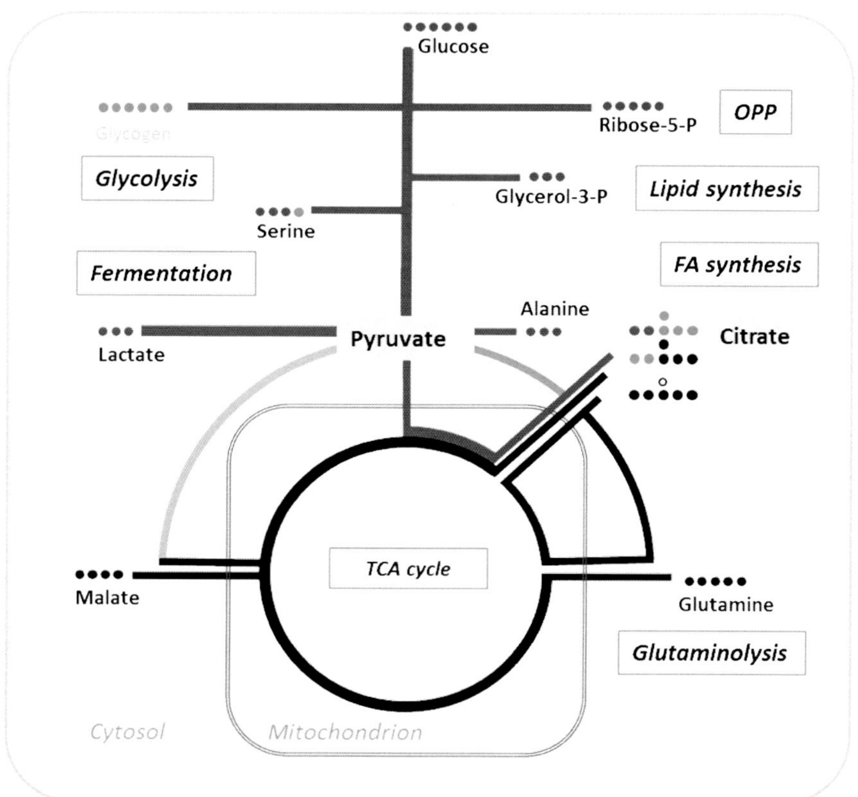

Figure 9.5 *CCM including the carbon flow through glycolysis or glutaminolysis as observed in many cancer cells in vitro.* The TCA cycle is a central hub in the metabolism of the cell and can be replenished from different entry points. The scheme represents the metabolic map of a fast proliferating cancer cell with a high glutaminolytic activity. (See the color plate.)

wiring of metabolic pathways make it hard to predict the activity or usage of distinct pathways.

The ultimate goal is the quantitative description of all metabolic fluxes in absolute numbers to learn more about the reaction velocity of every enzyme. A beautiful review is given in Wiechert and Noh (2013). Metabolic flux analysis (MFA) under nonstationary conditions is one of the most demanding approaches: It requires (i) quantitative information of the majority of metabolites, (ii) multiple sample time points to monitor the dynamics, (iii) a complete network of reactions, and (iv) huge computation power. It should be noted that even an incomplete model can sometimes produce good fits of the data to the output of the model (van Winden, Verheijen, &

Heijnen, 2001). Often, it is not possible to fulfill all the needs for comprehensive modeling (e.g., taking biopsies from human patients only derives one time point). As demonstrated, mathematical modeling is not always required to gain functional information from instationary isotope labeling studies.

4. OUTLOOK

The main interest is to understand when and how the metabolism of the cancer cell can be used as a therapeutic target. The application of pSIRM to decode the dynamics and regulation of cellular metabolism is very powerful. Even more, the use of such techniques to analyze *in vivo* metabolism may have an enormous potential and great advantages.

Such *in vivo* studies were already performed and allowed new insights in human metabolism. Dusick co-workers reported the increased activity of PPP after brain injuries. They injected 1,2-$^{13}C_2$-glucose into human patients and measured subsequently the labeling pattern in serum lactate (Dusick et al., 2007). Fan et al. (2009) injected ^{13}C glucose in humans with lung tumors shortly before their resection and analyzed the alterations in metabolism. Already these few examples may demonstrate how stable isotope labeling can be used in humans to understand *in vivo* metabolic processes and may outline the direction of isotope-resolved metabolomics for the future.

REFERENCES

Bak, L. K., Schousboe, A., Sonnewald, U., & Waagepetersen, H. S. (2006). Glucose is necessary to maintain neurotransmitter homeostasis during synaptic activity in cultured glutamatergic neurons. *Journal of Cerebral Blood Flow and Metabolism*, *26*(10), 1285–1297.

Calvin, M., & Benson, A. A. (1949). The path of carbon in photosynthesis IV: The identity and sequence of the intermediates in sucrose synthesis. *Science*, *109*(2824), 140–142.

Dietmair, S., Timmins, N. E., Gray, P. P., Nielsen, L. K., & Kromer, J. O. (2010). Towards quantitative metabolomics of mammalian cells: Development of a metabolite extraction protocol. *Analytical Biochemistry*, *404*(2), 155–164.

Dorr, J. R., Yu, Y., Milanovic, M., Beuster, G., Zasada, C., Dabritz, J. H., et al. (2013). Synthetic lethal metabolic targeting of cellular senescence in cancer therapy. *Nature*, *501*(7467), 421–425.

Dusick, J. R., Glenn, T. C., Lee, W. N., Vespa, P. M., Kelly, D. F., Lee, S. M., et al. (2007). Increased pentose phosphate pathway flux after clinical traumatic brain injury: A [1,2-13C2]glucose labeling study in humans. *Journal of Cerebral Blood Flow and Metabolism*, *27*(9), 1593–1602.

Entner, N., & Doudoroff, M. (1952). Glucose and gluconic acid oxidation of *Pseudomonas saccharophila*. *Journal of Biological Chemistry*, *196*(2), 853–862.

Evans, M. C., Buchanan, B. B., & Arnon, D. I. (1966). New cyclic process for carbon assimilation by a photosynthetic bacterium. *Science*, *152*(3722), 673.

Fan, T. W., Lane, A. N., Higashi, R. M., Farag, M. A., Gao, H., Bousamra, M., et al. (2009). Altered regulation of metabolic pathways in human lung cancer discerned by (13)C stable isotope-resolved metabolomics (SIRM). *Molecular Cancer*, *8*, 41.

Fan, T. W., Lorkiewicz, P. K., Sellers, K., Moseley, H. N., Higashi, R. M., & Lane, A. N. (2012). Stable isotope-resolved metabolomics and applications for drug development. *Pharmacology & Therapeutics*, *133*(3), 366–391.

Farquhar, G. D., Ehleringer, J. R., & Hubick, K. T. (1989). Carbon isotope discrimination and photosynthesis. *Annual Review of Plant Physiology and Plant Molecular Biology*, *40*(1), 503–537.

Fernandez, C. A., Des Rosiers, C., Previs, S. F., David, F., & Brunengraber, H. (1996). Correction of 13C mass isotopomer distributions for natural stable isotope abundance. *Journal of Mass Spectrometry*, *31*(3), 255–262.

Gambhir, S. S. (2002). Molecular imaging of cancer with positron emission tomography. *Nature Reviews. Cancer*, *2*(9), 683–693.

Garber, K. (2004). Energy boost: The Warburg effect returns in a new theory of cancer. *Journal of the National Cancer Institute*, *96*(24), 1805–1806.

Groheux, D., Espie, M., Giacchetti, S., & Hindie, E. (2013). Performance of FDG PET/CT in the clinical management of breast cancer. *Radiology*, *266*(2), 388–405.

Halket, J. M., & Zaikin, V. G. (2003). Derivatization in mass spectrometry—1. Silylation. *European Journal of Mass Spectrometry (Chichester, England)*, *9*(1), 1–21.

Hellerstein, M. K., & Neese, R. A. (1999). Mass isotopomer distribution analysis at eight years: Theoretical, analytic, and experimental considerations. *American Journal of Physiology*, *276*(6 Pt 1), E1146–E1170.

Hiller, K., Wegner, A., Weindl, D., Cordes, T., Metallo, C. M., Kelleher, J. K., et al. (2013). NTFD–a stand-alone application for the non-targeted detection of stable isotope-labeled compounds in GC/MS data. *Bioinformatics*, *29*(9), 1226–1228.

Hsu, P. P., & Sabatini, D. M. (2008). Cancer cell metabolism: Warburg and beyond. *Cell*, *134*(5), 703–707.

Jennings, M. E., 2nd., & Matthews, D. E. (2005). Determination of complex isotopomer patterns in isotopically labeled compounds by mass spectrometry. *Analytical Chemistry*, *77*(19), 6435–6444.

Kempa, S., Hummel, J., Schwemmer, T., Pietzke, M., Strehmel, N., Wienkoop, S., et al. (2009). An automated GCxGC-TOF-MS protocol for batch-wise extraction and alignment of mass isotopomer matrixes from differential 13C-labelling experiments: a case study for photoautotrophic-mixotrophic grown *Chlamydomonas reinhardtii* cells. *Journal of Basic Microbiology*, *49*(1), 82–91.

Lane, A. N., Fan, T. W., Higashi, R. M., Tan, J., Bousamra, M., & Miller, D. M. (2009). Prospects for clinical cancer metabolomics using stable isotope tracers. *Experimental and Molecular Pathology*, *86*(3), 165–173.

Linehan, W. M., & Rouault, T. A. (2013). Molecular pathways: Fumarate hydratase-deficient kidney cancer—Targeting the Warburg effect in cancer. *Clinical Cancer Research*, *19*(13), 3345–3352.

Liu, L., Ulbrich, J., Muller, J., Wustefeld, T., Aeberhard, L., Kress, T. R., et al. (2012). Deregulated MYC expression induces dependence upon AMPK-related kinase 5. *Nature*, *483*(7391), 608–612.

Locasale, J. W., Grassian, A. R., Melman, T., Lyssiotis, C. A., Mattaini, K. R., Bass, A. J., et al. (2011). Phosphoglycerate dehydrogenase diverts glycolytic flux and contributes to oncogenesis. *Nature Genetics*, *43*(9), 869–874.

MacNaught, A. D., & Wilkinson, A. (1997). *IUPAC. Compendium of chemical terminology—The "Gold Book"* (2nd ed.). Oxford: Blackwell Scientific Publications.

Nanchen, A., Fuhrer, T., & Sauer, U. (2007). Determination of metabolic flux ratios from 13C-experiments and gas chromatography–mass spectrometry data: Protocol and principles. *Methods in Molecular Biology, 358*, 177–197.

Noh, K., Gronke, K., Luo, B., Takors, R., Oldiges, M., & Wiechert, W. (2007). Metabolic flux analysis at ultra short time scale: Isotopically non-stationary 13C labeling experiments. *Journal of Biotechnology, 129*(2), 249–267.

Noh, K., & Wiechert, W. (2011). The benefits of being transient: Isotope-based metabolic flux analysis at the short time scale. *Applied Microbiology and Biotechnology, 91*(5), 1247–1265.

Pedersen, P. L. (2007). Warburg, me and Hexokinase 2: Multiple discoveries of key molecular events underlying one of cancers' most common phenotypes, the "Warburg Effect", i.e., elevated glycolysis in the presence of oxygen. *Journal of Bioenergetics and Biomembranes, 39*(3), 211–222.

Pelicano, H., Martin, D. S., Xu, R. H., & Huang, P. (2006). Glycolysis inhibition for anticancer treatment. *Oncogene, 25*(34), 4633–4646.

Rosman, K., & Taylor, P. (1998). Isotopic compositions of the elements 1997. *Journal of Physical and Chemical Reference Data, 27*, 1275.

Schoenheimer, R., & Rittenberg, D. (1935). Deuterium as an indicator in the study of intermediary metabolism. *Science, 82*(2120), 156–157.

Schoenheimer, R., & Rittenberg, D. (1938). The application of isotopes to the study of intermediary metabolism. *Science, 87*(2254), 221–226.

van Winden, W., Verheijen, P., & Heijnen, S. (2001). Possible pitfalls of flux calculations based on (13)C-labeling. *Metabolic Engineering, 3*(2), 151–162.

Warburg, O. (1956). On the origin of cancer cells. *Science, 123*(3191), 309–314.

Wasylenko, T. M., & Stephanopoulos, G. (2013). Kinetic isotope effects significantly influence intracellular metabolite 13C labeling patterns and flux determination. *Biotechnology Journal, 8*(9), 1088–1089.

Wiechert, W., & Noh, K. (2013). Isotopically non-stationary metabolic flux analysis: Complex yet highly informative. *Current Opinion in Biotechnology, 24*(6), 979–986.

Wise, D. R., Ward, P. S., Shay, J. E., Cross, J. R., Gruber, J. J., Sachdeva, U. M., et al. (2011). Hypoxia promotes isocitrate dehydrogenase-dependent carboxylation of alpha-ketoglutarate to citrate to support cell growth and viability. *Proceedings of the National Academy of Sciences of the United States of America, 108*(49), 19611–19616.

Wittmann, C., & Heinzle, E. (1999). Mass spectrometry for metabolic flux analysis. *Biotechnology and Bioengineering, 62*(6), 739–750.

Zamboni, N., Fendt, S. M., Ruhl, M., & Sauer, U. (2009). (13)C-based metabolic flux analysis. *Nature Protocols, 4*(6), 878–892.

Zamboni, N., Fischer, E., & Sauer, U. (2005). FiatFlux–a software for metabolic flux analysis from 13C-glucose experiments. *BMC Bioinformatics, 6*, 209.

Zwingmann, C., Richter-Landsberg, C., & Leibfritz, D. (2001). 13C isotopomer analysis of glucose and alanine metabolism reveals cytosolic pyruvate compartmentation as part of energy metabolism in astrocytes. *Glia, 34*(3), 200–212.

CHAPTER TEN

Single-Cell Imaging for the Study of Oncometabolism

Aleš Svatoš[*,1], **Alfredo J. Ibáñez**[†]

[*]Mass Spectrometry/Proteomic Research Group, Max Planck institute, Jena, Germany

[†]Department of Chemistry and Applied Biosciences, Eidgenössiche Technische Hochschule Zurich, Zurich, Switzerland

[1]Corresponding author: e-mail address: svatos@ice.mpg.de

Contents

Abstract

Metabolic profiling is commonly employed to investigate the global metabolic alterations of malignant cells or tissues. In the latter setting, neoplastic lesions are separated from adjacent, healthy tissues and their metabolites are quantified upon a chromatographic run coupled to mass spectrometry. Changes in the abundance of specific metabolites are then mapped on metabolic networks and the underlying metabolic circuitries are investigated as potential targets for the development of novel anticancer drugs. This approach, however, does not take into account the intrinsic heterogeneity of neoplastic lesions, which contain a large amount of non-transformed cells. To circumvent this issue, techniques have been developed that allow for the imaging of metabolites at the single-cell level. Here, we summarize established protocols that are suitable for imaging metabolites in animal cells (be them malignant or not) as well as in plant and prokaryotic cells. These methods are relevant for the study of the metabolic alterations that accompany oncogenesis and tumor progression.

Methods in Enzymology, Volume 543
ISSN 0076-6879
http://dx.doi.org/10.1016/B978-0-12-801329-8.00010-6

ABBREVIATIONS

AP-MALDI atmospheric pressure matrix-assisted laser desorption/ionization
LDI laser desorption/ionization
MALDI matrix-assisted laser desorption/ionization
MAMS microarrays for mass spectrometry
MS mass spectrometry
MSI mass spectrometry imaging
m/z mass over charge ratio

1. INTRODUCTION

Recently, technology in the "-omics" sciences has made substantial progress, especially in the areas of genomics and transcriptomics, allowing single-cell transcriptomic studies (Kalisky & Quake, 2011). Accordingly, metabolomics, the determination of a complete set of small molecules known as metabolites (<2500 Da with a typical molecular weight of 400 Da) in organisms or cells, has flourished (Fernie, Trethewey, Krotzky, & Willmitzer, 2004). The information metabolites offer is essential for understanding life because they are responsible for organisms' phenotypes. The structural diversity of metabolites is extraordinarily large in spite of their small size (Gunatilaka, 2006). They cover wide array of compound classes, including sugars, acids, bases, lipids, hormonal steroids, and many others (Koehn & Carter, 2005). Larger molecules that are made up of repeated building blocks, such as proteins and glycans, are no longer considered metabolites. Also, the structure of metabolites usually cannot be deduced by using genomic information. A notable exception represents polyketides. Traditionally, metabolites are divided into primary and secondary metabolites. Primary metabolites, directly involved in growth, development, and reproduction, are well investigated as the basic metabolic pathways and components are similar even between vastly different species. In contrast, secondary metabolites are often specific to a narrow set of species.

Typically, metabolomics (Saito & Matsuda, 2010) starts with harvesting the studied object. The metabolome is extracted from tissue with particular solvents, the tissue debris is removed by centrifugation or filtration, and metabolites are separated by chromatography and detected/identified with mass spectrometry (MS) or nuclear magnetic resonance. When sampling at cellular and subcellular levels, this protocol becomes impractical because the amount of contamination originating from solvents and the extraction apparatus may dramatically increase, and the signals of interest may be lost in the contaminant-related noise.

To overcome the dilution/contamination issue of typical metabolomic workflow, when working with cells novel methods combining sample preparation and metabolite detection were recently developed (Svatos, 2011). If the object is cellular, such as bacteria, yeast, or fungi, microfluidics seem to be the most appropriate way to sort, concentrate, disintegrate, and extract cells (Mellors, Jorabchi, Smith, & Ramsey, 2010; Szita, Polizzi, Jaccard, & Baganz, 2010; Wurm, Schopke, Lutz, Muller, & Zeng, 2010; Zare & Kim, 2010). In multicellular organisms, individual cells can be separated from each other by cellular matrix lysis before the microfluidics step, but it is important to show that such manipulation does not alter the metabolomic and developmental state of the studied cells. Probably, the best way of preventing stress is to keep the sample under ambient conditions. Combining sample preparation directly with the gasification and ionization of metabolites is the gentlest preparation method and has been recently used in cell-addressable laser-assisted ESI at ~30 μm resolution (Shrestha & Vertes, 2009) working well for plant cells. Because finding the proper position is problematic if micrometer scales are used, a blind method using mass spectrometry imaging (MSI) technology (Svatos, 2010) in which metabolites are desorbed and ionized from a predefined *x*,*y*-grid can simplify this issue.

The primary direct method for studying cell metabolites distribution would be secondary ion mass spectrometry (SIMS) introduced by Benninghoven in 1978 and later extended by Winograd (Winograd & Garrison, 2010). Here, a tightly focused primary beam of high-energy charged particles or ions eject and ionize molecules (secondary ions) of interest from the studied material with lateral resolution in range of 50 nm to 1 μm. However, under standard SIMS conditions, we were not able to see intact metabolites but rather chemical classes-specific fragments. Recently, by using C_{60}^{+}, SF_5^{+}, and ionic cluster (Au_n^{+}, Bi_n^{+}, SF_5^{+}) primary beams and treating samples with matrix or nanoparticles, small stable molecules have been imaged with excellent spatial resolution, allowing the internal contents of cells to be studied and 3D maps of their distribution constructed (Fletcher, Lockyer, Vaidyanathan, & Vickerman, 2007). In addition, cluster SIMS using a $C_{60}^{+\cdot}$ ion gun has been interfaced to hybrid Q-TOF instrumentation, providing excellent mass resolution and accuracy and enhancing the identification power of SIMS imaging.

The earliest studies of single cells were performed without sample preparation using the matrix-assisted laser desorption/ionization (MALDI)-TOF/MS method (Jimenez et al., 1998). Cells were deposited on the metallic target and treated with MALDI matrix solutions. Interestingly, peptides from neural cells were first analyzed concurrent with the development of the MALDI-MSI method (Caprioli, Farmer, & Gile, 1997). MALDI-based MSI is still a very attractive method and can perform cellular analysis in an imaging mode that allows different areas to be selectively irradiated with the laser foci bigger than the cell dimension using oversampling protocol. For single cell, the laser size should be focused close to 1 μm and effective ion extraction from the laser-evaporated ion plume should be achieved to overcome low amounts of metabolites in the cell volume. Those conditions have been recently met in Spengler's group working at University of Giessen in theirs SMALDI probe combined with tandem mass spectrometer (Römpp et al., 2010). System is working at ambient conditions and laser focus, and ion extraction is collinear and orthogonal to imaged sample (Figs. 10.1 and 10.2). Special resolution is for far limited by instrument sensitivity and 7 μm resolution has been demonstrated on MSI of HeLa cell lines (Römpp & Spengler, 2013).

Cell microarrays have been commonly paired with MS read-outs, when laser desorption/ionization (LDI) or MALDI setup is used (Ressine, Marko-Varga, & Laurell, 2007; Urban, Amantonico, & Zenobi, 2011; Urban, Schmidt, et al., 2011). One of the key reasons for their success is their ability to avoid the dilution of the analytical sample (by localizing it in a discrete micrometer-sized spot) for a more efficient laser ablation/ionization (Ahn et al., 2010; Ibáñez, Muck, Halim, & Svatoš, 2007). The microarrays for mass spectrometry (also known as MAMS) were first introduced by Urban et al. (2010). This particular type of microarrays was tailored for the study of cells grown in liquid medium using multiple-analytical read-outs, such as optical and MS. MAMS are capable of boosting the sensitivity of traditional LDI- or MALDI-based mass spectrometers to achieve single-cell level analysis by allowing the metabolic contents of the self-aliquoted cells to be localized within a discrete area (i.e., MAMS reservoir, Fig. 10.3; Ibánez et al., 2013; Urban, Amantonico, et al., 2011; Urban, Schmidt, et al., 2011). As typically, only one cell is harboring one MAMS microarray well no tight laser focus is needed and convention MALDI-TOF/MS instruments can be used. Increasing spatial resolution is associated with the obvious problem of available sensitivity. Performing cellular metabolomic measurements for major metabolites requires low femtomolar to high attomolar sensitivity, compatible with current technology; nevertheless, not sufficient to trace much less abundant yet important metabolites.

In this chapter, we have collected protocols for MALDI- and LDI-MS imaging of diverse samples ranging from intact plant tissue, immortal cell

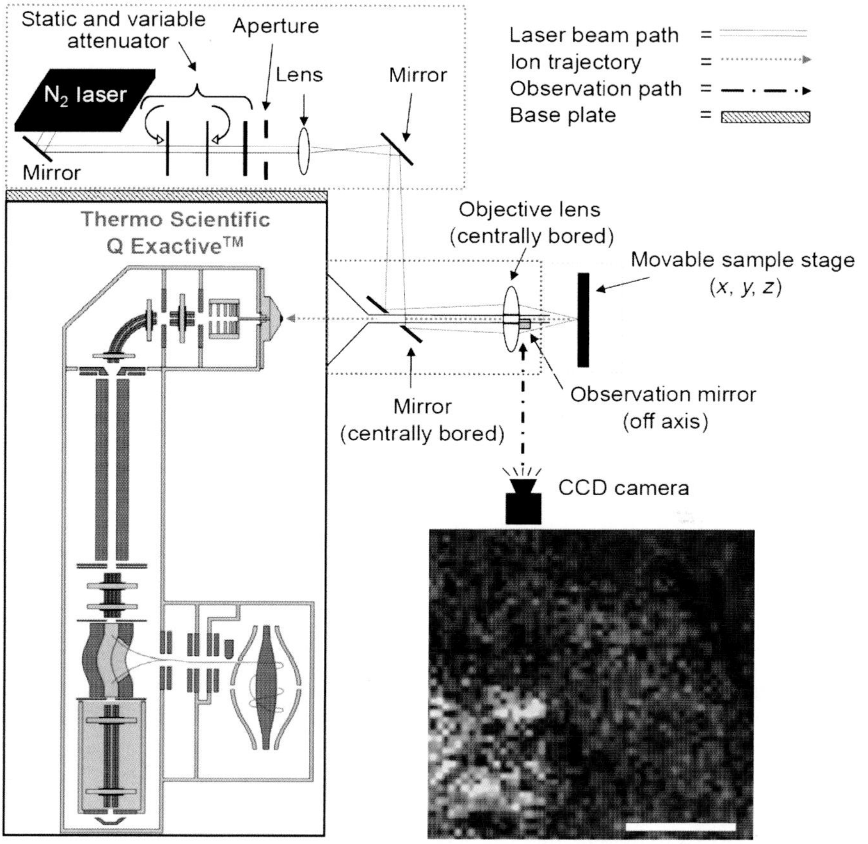

Figure 10.1 *Image10 AP-MALDI source (TransMIT, University Giessen, Germany) mounted on hybrid Q-Exactive mass spectrometer (Thermo Scientific, Bremen, Germany).* Inserts document lateral resolution of 10 μm, scale bar: 50 μm. (See the color plate.)

lines, and yeast cells. Detailed protocols for substrate handling, matrix application, data collection, and analysis are provided to assist community planning to implement MSI in their workflow.

2. METHODS

2.1. Sample preparation and handling

2.1.1 Intact tissues

1. Take the proper support for the imaging compatible to the MS instrument and the size and thickness of the imagined object. See the notes in the MS experiment section for details for the MS instruments used.

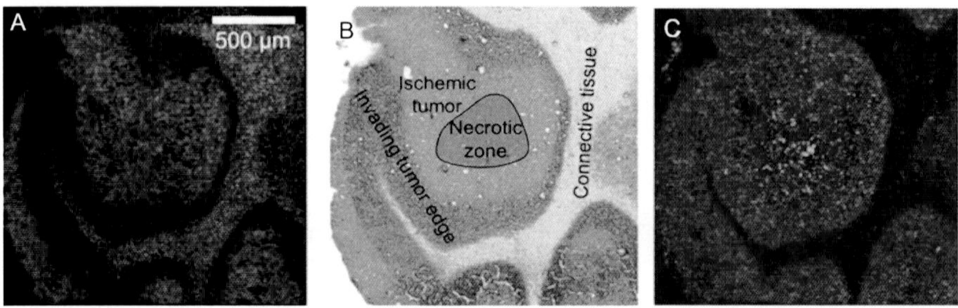

Figure 10.2 *Distribution of lipids in human non-small cell lung carcinoma.* Carcinoma was induced into a severe combined immunodeficiency (SCID) mouse model. (A) MS image, 10 µm pixel size, with false-color-coded distribution of lipids: red sphingomyelin SM (36:1), green cerebroside Cer (42:2), blue phosphatidylcholine PC (36:4). (B) H&E staining after measurement. (C) MS image, 10 µm pixel size, with false-color-coded distribution of lipids: green lysophosphatidylcholine LPC (16:1), red phosphatidylcholine PC (38:6). (See the color plate.)

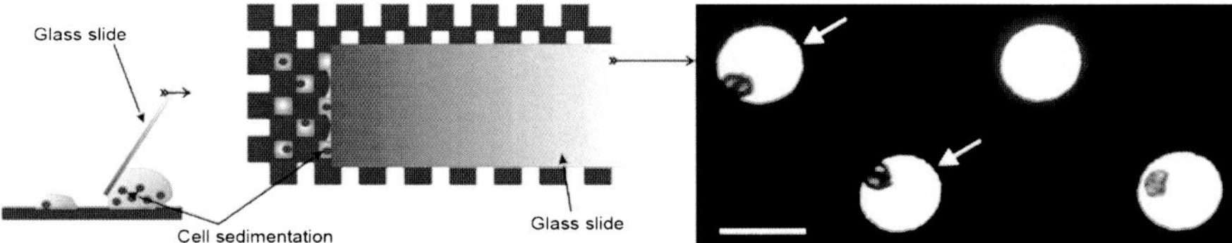

Figure 10.3 *MAMS array.* MAMS substrate has the self-aliquoting properties. Cell suspension deposited on the MAMS array is dragged over the array surface (left) and cells are self-aliquoted to individual hydrophilic cells. Arrows indicate recipient sites with the cells of *Cosmarium turpinii*. Scale bar: 100 µm. *Modified from Urban et al. (2010).* (See the color plate.)

2. Take a double side carbon conductive adhesive tape (Plano No. G3939, http://www.plano-em.de) and cut the tape ca. 2–3 mm bigger than the tissue and remove one glue cover.
3. Attach the tape by the sticky layer to glass microscopic slide or blank MALDI plate and after firm attachment remove the other cover. Using clean and sharp forceps are advisable, after having the one side of the tape firmly attached to the target.
4. Detach organ to be imaged (leaves, roots, flower petal, and sepal). The tissue should be flat and without big (max. 0.5–1 mm) trichomes or other imperfections. Difficult step, success depends on previous experiences with the object. The tissue removal must be done under controlled condition and fast. In some case, cooling the object will limit imposed stress. Reserve some time and material propreliminary experiments

and do not continue further before this step is not under full control. Use gloves and to prevent any leakage of internal cell content.

5. Hold the detached organ in a fine forceps and put on the sticky tape starting with the distal part from the forceps and finally attach the whole tissue. Make sure the tape is under whole tissue.
6. Finally, gently press the organ on the plate using dry paper tissue to ensure the firm attachment. When the organ (leave) have large number of trichomes, it is not easy. Instead double-sided tape fast setting epoxide glue (Hardman®, www.royaladhesives.com, Bellevile, NJ, USA) can be used with compromised results.
7. Apply the matrix or with start the measurement, if laser desorption will be employed to prevent tissue deformation and degradation.

2.1.2 Cryosectioning

1. Clean the sectioning blade or put new one into cryostat and adjust the cutting temperature. Depending on sample, ca. −15 to −20 °C is usually used.
2. Froze the sample in liquid nitrogen, transferred to a cryostat cooled to the desired temperature (−18 °C) and fix it to a sample plate via either a droplet of water (ice) or of OCT medium. The orientation should be observed to ensure if longitudinal or cross sections are desirable.
3. Firstly, set the cutting thickness large to remove the ice, OCT medium and when the desired part is already exposed, change the setting to desired thickness. Thickness below 15 μm is difficult to obtain, but for most of applications 20–25 μm are sufficient.
4. Transfer the fine sections immediately on indium tin oxide (ITO)-coated glass slides (Bruker Daltonics, Bremen, Germany) and fix the sections by the thaw-mounting where the ice in the section is melted by touching the opposite glass side by a finger for 1–3 s.
5. Transfer the fine sections into a dessicator for 0.5–1.0 h until completely dried. The time until complete dryness varied and was longest for the high water content tissues (roots). Alternatively, the matrix can be immediately applied (after taking photo) by spraying.
6. Use a Staedtler triplus gel-liner (silver, 0.4 mm; Staedtler-Mars, Nürnberg, Germany, http://www.staedler.com) to place marks close to the samples to define their position. The granular particles of this gel-roller marker are useful to alleviate the laser positioning during LDI-MSI.
7. Capture a microscopic image with a stereomicroscope (e.g., Leica MZ6, Wetzlar, Germany) connected to a digital camera (AxioCam ICc1, Zeiss, Jena, Germany) or similar microscopy system.

2.1.3 Preparing immortal HeLa cell lines on the glass slide

1. Suspend HeLa cell in serum free media (Quantum 101 for HeLa cells with L-glutamine, PAA, Pasching, Austria). (The cell line is categorized as biosafety level 1. Safety data sheet can be downloaded from: http://www.dsmz.de/catalogues/details/culture/ACC57.html?tx_dsmzresources_pi5%5BreturnPid%5D=192.)
2. Cut a 1.5 × 1.5 cm ITO glass (Bruker Daltonics, Bremen, Germany) using a laboratory glass cutter. Use firm gloves and take care to prevent injury.
3. Apply the cell suspension on ITO-coated glass slides (Bruker Daltonics, Bremen, Germany) and directly grown them on glass for 12 h in an incubator at 37 °C and an atmosphere of 5% CO_2.
4. Aspirate of media using Ependorph 100 μL tip and wash the cells two times with phosphate buffered saline (PBS) (Dulbeccoss̈ PBS, PAA, Pasching, Austria) followed by fixation in 0.25% glutaraldehyde (EM grade, Agar Scientific, Essex, UK) for 15 min. This fixation step leads to cross-linking of proteins in the cells. Kept buffer solutions at 37 °C in order to prevent disruption of cells before fixation.
5. Stain the cells with 3,3′-dihexyloxacarbocyanine iodide ($DIOC_6(3)$) (Invitrogen Live Technologies GmbH, Darmstadt, Germany) solution (c=0.001 μg/mL) in PBS for 5 min. $DIOC_6(3)$ is a fluorescence dye which stains cell membranes. Invitrogen product information DIOC6 (3), http://products.invitrogen.com/ivgn/product/D273 (accessed December 15, 2011).
6. Finally, rinse the cells with double-distilled water at room temperature in order to remove excessive $DIOC_6(3)$ from the sample.
7. Capture fluorescence images of the sample with an Olympus BX-40 microscope (Olympus Europa GmbH, Hamburg, Germany) at an emission wavelength of 501 nm prior to matrix application.
8. Attach the glass on the metallic target provided with image10 AP-LDI source using conventional double-sided tape (e.g., 3 *M*).

2.2. Matrix deposition

2.2.1 Paint brush

2.2.1.1 Matrix application for negatively charged molecules (glucosinolates, acids, etc.)

1. Prepare solution 9-aminoacridine free base at 15 mg/mL in HPLC grade methanol.
2. Mounted leaves on a MALDI stainless steel target plate (Fig. 10.2B, LM; Waters, UK) using a double-sided adhesive tape (Plano No. G3939, http://www.plano-em.de) with the abaxial surface of the leaf facing up.

3. Add ca. 0.3 μL of sulfated PEG 600 (1 mg/mL methanol) and standards of analytes at appropriate concentration for checking its behavior under the MSI conditions. The analytes can be added on additional test leaf for adjusting laser intensities as the ionization efficiency strongly depends on surface morphology and chemistry.
4. The fixed leaves were spray-coated with the 9-aminoacridine solution prepared at step (1) using a commercial airbrush with a 0.15 mm diameter sprayer jet (Harder&Steenbeck, www.airbrushuniverse.com, Norderstedt, Germany).
5. Keep the target plate was at a 45° angle against a plastic support and the sprayer held at a distance of 13 cm from the plate and in the axis to the canter of the leaf. This insured that the cone of the spray reaching the target covered the entire leaf; adjust for bigger leaves if needed.
6. Spray for 20–22 s, following by 5 min of drying. This process is typically carried out 15 times to give maximal signal strength. The interval between consecutive sprays insured negligible solvent build-up on the leaf, which was essential to prevent analyte delocalization.
7. After the last coat is applied keep the preparation at dessicator at reduced pressure to dry it. It will shorten pumping time for the MALDI-TOF instrument and prevent MALDI source contamination. Take care that matrix coat is not pealing out after the drying. Samples with coat damaged should not be used. For *Arabidopsis thaliana*, the abaxial side of the leaf up as this side offered a better surface for uniform matrix deposition owing to fewer trichomes; however, with appropriate care during the whole process MSI could be carry out also on the adaxial side.
8. Acquire MSI in negative ion mode.

2.2.2 Using ImagePrep automatic instrument

1. Prepare dilutions of particular matrices were prepared after adaptation to the various tissues. 2,5-Dihydroxybenzoic acid (DHB) was diluted to 30 mg/mL in 50% (v/v) methanol, 0.2% (v/v) trifluoroacetic acid (TFA) and HCCA to 7 mg/mLl in 60% (v/v) acetonitrile, 0.2% (v/v) TFA.
2. Insert the fixed sample into ImagePrep instrument.
3. Apply matrix solutions via sensor-controlled vibrational vaporization, utilizing the ImagePrep instrument (Bruker Daltonics) software control as per the manufacturer's instructions.
4. Use manufacture-provided spray protocols specific for the particular matrix substance.

2.2.3 Matrix deposition with rotating-stage sprayer

This device is now commercially available and will be part of the image10 atmospheric pressure matrix-assisted laser desorption/ionization (AP-MALDI) high-resolution imaging source (TransMIT GmbH, Giessen, Germany). For detailed description, see this reference Bouschen, Schulz, Eikely, and Spengler (2010).

1. Prepare DHB solution (30 mg/mL in 50:50 acetone/water/0.5% TFA).
2. Fill the syringe in the pneumatic sprayer with the matrix solution. Turn on nitrogen gas and add dummy target on the rotation stage.
3. Set the syringe pump to deliver 10–20 μL/min of the matrix solution, start the rotation of the sample stage and spray for ca 5 min.
4. Stop rotation, gas, and the pump and observe the crystal size and coverage under magnification.
5. If expected matrix coverage is observed, put the HeLa on ITO glass prepared above and start the rotation, nitrogen gas and syringe. Use optimized flow and time determined on dummy slide.
6. Inspect the final cover under microscope, using fluorescence microscope (e.g., Olympus BX-40, Olympus Europa GmbH, Hamburg, Germany) and take the image.
7. Use image10 AP-MALDI source for imaging.

2.2.4 Sublimation

Sublimation is very effective method for matrix deposition on single cell. Very fine crystals are typically formed and the surface is uniformly covered. However, the metabolite extraction is limited and if signal intensity is low an additional step cold be applied to improve the metabolite extraction. It may include treatment of the sublimed sample with solvent vapours compatible with the target material and additionally dissolving the compounds of interest. For positive mode DHB matrix is very suitable, ACTH could be used in some cases.

Caution: glasses should be worn during sublimation and vacuum operations!

1. Take sublimation Apparatus; micro no. 137022001 from Rettberg (Gottingen, Germany, http://www.rettberg.biz/). Cover the low part of the cooling finger with thick aluminum folia to ensure a good contact of the MALDI plate (Fig. 10.4A). This size of apparatus can be used for MALDI plate up to 5.4 × 4.1 cm (e.g., Waters blank MALDI Plate No. M881099BD1, http://www.waters.com). Also, microscopic glass plates could be used when the tissue is fixed on carbon conductive

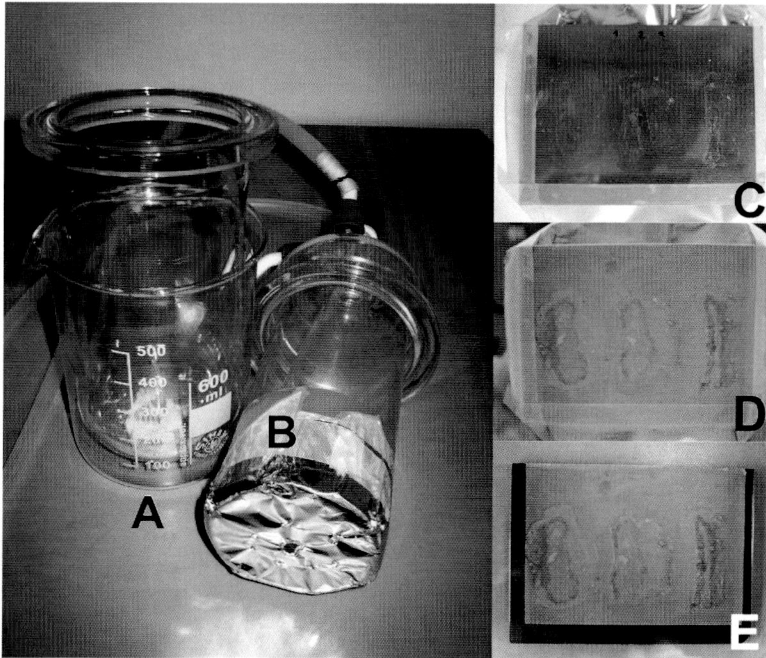

Figure 10.4 *Sublimation apparatus.* Disassembled apparatus (micro no. 137022001, Rettberg) consist of outer vessel (A) to which the matrix used for sublimation in deposited and inner finger that is covered with a preferred alumina folia for better heat dissipation (B). MALDI plate is fixed by adhesive tape onto the folia (C). After sublimation (D) the plate is detached and visually inspected (E).

adhesive tape (Plano No. G3939, http://www.plano-em.de) on an ITO slide (Bruker Daltonic, http://www.bdal.com) is utilize to prevent mass shift due to matrix charging.

2. Prepare the sample using one of the possibilities above (cryosectioned or intact flat tissue) and fix it on the blank MALDI plate. Leave place on the plate edges for the plate fixation to the cooling finger and adding calibrant/analytes to the plate after sublimation.
3. Disassembly the sublimation unit. Add the matrix (ca. 1 g) in the outer tube bottom.
4. Fix the plate on the aluminum folio using a Scotch tape, cut in folds to ensure firm attachment (Fig. 10.4B). Two or three sides are sufficient; make sure the plate is in axis and not hindering apparatus reassembling.
5. Add the calibration mixture (PEG oligomers) and optional a standards directly on the layer of matrix well apart the imaging tissue (ca. 0.3–0.5 μL).

6. Prepare a silicon oil bath, heating/stirring plate with temperature regulation and rotary vacuum pump allowing ca 10^{-3} Torr vacuum. Preheat the bath to desired temperature (e.g., 140 °C at 1×10^{-3} Torr for DHB).
7. Fix the assembled apparatus with matrix and MALDI plate in place, attach to vacuum (wear glasses) and cooling water. The water entrance must be on the side, which is connected to a glass tube protruding to the bottom of cooling finger.
8. Carefully immerse the apparatus into the oil bath such that only ca. 1/3 of the distance from the outer part holding matrix and the MALDI plate is immersed.
9. Sublime for ca. 10 min. The time and temperature should be optimized for every matrix/tissue to achieve the proper sample coverage and sufficient MALDI signal intensity.
10. After 10 min rise the apparatus out of oil bath, stop heating and if the outer glass temperature will be ca. 30 °C break vacuum, and stop cooling water. Went apparatus gently into air or inert gas atmosphere.
11. Disassemble the apparatus and remove the target plate (Fig. 10.4D).
12. Perform MSI experiment as described later.

2.3. MALDI-MS imaging measurement

Caution: If operating lasers wear safety glasses and make sure that unauthorized persons did not enter the room during the time the laser is on.

2.3.1 Bruker Ultraflex III instruments

1. Fit the above-prepared ITO slides into a Slide Adapter II MALDI target (Bruker Daltonics) and introduce the adapter into source.
2. Select the appropriate tune file and polarity for measurement.
3. Perform mass calibrations using standards deposited on the slide using the common procedure (see to instrument manual) using a table of theoretical mass over charge ratio (m/z) values.
4. Adjust laser power using standards for a class of your compounds and check the signal intensity to obtain a relative signal intensity of 25,000 based on measurements from the tissue section prior to the automated MSI measurement, and chose the number of laser shots needed.
5. Create file for MSI measurements using taken photograph before matrix applications using Flex Imaging software v2.1 (Bruker Daltonics), and autoexecute sequences, as per the manufacturer's instructions. Select desired spatial resolution and laser frequency. For very small sections

(such as young seeds, cross section), use 10–15 μm laser raster was used in order to get sufficient representation of the various tissues. Select an m/z range of 80–1300 for small metabolites and set sample rate to 0.5 Gs/s. Make sure there is enough space on HD.

6. Performed MSI using available instrument (e.g., ultrafleXtreme MALDI-TOF/TOF device (Bruker Daltonics), equipped with a smart beam-II laser with a repetition rate of 1000 Hz). Carry out the measurement of the spots in random order to eliminate influences of measurement order.

2.3.2 MS imaging of individual HeLa cells with high lateral and mass resolution

For MALDI-MSI measurements, a Fourier transform orbital trapping mass spectrometer Q-Exactive (Thermo Fisher Scientific GmbH, Bremen, Germany) equipped with the atmospheric pressure imaging ion source imagine10 (TransMIT GmbH, Giessen) MALDI source was used. MS/MS measurements in ion trap mode were performed manually from selected ions. The isolation window for precursor ions was 2 mass units (Da). Fragmentation mode was collision-induced dissociation.

1. Introduce above-prepared cells on HeLa-ITO glass target (AP-MALDI) high-resolution imaging source (TransMIT GmbH, Giessen, Germany) attached to a Fourier transform orbital trapping mass spectrometer (Exactive Orbitrap, Thermo Fisher Scientific GmbH, Bremen, Germany). The ion optical and laser setup as well as details on the measurement procedure are described in instruction materials.
2. Set the step size of the sample stage to 7 μm in both x- and y-direction. The ablation area caused by the laser will be about 5 μm in diameter. Summe 30 laser pulses for each measured mass spectrum. Set Q-Exactive in positive-ion mode and in the mass to charge range m/z 100–1000. Use maximal mass resolution ($R = 100\,000$ at m/z 200). Use the lock mass feature of the instrument to maximize the mass data precision.
3. Acquire the selected area of interest, start the measurements and observe performance and stability of the instrument.
4. The mass data software provided by the producer can evaluate obtained data.

3. MAMS ARRAYS

The MAMS can be purchased from Sigma-Aldrich (order no. Fluka-50757). Follow all good-laboratory-practice procedures while growing and handling the cell cultures. We strongly recommend that all solutions to quench and process the cells, as well as those used for the MALDI matrix (if needed) to be prepare using ultrapure water and MS compatible reagents (e.g., LC–MS grade).

3.1. Preparing the MAMS substrates

1. Place the MAMS substrate in a glass or plastic container. Fill the container with acetone, and ultrasonicate it for 15 min at room temperature.
2. Flush the MAMS substrate first with 50–100 mL of acetone, then with an equal volume of ethanol and finally rinse the MAMS substrate with ultrapure water.
3. Dry the MAMS substrate under a flow of nitrogen. To avoid introducing new dust particles onto the MAMS substrate filter the nitrogen gas stream with a sieve.
4. Observe the MAMS substrate under a microscope to take notice of any impurity that can later be confused with the cells to be studied.

3.2. Applying the cell cultures onto the MAMS substrate for analysis

Cell handling is a very critical step in a metabolomic study. The following procedures must be carried out at low temperatures—preferably lower than −10 °C or −20 °C to keep the cells quenched—unless otherwise specified (Sellick, Hansen, Stephens, Goodacre, & Dickson, 2011).

Due to the omniphobic properties of the polymer coating used in the MAMS fabrication, the MAMS substrate can fractionate (self-aliquot) a larger liquid volume into discrete droplets (Fig. 10.3; Urban et al., 2010). This can be accomplished by pulling/dragging the liquid over the MAMS surface. As a consequence, cells can be easily distributed through the MAMS substrate by dragging a cell suspension over the surface. It is also possible to apply the cells by using other cell handling methods such as a microspotter and microfluidic devices. The distribution of the number of cells within one reservoir of the MAMS substrate is stochastic (i.e., Poisson distribution). Thus, the density of the cell suspension applied on the MAMS substrate is directly correlated with the mean number of cells trapped in the reservoirs. Typically, the original cell culture must be preconcentrated to a cell density of 10^7 cells per milliliter. Independently of the method used for applying the cells onto the MAMS substrate, due to the low temperature requirements, cells will be more commonly suspend in organic solvents. Unfortunately, the used of organic solvents make cells more leaky, which translate in the loss of sensitivity for some metabolites. It is strongly suggested that prior running this type of single-cell analysis, the researchers would have a good understanding of what are the best conditions to quench, and handle their cells to avoid any analytical or biological artifacts (Urban, Amantonico, et al., 2011; Urban, Schmidt, et al., 2011).

1. The counting of the cells can be achieved using any microscope. Once more, great care must be taken to avoid changes in the metabolite levels of the cells trap on the reservoirs (i.e., the cells must be quenched or free of any abiotic-related stress during the cell counting procedure).
2. Once the cell counting is completed, it is strongly suggested to freeze-dry the cells on the MAMS substrate to ensure no further enzymatic activity will occur.

3.3. MS measurement of the cells located on the MAMS substrate

Once the cells are on the MAMS substrate, the following steps resemble very much to those previously describe in this chapter for MSI using either a LDI or a MALDI setup in combination with a MS read-out.

1. In the case of MALDI-based ionization method is used, an exogenous light-absorbing molecule has to be applied onto the sample (i.e., MALDI matrix). The most common methods of MALDI matrix application are: (i) airbrush systems (Section 2.2.1), (ii) ultrasonic scaler (Section 2.2.2), and (iii) sublimation chambers (Section 2.2.4).
2. Acquire the MS spectrum for each individual reservoir.

3.4. Data processing and data analysis

To avoid erroneous molecular formula assignments, researchers are advice to follow the "Seven Golden Rules for Metabolomic Studies" that were introduced by the Fiehn Lab (Kind & Fiehn, 2007). We will briefly describe here some of the key steps associated with the seven golden rules for chemical assignments (identities) of metabolites. The identity of a metabolite must be made on the basis of accurate mass measurements, their isotope pattern, and comparing their MS/MS spectra to commercially available standards. Additional steps can be corroborating the presence of the metabolite in metabolic databases for the specific cell organism under studied or monitoring the mass shift due to growing the cells on isotope labeled medium (^{13}C, ^{15}N, ^{18}O, etc.).

REFERENCES

Ahn, J. Y., Lee, S. W., Kang, H. S., Jo, M., Lee, D. K., Laurell, T., et al. (2010). Aptamer microarray mediated capture and mass spectrometry identification of biomarker in serum samples. *Journal of Proteome Research*, *9*(11), 5568–5573.

Bouschen, W., Schulz, O., Eikely, D., & Spengler, B. (2010). Matrix vapor deposition/recrystallization and dedicated spray preparation for high-resolution scanning microprobe matrix-assisted laser desorption/ionization imaging mass spectrometry (SMALDI-MS) of tissue and single cells. *Rapid Communications in Mass Spectrometry*, *24*(3), 355–364.

Caprioli, R. M., Farmer, T. B., & Gile, J. (1997). Molecular imaging of biological samples: Localization of peptides and proteins using MALDI-TOF MS. *Analytical Chemistry*, *69*(23), 4751–4760.

Fernie, A. R., Trethewey, R. N., Krotzky, A. J., & Willmitzer, L. (2004). Innovation—Metabolite profiling: From diagnostics to systems biology. *Nature Reviews Molecular Cell Biology*, *5*(9), 763–769.

Fletcher, J. S., Lockyer, N. P., Vaidyanathan, S., & Vickerman, J. C. (2007). TOF-SIMS 3D biomolecular imaging of Xenopus laevis oocytes using buckminsterfullerene (C60) primary ions. *Analytical Chemistry*, *79*(6), 2199–2206.

Gunatilaka, A. A. L. (2006). Natural products from plant-associated microorganisms: Distribution, structural diversity, bioactivity, and implications of their occurrence. *Journal of Natural Products*, *69*(3), 509–526.

Ibánez, A. J., Fagerer, S. R., Schmidt, A. M., Urban, P. L., Jefimovs, K., Geiger, P., et al. (2013). Mass spectrometry-based metabolomics of single yeast cells. *Proceedings of the National Academy of Sciences of the United States of America*, *110*(22), 8790–8794.

Ibáñez, A. J., Muck, A., Halim, V., & Svatoš, A. (2007). Trypsin-linked copolymer MALDI chips for fast protein identification. *Journal of Proteome Research*, *6*(3), 1183–1189.

Jimenez, C. R., Li, K. W., Dreisewerd, K., Spijker, S., Kingston, R., Bateman, R. H., et al. (1998). Direct mass spectrometric peptide profiling and sequencing of single neurons reveals differential peptide patterns in a small neuronal network. *Biochemistry*, *37*(7), 2070–2076.

Kalisky, T., & Quake, S. R. (2011). Single-cell genomics. *Nature Methods*, *8*(4), 311–314.

Kind, T., & Fiehn, O. (2007). Seven Golden Rules for heuristic filtering of molecular formulas obtained by accurate mass spectrometry. *BMC Bioinformatics*, *8*, 105.

Koehn, F. E., & Carter, G. T. (2005). The evolving role of natural products in drug discovery. *Nature Reviews Drug Discovery*, *4*(3), 206–220.

Mellors, J. S., Jorabchi, K., Smith, L. M., & Ramsey, J. M. (2010). Integrated microfluidic device for automated single cell analysis using electrophoretic separation and electrospray ionization mass spectrometry. *Analytical Chemistry*, *82*(3), 967–973.

Ressine, A., Marko-Varga, G., & Laurell, T. (2007). Porous silicon protein microarray technology and ultra-/superhydrophobic states for improved bioanalytical readout. *Biotechnology Annual Review*, *13*, 149–200.

Römpp, A., Guenther, S., Schober, Y., Schulz, O., Takats, Z., Kummer, W., et al. (2010). Histology by mass spectrometry: Label-free tissue characterization obtained from high-accuracy bioanalytical imaging. *Angewandte Chemie International Edition*, *49*(22), 3834–3838.

Römpp, A., & Spengler, B. (2013). Mass spectrometry imaging with high resolution in mass and space. *Histochemistry and Cell Biology*, *139*(6), 759–783.

Saito, K., & Matsuda, F. (2010). Metabolomics for functional genomics, systems biology, and biotechnology. *Annual Review of Plant Biology*, *61*, 463–489.

Sellick, C. A., Hansen, R., Stephens, G. M., Goodacre, R., & Dickson, A. J. (2011). Metabolite extraction from suspension-cultured mammalian cells for global metabolite profiling. *Nature Protocols*, *6*(8), 1241–1249.

Shrestha, B., & Vertes, A. (2009). In situ metabolic profiling of single cells by laser ablation electrospray ionization mass spectrometry. *Analytical Chemistry*, *81*(20), 8265–8271.

Svatos, A. (2010). Mass spectrometric imaging of small molecules. *Trends in Biotechnology*, *28*(8), 425–434.

Svatos, A. (2011). Single-cell metabolomics comes of age: new developments in mass spectrometry profiling and imaging. *Analytical Chemistry*, *83*, 5037–5044.

Szita, N., Polizzi, K., Jaccard, N., & Baganz, F. (2010). Microfluidic approaches for systems and synthetic biology. *Current Opinion in Biotechnology*, *21*(4), 517–523.

Urban, P. L., Amantonico, A., & Zenobi, R. (2011). Lab-on-a-plate: Extending the functionality of MALDI-MS and LDI-MS targets. *Mass Spectrometry Reviews*, *30*(3), 435–478.

Urban, P. L., Jefimovs, K., Amantonico, A., Fagerer, S. R., Schmid, T., Madler, S., et al. (2010). High-density micro-arrays for mass spectrometry. *Lab on a Chip*, *10*(23), 3206–3209.

Urban, P. L., Schmidt, A. M., Fagerer, S. R., Amantonico, A., Ibañez, A., Jefimovs, K., et al. (2011). Carbon-13 labelling strategy for studying the ATP metabolism in individual yeast cells by micro-arrays for mass spectrometry. *Molecular BioSystems*, 7(10), 2837–2840.

Winograd, N., & Garrison, B. J. (2010). Biological cluster mass spectrometry. *Annual Review of Physical Chemistry*, *61*, 305–322.

Wurm, M., Schopke, B., Lutz, D., Muller, J., & Zeng, A. P. (2010). Microtechnology meets systems biology: The small molecules of metabolome as next big targets. *Journal of Biotechnology*, *149*(1–2), 33–51.

Zare, R. N., & Kim, S. (2010). Microfluidic platforms for single-cell analysis. *Annual Review of Biomedical Engineering*, *12*, 187–201.

CHAPTER ELEVEN

Study of Cellular Oncometabolism via Multidimensional Protein Identification Technology

Claire Aukim-Hastie[*,†,1,2], **Spiros D. Garbis**[†,3]

[*]Faculty of Health & Medical Sciences, University of Surrey, Guildford, United Kingdom
[†]Faculty of Medicine, Cancer Sciences and CES Units, Institute for Life Sciences, University of Southampton, Southampton, United Kingdom
[1]Corresponding author: e-mail address: c.aukim-hastie@surrey.ac.uk
[2]Senior author.
[3]Co-senior author.

Contents

Abstract

Cellular proteomics is becoming a widespread clinical application, matching the definition of bench-to-bedside translation. Among various fields of investigation, this approach can be applied to the study of the metabolic alterations that accompany oncogenesis and tumor progression, which are globally referred to as oncometabolism. Here, we describe a multidimensional protein identification technology (MuDPIT)-based

Methods in Enzymology, Volume 543
ISSN 0076-6879
http://dx.doi.org/10.1016/B978-0-12-801329-8.00011-8

strategy that can be employed to study the cellular proteome of malignant cells and tissues. This method has previously been shown to be compatible with the reproducible, in-depth analysis of up to a thousand proteins in clinical samples. The possibility to employ this technique to study clinical specimens demonstrates its robustness. MuDPIT is advantageous as compared to other approaches because it is direct, highly sensitive, and reproducible, it provides high resolution with ultra-high mass accuracy, it allows for relative quantifications, and it is compatible with multiplexing (thus limiting costs).This method enables the direct assessment of the proteomic profile of neoplastic cells and tissues and could be employed in the near future as a high-throughput, rapid, quantitative, and cost-effective screening platform for clinical samples.

1. INTRODUCTION

Metabolism is a highly integrated process of chemical reactions allowing living organisms to obtain energy from their environment to fuel the synthesis of macromolecules. Primarily found in mitochondria in eukaryotes, these metabolic pathways when dysregulated often lead to diseases such as cancer. Cancer cells grow and divide at an increased rate and as such, have a unique metabolism. The process of carcinogenesis involves changes in multiple cellular pathways and for these to be studied, a global strategy is required which can interrogate this complexity simultaneously. Mass spectrometry-based proteomics has the analytical power to perform such analyses and give insights into complex signaling pathways such as those involved in oncometabolism.

1.1. Proteomics

The term proteome was coined in 1994 to describe the linguistic analogue of the genome, being the entire protein complement expressed by a genome or a cell/tissue type (Cai, Chiu, & He, 2004; Naaby-Hansen, Nagano, Gaffney, Masters, & Cramer, 2003). Proteomics can be defined as the study of all proteins at the subcellular, cellular, tissue, or even organism level (Cai et al., 2004). Proteomics is the large scale study of proteins, ideally proteomic studies seek to characterize and quantitate all proteins within a sample at a particular time under a particular set of environmental conditions (Larsen, Sørensen, Fey, Larsen, & Roepstorff, 2001). However, proteomics is not solely used to obtain a snapshot of proteins at a given time, it can also be used to identify protein interactions with other proteins, DNA, or RNA, allowing analysis of complex biochemical pathways such as those involved in cellular metabolism (Emili & Cagney, 2000).

Studies of protein expression and interaction are not mutually exclusive, as one gives functional significance to the other. In proteomics, however, these studies are split into two disciplines "expression proteomics" and "functional or cell map proteomics" (Blackstock & Rowley, 1999). The difference between the two disciplines lies mainly in the techniques used and the number of proteins studied. The term "expression proteomics" is associated with the description of the total proteome of the subject, the differential expression of proteins between subjects and under different treatment conditions. This is sometimes denoted "discovery proteomics" as the researcher is seeking to identify both known and unknown proteins (Cai et al., 2004). Global analyses such as these require the separation of thousands of proteins from the original sample. The proteome unlike the genome is in a constant state of dynamic flux producing a mixture of heterogeneous proteins with differing biochemical properties and expression levels. It is thought that proteins with low abundance may be masked by other "housekeeping proteins" present at high levels within the cell, this can be rectified by selective removal of abundant proteins. This as with all prefractionation techniques has the associated risk of protein loss (Zuo & Speicher, 2000).

Considering the above, analysis of the prokaryote proteome with its genome complement of less than 3000 genes appears daunting and analysis of eukaryote proteome with greater than 10,000 proteins and the added complication of mRNA processing and complex posttranslational modification (PTM), impossible (Zuo & Speicher, 2000). There is however, a technique commonly used in the global expression analysis of complex genomes which is both sensitive enough to give single protein resolution and discrimination of PTM. Proteomic methods such as MuDPIT are the only techniques available that allow the analysis of splice variants and PTM, as well as more conventional expression level quantification. Such modifications may be functionally critical and determine a disease, that is, phosphorylation and glycosylation, as is the case of the prion diseases CJD and Alzheimer's, respectively (Banks et al., 2000). Novel PTM, mostly phosphorylation, of several oncogene products and cell cycle components (including p53), have been identified in transformed liver cells (Yan, Pan, Olman, Hettich, & Xu, 2004). In addition to this aberrant glycosylation of several cancer-associated proteins has also been reported, and proteomics-based approaches are ideally placed to characterize such aberrations (Banks et al., 2000).

1.2. Cellular proteomics

The process of carcinogenesis leads to changes in the array of proteins present in the cellular metabolome. Historically, proteomic analysis has been centered on the plasma membrane where changes may be seen in the number or activity of growth factor receptors heightening growth factor responsiveness, in the number or type of cell adhesion molecules expressed, for example, integrins, or in the number/activity of proteases present on the cell surface. Analysis of differential expression of such proteins in cancer is extremely important, as due to their position on the cell surface they have potential for use as diagnostic and/or prognostic markers and therapeutic targets. In the latter capacity, it may be possible to tailor treatments more specifically to the cancer cells and reduce the harmful side effects of anticancer treatments. Research has led to the identification of a number of cancer-associated plasma membrane proteins, which currently account for 70% of all-known pharmaceutical drug targets.

The basic structure of the plasma membrane is a lipid bilayer, studded with integrally and peripherally attached proteins. These membrane-associated proteins confer the majority of the plasma membrane's unique functions. Within the plasma membrane, further complexity is present in the form of subdomains called lipid rafts, enriched in cholesterol, glycosphingolipids, and signaling molecules. These structures appear small in size but can represent a relatively large proportion of the PM and are thought to be involved in the regulation of signal transduction (Pike & Sadler, 2004). It has been suggested that these subdomains represent areas of increased order, with a corresponding decrease in fluidity, likely to be due to the lipid/cholesterol composition and abundance of transmembrane proteins (Pike & Sadler, 2004). Further plasma membrane complexity is conferred by the aggregation of these lipid rafts into membrane invaginations or caveolae. This process clusters proteins into a plasma membrane microenvironment conducive to the initialisation of signaling cascades such as those involved in oncometabolism (Simons & Toomre, 2000).

Protein-based approaches have been used in the discovery of tumor markers for many years, from the discovery of Bence Jones proteins in the 1800s, to the generation of tumor-specific antibodies in epithelial cancer (Banks et al., 2000). Recent technological advances in the field of proteomics have led to their application in the relatively new field of oncometabolomics. The sensitivity and specificity of techniques such as MuDPIT in conjunction with the relative low per sample cost makes proteomics the ideal tool for the investigation of metabolomic biomarkers. It is known that the process of carcinogenesis leads to unusual stress on the cell and culminates in a shift in energy production and utilization. High-resolution proteomics can be used to study the biochemical intermediates and end products of this process directly from tumor tissue samples. Metabolomic profiling has been successfully used in the analysis of sarcoma tissue, which detected substantial numbers of polar metabolites despite the formalin-fixed nature of the starting sample (Kelly et al., 2011). Proteomic techniques have also been utilised in the study of castrate-resistant prostate cancer, which mapped altered metabolites to a pathway subsequently found to be significantly associated with time to treatment failure in clinical datasets (Kaushik et al., 2013). Such techniques have also been used to characterise cysteine catabolism in glioblastoma which was found to be a positive regulator of tumor growth and a powerful target for growth inhibition (Prabhu et al., 2013). Oncometabolomics holds great promise in the discovery of proteins involved in cancer progression that could be used as new biomarkers or therapeutic targets.

1.3. MuDPIT

MuDPIT is a quantitative proteomic method allowing multiplex analysis of several samples. This allows samples obtained from different stages of cancerous disease to be compared simultaneously resulting in absolute quantification of proteins of interest within the metabolome. The advantages of this method are that, as samples are run concurrently the impact of experimental error is reduced, electronspray ionization efficiency is improved and sensitivity enhanced. The use of tandem mass spectrometric evidence for quantification increases the selectivity, specificity, and confirmatory power in the identification and quantification of identified proteins (Garbis et al., 2008). The MuDPIT technique has also been shown to vastly improve the dynamic range of proteins present within a complex sample. Garbis et al. (2011) demonstrated that this approach could lead to the identification and quantification of proteins with orders of magnitude of twelve less than the most abundant protein, representing a novel level of in-depth proteome coverage (Garbis et al., 2011).

Current literature suggests that carcinogenesis is a cooperative process involving the tissue as a whole (including normal and transformed cells) and not as a result of processes within individual cells alone. MuDPIT has been used to identify changes in protein expression associated with prostate carcinogenesis from whole tissue sections allowing the inclusion of analysis of the heterogeneous tumor microenvironment (Garbis et al., 2008). The enhanced resolving power of this technique is also able to reliably distinguish protein isoforms and expression differences between them. Such was the case with human papillomavirus (HPV) proteins analyzed in ThinPrep cervical smear specimens along with their phosphorylation status despite the use of positive ion electrospray ionization and without recourse to targeted enrichment (Papachristou et al., 2013) The combination of good multidimensional chromatographic technique combined with the use of both He-based collision-induced dissociation (CID) and Nitrogen C-trap high energy dissociation (HCD) at ultra-high MS detection afforded by the FT-Orbitrap system provided the necessary selectivity, specificity, and sensitivity to allow the protein isoform and their phosphorylated variants to be detected.

2. SAMPLE PROTOCOL

There follows a sample protocol for MuDPIT analysis of samples from cells or tissues, this protocol can be adapted for clinical samples and examples of this can be observed in Section 4.

2.1. For cell pellets

2.1.1. To each sample tube-containing cell pellets, add 100 μL Triethylammonium bicarbonate, 0.5 *M*

2.1.2. Sonicate for 3 min.

2.1.3. Vortex to mix.

2.1.4. Sonicate for 3 min.

2.1.5. Add 5 μL of the Denaturant (SDS 2%). For more than 10 × 106 cells, add 200 μL Buffer and 10 μL Denaturant.

2.1.6. Sonicate for 3 min.

2.1.7. Vortex to mix.

2.1.8. Sonicate for 3 min.

2.1.9. Room temperature for 1 h (solution A).

2.1.10. Centrifuge 10 min/10,000 rpm.

2.1.11. Transfer the supernatant liquid into numbered clean tubes.

2.2. For tissue specimens

2.2.1. Transfer the tissue specimen in a mortar and add liquid nitrogen. After the evaporation of liquid nitrogen, directly homogenize the frozen tissue with a pestle. Repeat the procedure until homogenization and transfer the powder quantitatively into a clean Eppendorf tube.

2.2.2. To each sample tube-containing homogenized tissue, add 250 μL triethylammonium bicarbonate, 0.5 *M*.

2.2.3. Sonicate for 10 min.

2.2.4. Vortex to mix.
2.2.5. Sonicate for 10 min.
2.2.6. Add 12.5 μL SDS 2%.
2.2.7. Sonicate for 10 min.
2.2.8. Vortex to mix.
2.2.9. Sonicate for 10 min
2.2.10. Incubate at Room temperature for 1 h.
2.2.11. Centrifuge 10 min/13,000 rpm.
2.2.12. Transfer the supernatant liquid into numbered clean tubes.

2.3. Spectrophotometer calibration with Bradford standard protein

2.3.1. Create standard bovine serum albumin solutions: 0, 0.2, 0.4, 0.6, 0.8, and 1.0 mg/mL in an appropriate buffer.
2.3.2. Dilute 10 μL of each standard solution in 990 μL Bradford Reagent, mix into a photometric cuvette with a pipette, and calibrate the spectrophotometer according to manufacturer instructions.
2.3.3. Dilute 10 μL of your sample in 190 μL ultra pure water.
2.3.4. Further dilute 10 μL of this solution in 990 μL Bradford Reagent and mix into a photometric cuvette with a pipette and measure sample.
2.3.5. If any measurement is out of the Bradford dynamic range then make a new dilution for the particular sample.
2.3.6. Calculate the appropriate volumes for each of your samples to create a final content of 100 μg total protein for each sample.
Example: If sample A = 659.5 μg/mL then
total protein in 200 μL = (659.5 × 200)/1000 = 131.9 μg
therefore, 10 μL of the sample contains 131.9 μg protein
and (10 × 100)/131.9 = 7.58–7.6 μL of solution A contain 100 μg protein.
2.3.7. Dilute each calculated volume of solution A into ultra pure water up to final volume 20 μL. If any calculation exceeds the volume of 20 μL by 2–3 μL then dilute every sample up to final volume 22–23 μL. If any calculation exceeds the volume of 20 μL by more than 3 μL then repeat calculations so as to take less than 100 μg protein of each sample and the volume of the most diluted sample does not exceed 23 μL.

2.4. Reducing the proteins and blocking cysteine

2.4.1. Add 2 μL 50 m*M* carboxymethyl phosphine (TCEP), to each sample
2.4.2. Incubate in a heat block at 60 °C for 1 h.

2.4.3. Add 1 μL 200 m*M* methyl methanethiosulfonate to each tube and directly vortex to mix and then spin down at 8000 rpm for 1 min.
2.4.4. Incubate at room temperature for 10 min.

2.5. Digesting the proteins with trypsin

2.5.1. Add 4 μL trypsin solution (500 ng/μL) and 26 μL H_2O (protein/trypsin ~ 50:1) to each tube.
2.5.2. Incubate overnight at room temperature.

2.6. Labeling the protein digests with the iTRAQ reagents

2.6.1. Add 70 μL of ethanol to each room temperature iTRAQ Reagent vial and then vortex to mix.
2.6.2. Transfer the contents of one iTRAQ Reagent vial to one sample tube.
2.6.3. Incubate at room temperature for 1 h.
2.6.4. Mix the content of the sample tubes together into a clean tube.
2.6.5. Place samples in a centrifugal vacuum concentrator for 30 min until final volume reaches 100 μL.

2.7. Strong cation exchange chromatographic fractionation of peptides

The HPLC system used in this protocol is a Dionex, P680 HPLC Pump, PDA-100 photodiode Array Detector. The columns used where poly-sulfoethyl A, 200 × 4.6 mm i.d., 5 μm, 1000 Ao. The protocol for separation as programmed into this device is included in Table 11.1.

2.7.1. Mobile phases

2.7.1.1. Prepare 1 L of mobile phase (A) composed of: 25% acetonitrile (ACN, Lab-Scan, 99.9%), 10 m*M* (1.36 g/L) KH2PO4, and then add H_3PO_4 (17%) dropwise until pH 3 using electronic pH meter calibrated with buffers at pH 2, 4, 7, and 10.

Table 11.1 Chromatographic separation method

Time	Mobile phase flow (1 mL/min)
For 10 min	100% A
For 45 min	Gradient up to 25% B
For 10 min	Gradient up to 60% B
For 5 min	Gradient up to 100% B
For 5 min	100% B

2.7.1.2. To prepare mobile phase (B) composed of: 25% ACN, 10 m*M* KH_2PO_4, 1 *M* KCl, pH 3, take 0.5 L of solution (A) and dissolve 37.28 g KCl (Panreac 99.5%).

2.7.1.3. To prepare mobile phase (C) composed of 0.3 *M* CH_3COONa/0.2 *M* NaH_2PO_4, dissolve 12.3045 g CH_3COONa and 13.799 g $NaH_2PO_4 \cdot H_2O$ into 0.5 L H_2O and then add H_3PO_4 (17%) dropwise until pH 5.7.

2.7.1.4. Mobile phase (D) 50% Methanol.

2.7.2. Purge channels A, B, C, and D.

2.7.3. Column conditioning: Purge column with (A) for 45 min at 1 mL/min and $T = 30$ °C. If the column has been stored for a long time without being used then first purge column with (C) for 45 min.

2.7.4. Prepare solution E for SCX fractionation: In the tube-containing solution E, add mobile phase (A) up to final volume 700 μL and then filter sample with PVDF 0.22 μm.

2.7.5. Remove bubbles from the syringe and inject sample into the 500 μL sample loop with the valve at load position.

2.7.6. Using the appropriate software (Chromeleon in this case) autozero the lamp and if intensity is larger than 8 × 106 then set the valve at inject position and separations starts.

2.7.7. After 10 min, start collecting fractions for every 1 min (no more than 30 fractions).

2.7.8. Column cleaning and storage: Purge column with (D) 50 min at 0.5 mL/min and then store it at 2 °C.

2.7.9. Dry fractions with centrifugal vacuum concentrator for 5–7 h.

2.8. Desalting fractions

2.8.1. Prepare fractions: Add 200 μL H_2O/0.1% formic acid to each fraction and vortex to mix.

2.8.2. Conditioning silica columns.

2.8.2.1. Load 2 × 200 μL ACN and centrifuge 100 × *g* without drying the column for 1 min each time using numbered 2 mL tubes.

2.8.2.2. Load 2 × 200 μL H_2O and centrifuge 100 × *g* without drying the column for 1 min each time.

2.8.3. Load and desalt fractions.

2.8.3.1. Empty tubes and load 200 μL of each fraction and centrifuge 100 × *g* without drying the column for 1 min.

2.8.3.2. Load 3 × 200 μL H_2O and centrifuge 100 × *g* without drying the column for 1 min each time.

2.8.4. Eluting peptides.

2.8.4.1. Replace dirty tubes with new clean numbered tubes and load 200 μL 50% ACN/0.1% formic acid and centrifuge 100 × *g* without drying the column for 1 min.

2.8.4.2. Load 200 μL 80% ACN/0.1% formic acid and centrifuge 100 × *g* without drying the column for 1 min.

2.8.4.3. Load 200 μL 100% ACN/0.1% formic acid and centrifuge 100 × *g* until drying the column for 3 or 4 min.

2.8.5. Filter desalted fractions using 1 mL syringe and microfilters (PVDF 0.22 μm). Purge every microfilter with 1 mL ACN before use and then filter the sample dropwise.

2.8.6. Dry desalted fractions with centrifugal vacuum concentrator (5–7 h).

2.9. LC–MS/MS analysis

2.9.1. Tuning and calibration of with direct infusion.

2.9.2. For tuning and calibration, fill an ESI emitter with 4 μL of or Renin substrate standard 10 pmol/μL so the solution reaches the intact capillary tip of the emitter.

2.9.3. Fit the emitter on the ESI source. Press and immobilize the plug at a stable pressure. Move carefully the emitter towards the curtain plate (close but off axis to the orifice plate hole) and notice if there is any droplet on the capillary tip of the emitter. If not, then bring the capillary tip onto the curtain plate and press it smoothly up and down or front and back, until a droplet appears.

2.9.4. Software settings:

2.9.4.1. Before start: Ensure the MS, autosampler, and nanopump are active. In order to maintain a stable flow for the LC system, apply the following settings: 0.15 μL/min, 90% mobile phase A (98% H_2O–2%ACN–0.1% formic acid), 10% mobile phase B (98% ACN–2% H_2O–0.1% formic acid), time 30 min.

2.9.4.2. Select MS method.

2.9.4.3. Acquire spectrum for 1 min with the following settings:
Mass range = 100–1000
IonSpray voltage = 1000

2.9.4.4. Optimize TIC and TOF-MS signal for *m/z* 880 on TOF-MS mode by changing the position of the emitter manually (intensity > 250 cps, $R > 5000$).

2.9.4.5. Aqcuire TOF-MS spectrum for 1 min with the following settings:
Mass range = 100–1000
IonSpray voltage = 1000
Collision energy = 45–50
Resolution = low resolution
Data Filename = "example" RENIN_15APR07
Sample name = "example" RENIN_15APR07_TOF_AT_BC

2.9.4.6. Optimize TIC and Product ion signal for *m/z* 880 on Product ion mode for 60 MCA cycles (intensity > 250 cps, $R > 4000$).

2.9.4.7. Aqcuire Product ion spectrum for 1 min on MCA mode with the following settings:
Mass range = 100–1000
IonSpray voltage = 1000
Collision energy = 45–50
Resolution = low resolution

2.9.4.8. Calibrate the spectrum by using the renin mass table and typing the true experimental values for *m/z* 879,..., 110,... as shown on the spectrum.

2.9.4.9. Aqcuire Product ion spectrum for 1 min on MCA mode with the following settings:
Mass range = 100–1000
IonSpray voltage = 1000
Collision energy = 45–50
Resolution = low resolution

2.9.4.10. Repeat steps 2.9.4.7–2.9.4.9

3. DATA PROCESSING

The HCD tandem mass spectra can be submitted to Sequest search engine implemented on the Proteome Discoverer software version 1.3.0.339 for peptide and protein identifications. All spectra can be searched against a UniProt Fasta file-containing 20,200 human reviewed entries. An ideal Sequest node would include the following parameters: precursor mass tolerance 10 ppm, fragment mass tolerance 20 mmu. Dynamic modifications based on oxidation of M (+15.995 Da), deamidation of N,

Q (+0.984 Da), phosphorylation of S (+79.966 Da), and static modifications based on iTRAQ8plex at any N-Terminus, K, Y (+304.205 Da) and methylthio at C (+45.988 Da) can be made. The level of confidence for peptide identifications can be estimated using the Percolator algorithm with decoy database searching. Normally, the Strict FDR is set to 0.01 and relaxed FDR set to 0.05. Validation can be based on the *q* value. The Reporter Ion Quantifier node can include a custom iTRAQ 8plex Quantification Method, integration window tolerance 20 ppm and integration method Most Confident Centroid. Protein ratios need to be normalized to protein median concentration and phosphorylation localization probability can be estimated with the phosphoRS node.

4. EXAMPLE APPLICATIONS OF MuDPIT

There follows two examples of the application of the MuDPIT method to the global proteome analysis of clinical samples, further analysis of the altered pathways discovered in these studies led to new insights into the oncometabolism of cells derived from the respective diseases.

4.1. Prostate cancer

The global and in-depth proteome analysis of cancer samples has proven to be a daunting task. Sample enrichment, involving the immunodepletion-based removal of high abundance, unwanted proteins often results in the co-removal of target, lower abundance proteins, as well as the potential for nonselective loss of enriched proteins (Garbis et al., 2011; Granger, Siddiqui, Copeland, & Remick, 2005; Gundry, Fu, Jelinek, Van Eyk, & Cotter, 2007; Kulasingam & Diamandis, 2008; Yocum, Yu, Oe, & Blair, 2005; Zolla, 2008; Zolotarjova et al., 2005). In order to bypass these limitations, MuDPIT-based method was developed for an in-depth proteome analysis of clinical samples derived from prostate cancer patients which had been previously studied at the tissue level. This method successfully captured a diverse range of proteins and phosphoproteins spanning >12 orders of magnitude native abundance levels from only 100 μL of unprocessed serum volume without recourse to immunodepletion strategies (Garbis et al., 2011). Additionally, this method also allowed the capturing of proteins encompassed in exosomes and lipid microvesicles specifically secreted by the

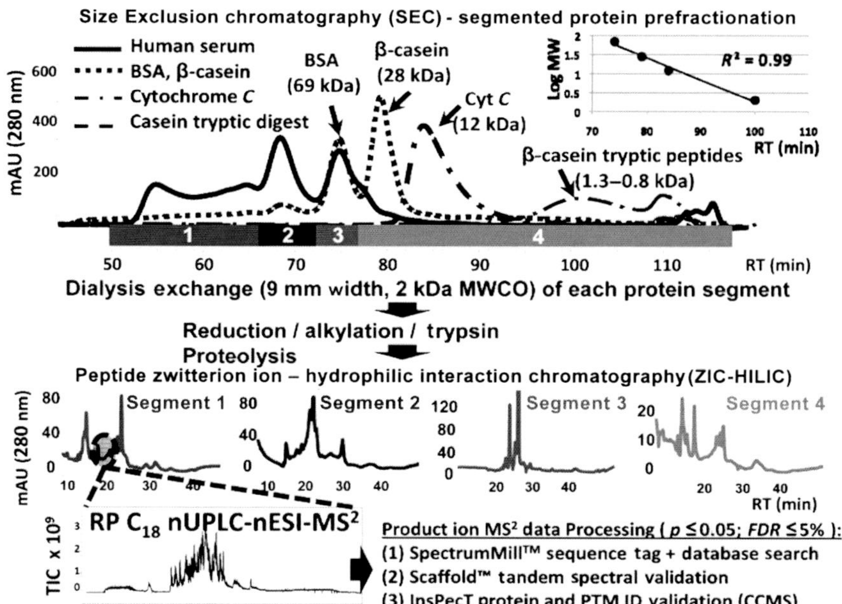

Figure 11.1 *iTRAQ 3D LC–MS method for the analysis of the serum samples from patients with benign prostate hyperplasia.* High performance size-exclusion chromatography was applied to 100 μL serum volumes from each specimen (6×) for the prefractionation of their protein content followed by their dialysis exchange purification, enrichment, reduction, and proteolysis. The resulting tryptic peptides are amenable to iTRAQ labeling and analysis to LC–MS analysis with reverse-phase nano-UPLC (0.75 mm × 25 cm × 1.9 μm particle) with nanoelectrospray ionization-tandem ultra-high-resolution mass spectrometry. *Reprinted with permission from Garbis et al. (2011), Copyright 2013 American Chemical Society.* (See the color plate.)

cancer cells (Emmanouilidou et al., 2010). An additional performance characteristic of the 3D study method was its accuracy and sensitivity in identifying the phosphorylated variant to a potential protein marker thus imparting an additional molecular feature in the more precise capturing of unique chemical signatures of disease and treatment response. Further refinements were made to the MuDPIT method (Fig. 11.1) in order to improve its performance characteristics and analysis throughput, while at the same time substantially reduce its running costs.

4.2. Cervical cancer

The ThinPrep cervical smear is an established biospecimen for the cytological and molecular screening against abnormal cells and HPV infection. In this particular application, an LC–MS method based on isobaric (iTRAQ)

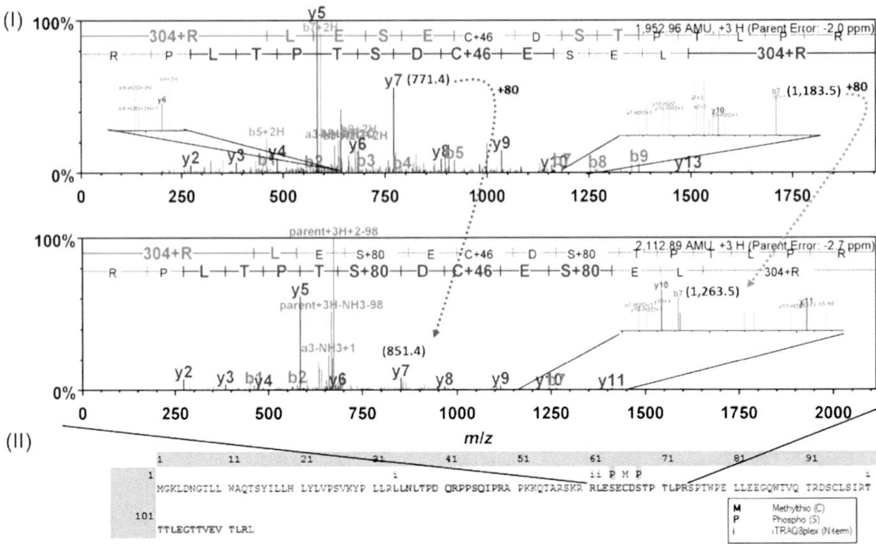

Figure 11.2 (I) Annotated ITMS CID MS/MS spectrum of the phosphopeptide R(iTRAQ8plex)LES(phospho)EC(methylthio)DS(phospho)TPTLPR with *m/z* 705.30310 Da (−2.61 ppm), *z* = +3, traceable to Protein E4 of HPV43 found in experiment A along with the annotated ITMS CID MS/MS spectrum of the native peptide. (II) The peptide sequence coverage of Protein E4 of HPV43. *Reprinted with permission from Garbis et al. (2011) and Papachristou et al. (2013), Copyright 2013 American Chemical Society.* (See the color plate.)

labeling and high-resolution FT-MS was used for the proteomic analysis of 23 human ThinPrep smear specimens. Tandem mass spectrometry analysis was performed with both nitrogen high collision dissociation (HCD MS/MS) and CID MS/MS peptide fragmentation modes. The overall analysis yielded the analysis of over 3200 unique proteins at FDR < 1%. The analysis also included six HPV-derived proteins including the high-risk HPV16 type in the specimens tested. A further testament to the sensitivity and selectivity of the proposed study method was the confident detection of a significant number of phosphopeptides in these specimens (Fig. 11.2). Classification of the whole proteome according to Gene Ontologies was performed with DAVID software and is shown in pie chart form in Fig. 11.3. A broad range of biological processes and molecular functions of proteins with diverse cellular component origin was determined. The advantages of whole cervical smear specimen analysis include the potential for the identification of extracellular proteins, the secretion of which may be controlled by disease-specific mechanisms.

(I) **Tissue expression**

Term		Count	%	p-value
Keratinocyte		10	16.1	4.0E-10
Platelet		11	17.7	1.2E-5
Cajal-Retzius cell		7	11.3	4.2E-5
Tongue		9	14.5	5.2E-5
Bile		4	6.5	5.3E-5
Pancreas		13	21.0	7.0E-5
Epithelium		21	33.9	1.7E-4

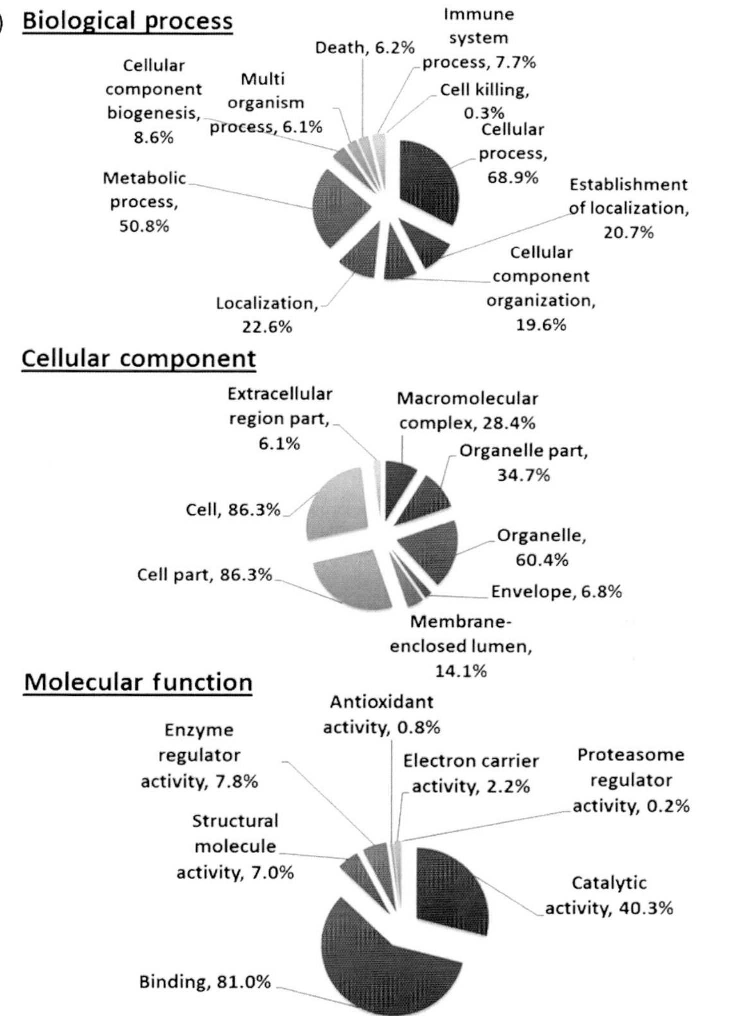

Figure 11.3 *Gene ontology (GO) annotation of the identified proteome as computed by the DAVID software.* (I) Classification of the top 50 high-scoring proteins per experiment based on their tissue expression consensus. (II) Classification of the total proteome based on biological process (BP), cellular component (CC), and molecular function (MF) terms. *Reprinted with permission from Papachristou et al. (2013), Copyright 2013 American Chemical Society.*

5. SUMMARY

The deregulation of cell proliferation in carcinogenesis requires adjustments of energy metabolism to fuel deregulated proliferation. The reprogramming of glucose metabolism in cancer cells leading to aerobic glycolysis requires a complex attenuation of cellular machinery e.g., glucose transporters, allowing greater glucose import into the cell. The neoplastic respiratory chain malfunction is an essential process to sustain ATP supply and is likely to be due to mitochondrial DNA mutations rease in glycolysis to sustain ATP supply (Antico Arciuch, Elguero, Poderoso, & Carreras, 2012).

Oncometabolism is an extremely complex process requiring an equally dynamic analysis tool to achieve highly sensitive and specific resolution of the pathways involved. Recent studies have utilized new advances in proteomic technology to elucidate clinically significant pathways which in time may prove to be of use as biomarkers of disease or therapeutic targets. In oncology, proteomics is fast becoming the most widespread clinical application of research, making the technique translatable from bench to bedside. A successful application of proteomics to oncology can be observed in the example of squamous cell carcinoma of the bladder, where several proteins unique to metastatic lesions were discovered. From this study, psoriacin was highlighted as a potential marker of the primary bladder carcinoma. An ELISA assay is currently being developed for psoriacin allowing its potential as a diagnostic marker to be evaluated. These researchers have now set up a comprehensive proteomic database for bladder carcinoma including profiles for both transitional and squamous cell tumors (Celis, Wolf, & Ostergaard, 1999; Chambers, Naumov, Vantyghem, & Tuck, 2000).

The MuDPIT method is proven to achieve in-depth analysis coverage of up to a thousand proteins reproducibly in replicate experiments using clinical samples. Its application to real clinical specimens demonstrates its robustness with the advantage over other approaches of being direct, highly sensitive, reproducible, high resolution with ultra-high mass accuracy, effective relative quantitative analysis, and economical due to its multiplexing capacity to coanalyze multiple biological specimens. This could lead to a future high throughput, rapid, quantitative, and economical screening method for patient samples.

REFERENCES

Antico Arciuch, V. G., Elguero, M. E., Poderoso, J. J., & Carreras, M. C. (2012). Mitochondrial regulation of cell cycle and proliferation. *Antioxid Redox Signal*, *16*(10), 1150–1180. http://dx.doi.org/10.1089/ars.2011.4085.

Banks, R. E., Dunn, M. J., Hochstrasser, D. F., Sanchez, J. C., Blackstock, W., Pappin, D. J., et al. (2000). Proteomics: New perspectives, new biomedical opportunities. *Lancet*, *356*(9243), 1749–1756. http://dx.doi.org/10.1016/S0140-6736(00)03214-1.

Blackstock, W., & Rowley, A. (1999). Probing cellular complexity with proteomics. *Current Opinion in Molecular Therapeutics*, *1*(6), 702–711.

Cai, Z., Chiu, J. F., & He, Q. Y. (2004). Application of proteomics in the study of tumor metastasis. *Genomics, Proteomics & Bioinformatics*, *2*(3), 152–166.

Celis, J. E., Wolf, H., & Ostergaard, M. (1999). Proteomic strategies in bladder cancer. *IUBMB Life*, *48*(1), 19–23. http://dx.doi.org/10.1080/713803472.

Chambers, A. F., Naumov, G. N., Vantyghem, S. A., & Tuck, A. B. (2000). Molecular biology of breast cancer metastasis. Clinical implications of experimental studies on metastatic inefficiency. *Breast Cancer Research*, *2*(6), 400–407.

Emili, A. Q., & Cagney, G. (2000). Large-scale functional analysis using peptide or protein arrays. *Nature Biotechnology*, *18*(4), 393–397. http://dx.doi.org/10.1038/74442.

Emmanouilidou, E., Melachroinou, K., Roumeliotis, T., Garbis, S. D., Ntzouni, M., Margaritis, L. H., et al. (2010). Cell-produced alpha-synuclein is secreted in a calcium-dependent manner by exosomes and impacts neuronal survival. *Journal of Neuroscience*, *30*(20), 6838–6851. http://dx.doi.org/10.1523/JNEUROSCI.5699-09.2010.

Garbis, S. D., Roumeliotis, T. I., Tyritzis, S. I., Zorpas, K. M., Pavlakis, K., & Constantinides, C. A. (2011). A novel multidimensional protein identification technology approach combining protein size exclusion prefractionation, peptide zwitterion-ion hydrophilic interaction chromatography, and nano-ultraperformance RP chromatography/nESI-MS2 for the in-depth analysis of the serum proteome and phosphoproteome: Application to clinical sera derived from humans with benign prostate hyperplasia. *Analytical Chemistry*, *83*(3), 708–718. http://dx.doi.org/10.1021/ac102075d.

Garbis, S. D., Tyritzis, S. I., Roumeliotis, T., Zerefos, P., Giannopoulou, E. G., Vlahou, A., et al. (2008). Search for potential markers for prostate cancer diagnosis, prognosis and treatment in clinical tissue specimens using amine-specific isobaric tagging (iTRAQ) with two-dimensional liquid chromatography and tandem mass spectrometry. *Journal of Proteome Research*, 7(8), 3146–3158. http://dx.doi.org/10.1021/pr800060r.

Granger, J., Siddiqui, J., Copeland, S., & Remick, D. (2005). Albumin depletion of human plasma also removes low abundance proteins including the cytokines. *Proteomics*, *5*(18), 4713–4718. http://dx.doi.org/10.1002/pmic.200401331.

Gundry, R. L., Fu, Q., Jelinek, C. A., Van Eyk, J. E., & Cotter, R. J. (2007). Investigation of an albumin-enriched fraction of human serum and its albuminome. *Proteomics Clinical Applications*, *1*(1), 73–88. http://dx.doi.org/10.1002/prca.200600276.

Kaushik, A., Vareed, S., Basu, S., Putluri, V., Putluri, N., Panzitt, K., et al. (2013). Metabolomic profiling identifies biochemical pathways associated with castration-resistant prostate cancer. *Journal of Proteome Research,* (Dec 20), *13*(2), pp. 1088–1100. http://dx.doi.org/10.1021/pr401106h.

Kelly, A., Breitkopf, S., Yuan, M., Goldsmith, J., Spentzos, D., & Asara, J. (2011). Metabolomic profiling from formalin-fixed, paraffin-embedded tumor tissue using targeted LC/MS/MS: Application in sarcoma. *PLoS One*, *6*(10).

Kulasingam, V., & Diamandis, E. P. (2008). Strategies for discovering novel cancer biomarkers through utilization of emerging technologies. *Nature Clinical Practice Oncology*, *5*(10), 588–599. http://dx.doi.org/10.1038/ncponc1187.

Larsen, M. R., Sørensen, G. L., Fey, S. J., Larsen, P. M., & Roepstorff, P. (2001). Phosphoproteomics: Evaluation of the use of enzymatic de-phosphorylation and differential mass spectrometric peptide mass mapping for site specific phosphorylation assignment in proteins separated by gel electrophoresis. *Proteomics*, *1*(2), 223–238. http://dx.doi.org/10.1002/1615-9861(200102)1:2<223::AID-PROT223>3.0.CO;2-B.

Naaby-Hansen, S., Nagano, K., Gaffney, P., Masters, J. R., & Cramer, R. (2003). Proteomics in the analysis of prostate cancer. *Methods in Molecular Medicine*, *81*, 277–297. http://dx.doi.org/10.1385/1-59259-372-0:277.

Papachristou, E. K., Roumeliotis, T. I., Chrysagi, A., Trigoni, C., Charvalos, E., Townsend, P. A., et al. (2013). The shotgun proteomic study of the human ThinPrep cervical smear using iTRAQ mass-tagging and 2D LC-FT-Orbitrap-MS: The detection of the human papillomavirus at the protein level. *Journal of Proteome Research*, *12*(5), 2078–2089. http://dx.doi.org/10.1021/pr301067r.

Pike, L. J., & Sadler, J. E. (2004). Proteomics, genomics and the future of medical education. *Missouri Medicine*, *101*(5), 496–499.

Prabhu, A., Sarcar, B., Kahali, S., Yuan, Z., Johnson, J. J., Adam, K. P., et al. (2013). Cysteine catabolism: A novel metabolic pathway contributing to glioblastoma growth. *Cancer Research: Vol. 74*, 787–796.

Simons, K., & Toomre, D. (2000). Lipid rafts and signal transduction. *Nature Reviews Molecular Cell Biology*, *1*(1), 31–39. http://dx.doi.org/10.1038/35036052.

Yan, B., Pan, C., Olman, V. N., Hettich, R. L., & Xu, Y. (2004). Separation of ion types in tandem mass spectrometry data interpretation—A graph-theoretic approach. *Proceedings of the 2004 IEEE Computational Systems Bioinformatics Conference*, 236–244.

Yocum, A. K., Yu, K., Oe, T., & Blair, I. A. (2005). Effect of immunoaffinity depletion of human serum during proteomic investigations. *Journal of Proteome Research*, *4*(5), 1722–1731. http://dx.doi.org/10.1021/pr0501721.

Zolla, L. (2008). Proteomics studies reveal important information on small molecule therapeutics: A case study on plasma proteins. *Drug Discovery Today*, *13*(23–24), 1042–1051. http://dx.doi.org/10.1016/j.drudis.2008.09.013.

Zolotarjova, N., Martosella, J., Nicol, G., Bailey, J., Boyes, B. E., & Barrett, W. C. (2005). Differences among techniques for high-abundant protein depletion. *Proteomics*, *5*(13), 3304–3313. http://dx.doi.org/10.1002/pmic.200402021.

Zuo, X., & Speicher, D. W. (2000). A method for global analysis of complex proteomes using sample prefractionation by solution isoelectrofocusing prior to two-dimensional electrophoresis. *Analytical Biochemistry*, *284*(2), 266–278. http://dx.doi.org/10.1006/abio.2000.4714.

CHAPTER TWELVE

In Vivo Quantitative Proteomics for the Study of Oncometabolism

Steven Reid, Juan Ramon Hernandez-Fernaud, Sara Zanivan[1]
Vascular Proteomics Group, Cancer Research UK Beatson Institute, Glasgow, United Kingdom
[1]Corresponding author: e-mail address: s.zanivan@beatson.gla.ac.uk

Contents

Abstract

The active reprograming of cellular metabolism is a primary driver of oncogenesis and a hallmark of established neoplastic lesions. Much of this reprogramming depends on the expression levels and posttranslational modifications (PTMs) of metabolic enzymes. Stable isotope labeling of amino acids in culture (SILAC) is an amino acid-based labeling technique that can be used both *in vitro* and *in vivo* to comparatively assess the levels and PTMs of proteins. To this aim, SILAC-labeled cell lysates can be spiked into each sample as a standard, followed by the analysis of specimens by mass spectrometry (MS). Combined with appropriate protocols for the lysis and preparation of samples for MS, this technique allows for the accurate and in-depth quantification of the proteome of a wide variety of cell and tissue samples. In particular, SILAC can be employed to infer the metabolic state of neoplastic lesions and obtain a profound understanding of the proteomic alterations that accompany oncogenesis and tumor progression. Here,

Methods in Enzymology, Volume 543
ISSN 0076-6879
http://dx.doi.org/10.1016/B978-0-12-801329-8.00012-X

we describe a proteomic approach based on SILAC, high-resolution chromatography and high-accuracy MS for comparing levels and phosphorylation status of proteins between the samples of interest. This method can be applied not only to the proteomic study of oncometabolism in murine tissues, but also to the study of cellular samples and human specimens.

1. INTRODUCTION

Cancer cell metabolism is vastly different to that of normal cells, and it has been established as a defined hallmark of cancer (Hanahan & Weinberg, 2011). During cancer progression, tumor cells may switch their metabolism to anaerobic glycolysis. This is partly driven by common mutations in cancer cells such as c-Myc and p53, which enhance glycolysis by affecting proteins such as glucose transporter-1, lactate dehydrogenase A (LDHA), and pyruvate kinase M2 (PKM2) (Semenza, 2010; Yeung, Pan, & Lee, 2008). PKM2 catalyzes the formation of pyruvate and is key in controlling whether glycolysis fuels ATP generation or biosynthetic processes, whereas LDHA catalyzes the subsequent step from pyruvate to lactate. This oncometabolism is functional in cancer progression as the loss of LDHA inhibits transformation and xenograft growth (Dang, Le, & Gao, 2009). Furthermore, it has become apparent that there is an intricate interplay between the cellular proteome and the metabolism. For example, tumors are commonly hypoxic resulting in the increased expression of the transcription factor hypoxia-induced factor 1 (HIF-1). Downstream targets of HIF-1 include pyruvate dehydrogenase kinase 1, which phosphorylates and inhibits the pyruvate dehydrogenase, a key enzyme that links glycolysis to the tricarboxylic acid cycle (Kim, Tchernyshyov, Semenza, & Dang, 2006). HIF-1 can also induce LDH expression, which reinforces the HIF-1 induced increase in glycolysis while inhibiting oxidative phosphorylation (Firth, Ebert, & Ratcliffe, 1995). Proteins are not only regulated at the expression level, their stability and activity can also be regulated by posttranslational modifications (PTMs). The advantage of altering a reversible modification such as phosphorylation or acetylation allows the activity of key metabolic enzymes to rapidly respond to the cells energy needs (Wang et al., 2010). For example, PKM2 is controlled at the expression level by c-Myc and HIF-1 and also by multiple modifications (Chaneton & Gottlieb, 2012; Hitosugi et al., 2009). Therefore, by quantitatively studying protein and PTMs, we can infer the metabolic state of the cell and gain a deeper understanding of metabolic alteration in cancer. For example, quantitative mass spectrometry (MS)-based analysis of the proteome and phosphoproteome of normal mouse skin and tumor tissues showed that Ldha and Pkm2 were increased in the tumor compared to normal tissue. However, despite the increase in Pkm2 at the protein level, there was a decrease in phosphorylation level at an uncharacterized serine at position 127 of Pkm2 (Zanivan et al., 2013).

In the past decade, MS technology combined with stable isotope-based approaches have dramatically improved and been successfully used to measure metabolic states and in-depth proteomes of tumor cells in culture (Chaneton et al., 2012; Geiger, Madden, Gallagher, Cox, & Mann, 2012). However, studying cells in culture has its limitations in how representative this is of the human disease. Conversely, murine models can recapitulate cancers and be exploited to explore metabolism *in vivo* (Marcotte & Muller, 2008). While *in vivo* metabolic studies are still at their infancy (Marin-Valencia et al., 2012), quantitative MS proteomics is supported by an established and robust workflow (Lamond et al., 2012).

The most widely used MS approach to analyze complex protein mixtures is shotgun proteomics, which relies on converting all proteins into smaller peptides by protease digestion. Trypsin is the commonly used enzyme because it cleaves the C-terminus of arginine and lysine yielding peptides of 4–40 amino acids in length with a basic C-terminus, which is the optimal mass range for efficient MS fragmentation and makes them easily ionizable at low pH. Peptides are separated using high-resolution liquid chromatography (LC) and analyzed with high mass accuracy, speed, and sensitivity mass analyzers. To obtain accurate quantitative comparisons of protein levels, a widely used approach is stable isotope labeling of amino acids in culture (SILAC) (Ong et al., 2002). SILAC is achieved by labeling proteins metabolically, by substituting (semi)essential amino acids with an exogenous supply of their labeled counterparts. Lysine- and arginine-holding non-radioactive heavy isotopes are commonly used: $^{13}C_6{}^{15}N_4$ L-arginine (Arg10) and $^{13}C_6{}^{15}N_2$ L-lysine (Lys8), referred to as heavy amino acids. Cells use the heavy amino acids in all new protein synthesis and become fully labeled over several divisions. This ensures all peptides are labeled at their C-terminus following trypsin digestion and therefore usable for quantification. Differently labeled cells can be subjected to a treatment and then lysed to obtain two proteomes, which are directly mixed together and remain distinguishable by MS. Early mixing increases the quantification accuracy by limiting variation during sample preparation (e.g., from differences in digestion efficiency or fractionation). The same concept can be applied using a labeled standard that is spiked into unlabeled samples and used as a reference (Geiger et al., 2011; Monetti, Nagaraj, Sharma, & Mann, 2011; Rayavarapu et al., 2013; Zanivan et al., 2013; Fig. 12.1). Spike-in SILAC is a versatile approach at low cost while maintaining the high accuracy of quantification. It has clear benefits as it can be applied to samples that cannot be labeled and permits comparison between numerous samples that can be separated in time, assuming the same standard mix is used. Spike-in SILAC combined to shotgun MS has already been successfully used *in vivo* to accurately measure normal and tumor mouse tissues to a depth of more than 5000 proteins and 10,000 phosphorylation sites (Monetti et al., 2011; Zanivan et al., 2013).

There are many protocols to prepare samples for quantitative MS proteomics, depending on the chromatography and MS setup, sample amount, whether the sample can be metabolically labeled or there is a standard available (Bantscheff, Lemeer, Savitski, & Kuster, 2012; Fig. 12.2). Here, we present a robust methodology used in our lab, which combines SILAC, high-resolution LC, and mass analyzer to measure in-depth proteome and phosphoproteome of mouse tissues and which can be applied to investigate metabolism in cancer (Fig. 12.1).

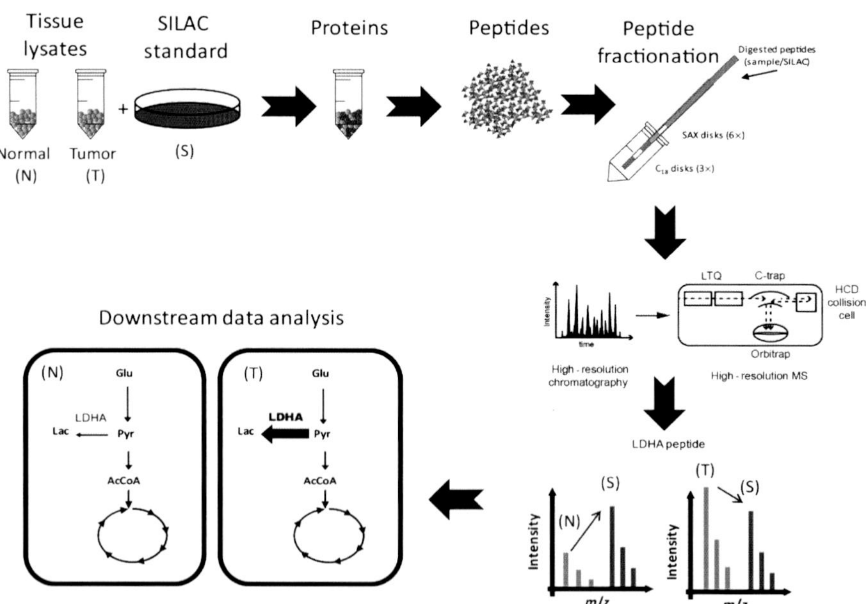

Figure 12.1 *MS shotgun proteomics workflow: from mouse tissue to metabolic pathway analysis.* (See the color plate.)

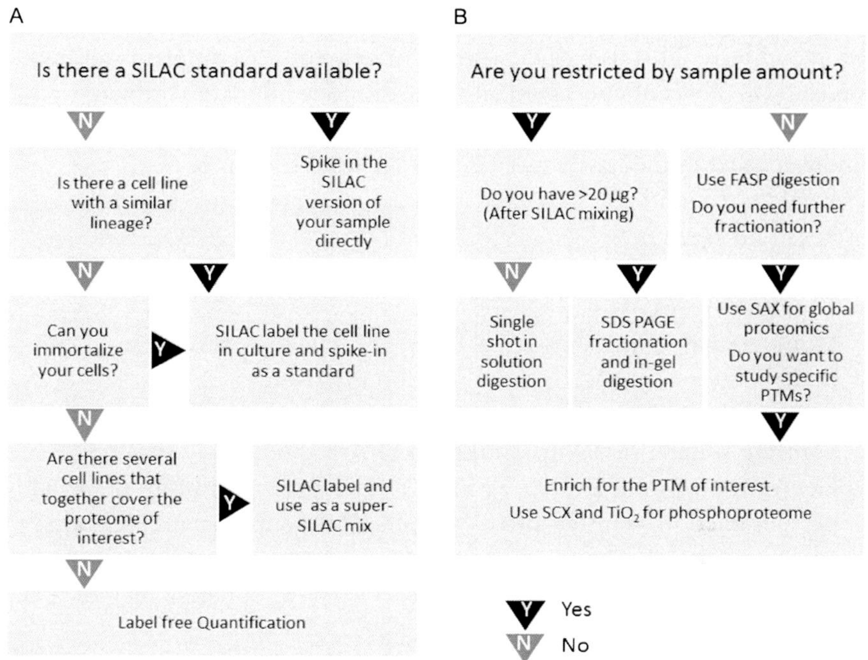

Figure 12.2 *Workflow for sample preparation.* Flow chart recommending which MS approaches to take depending on your experimental needs, answering the questions, follow (Y/N) for yes/no, respectively. (A) Helps readers choose between SILAC spike-in and label-free approach, (B) helps readers choose which sample preparation method to use according to sample quantity. Single shot, no fractionation at protein or peptide level before MS analysis. Super-SILAC mix, more than one SILAC-labeled cell lines mixed together to recapitulate the proteome to be analyzed at the MS. (See the color plate.)

2. SAMPLE PREPARATION

2.1. Tissue sample preparation

Materials required

- Liquid nitrogen (optional)
- Sodium dodecyl sulfate (SDS)
- Dithiothreitol (DTT)
- Tris base pH 7.4
- Bradford reagent (Abcam # ab119216)

Solutions and buffers

SDS lysis buffer: 4% SDS, 100 m*M* DTT, and 100 m*M* Tris base pH 7.4 (stored at −20 °C).

Once the target tissue or organ has been isolated, it undergoes homogenization and lysis. This uses SDS, a strong detergent for complete solubilization, and DTT, a reducing agent ensuring disulfide bonds are broken. Lysis is performed at room temperature (RT) to avoid SDS precipitation. If the sample is lysed immediately after isolation, jump to step 3.

1. Rapidly, transfer the extracted samples into sterile tubes and freeze in liquid nitrogen.
2. Grind the frozen tissue to a powder (e.g., using a mortar containing liquid nitrogen).
3. Homogenize (Ultra-Turrax T8 homogenizer, IKA Works) in SDS lysis buffer (use ~1 ml SDS lysis buffer/100 mg tissue).
4. Boil, 5 min 95 °C.
5. Sonicate using a metal tip until the DNA is completely sheared.
6. Centrifuge 10 min, 14,000 × *g*. The clear supernatant holds the soluble protein fraction, transfer into clean tube. If the supernatant is clear go to step 8.
7. Ultracentrifuge, 30 min 540,000 × *g*. The clear supernatant holds the soluble protein fraction.
8. Quantify the protein content by Bradford reagent, snap-freeze, and store the lysate at −20/−80 °C until preparation for MS analysis.

2.2. SILAC standard preparation

Materials required

SILAC medium (custom made, depleted for arginine and lysine)

Light and heavy aminoacids:

- L-Arginine:HCl light (Arg0; Sigma-Aldrich # A6969)
- L-Lysine:HCl light (Lys0; Sigma-Aldrich # L8662)
- L-Arginine:HCl heavy (Arg10; Cambridge Isotope Laboratories, Inc. # CNLM-539-H)
- L-Lysine:2HCl heavy (Lys8; Cambridge Isotope Laboratories, Inc. # CNLM-291-H)
- L-Proline light (Pro; Sigma-Aldrich # P5607 optional)

Phosphate buffered saline (PBS)

Normal and 10 kDa cutoff dialysed fetal bovine serum (FBS and 10 kDa FBS, respectively)

Medium supplements: penicillin–streptomycin, L-glutamine, growth factors (as required)

Trypsin (cell dissociation enzyme, optional)

Urea
Sodium chloride (NaCl)

Solutions and buffers

Arginine stock solution: 0.4 *M* arginine dissolved in PBS (stored at 4 °C).
Lysine and proline stock solutions: 0.8 *M* of lysine or proline dissolved in PBS (stored at 4 °C).
Urea buffer: 8 *M* urea, 75 m*M* NaCl, 50 m*M* Tris–HCl, pH 8.0 (stored at −20 °C).
SDS lysis buffer (see Section 2.1).

To perform accurate quantification, a SILAC spike-in standard should be qualitatively and quantitatively similar, though not identical, to the proteome of the sample to be analyzed. While for simpler organs, a single cell type can accurately mirror the tissue proteome, for example, hepatocyte cell line for the liver (Monetti et al., 2011), for complex tissue, a SILAC standard can be generated by mixing different SILAC-labeled cell lines together, referred to as Super-SILAC (Geiger, Cox, Ostasiewicz, Wisniewski, & Mann, 2010), for example, several tumor cell lines for tumor tissue. To define the best mixture for a Super-SILAC standard, we refer the reader to see Deeb, D'Souza, Cox, Schmidt-Supprian, & Mann (2012).

Once you have identified the cell line(s) to use as SILAC standard, you need to SILAC-label them so that at least 95% of the proteome is labeled to avoid interference with the sample quantification at the MS and then check how well the standard represents the proteome of interest.

We describe here the procedure to label the SILAC standard and check the labeling efficiency

1. When cells are 50–60% confluent, replace normal medium with SILAC medium supplemented with:
   a. Heavy amino acids: 1:1000 dilution from stock solution. The concentrations of Arg10 must be corrected when conversion of Arg10 to heavy proline (Pro6) is detected (if >5% of identified peptides contain Pro6 modification): either reduce Arg10 concentration or add some proline.
   b. 10 kDa FBS: as required (if cells do not grow well in 10 kDa FBS, use 80% 10 kDa FBS and 20% normal FBS; alternatively, use 100% normal FBS until 85% of incorporation is reached then swap to 80% 10 kDa FBS and 20% normal FBS until complete labeling).
   c. Penicillin–streptomycin, L-glutamine, and growth factors: as required.
2. Change medium every second day and split cells when confluent (avoid high split ratio because sparse cells might not grow well with 10 kDa

FBS; remove the dissociation enzyme (e.g., trypsin) from the medium because it is a source of nonlabeled amino acids and can impact the efficiency of the labeling). At each passage, lyse some cells in urea buffer for in solution digestion (Section 3.1.1) to monitor the incorporation of the SILAC amino acids.

3. Repeat step 2 until the incorporation of Arg10 and Lys8 is >95%. The incorporation efficiency is calculated separately for arginine and lysine containing peptides as

$$\left(1-\frac{1}{\text{ratio}\frac{H}{L}}\right)\times 100,$$

where "ratio H/L" = median of the ratio H/L of the quantified peptides in the "peptides.txt" output file of MaxQuant (Section 5), after removal of reverse and contaminant peptides.

4. Expand SILAC-labeled cells according to the amount of standard needed (always prepare in excess. The SILAC standard cannot be different between experiments that will be compared).
5. Lyse fully labeled SILAC cells with SDS lysis buffer (see Section 2.1).
6. For Super-SILAC only: mix appropriate amount of SILAC lysates from different cell lines.
7. Make aliquots of the SILAC standard, snap-freeze, and store at −80 °C.

2.3. Test suitability of SILAC standard

Materials required

Methanol

Chloroform

Solutions and buffers

SDS lysis buffer (see Section 2.1)

Urea buffer (see Section 2.2)

To test the suitability of the SILAC standard, the proteome of interest is compared to the SILAC standard as follows:

1. Mix 5–10 μg each of the sample and SILAC standard.
2. Precipitate the protein mixture with methanol–chloroform (to remove SDS, which is incompatible with the MS).
3. Resuspend the protein in 20 μl of urea buffer.
4. Digest in solution (see Section 3.1.1) and use 2 μg of peptides for nano-LC (nLC)–MS (4 h gradient, Table 12.1) and analyze data with

Table 12.1 Nanoliquid chromatography gradient

	Time (min)		Flow (nl/min)	%B	Temperature (°C)
Column equilibration	5		200	0	35
Sample loading	We use maximum flow at 280 bar (6 μl in approximately 17 min)			0	35
Gradient	**2 h**	**4 h**			
	0–5	0–5	200	5	35
	5–95	5–205	200	5–30	35
	95–110	205–220	200–250	30–60	35
	110–115	220–225	250	60–80	50
	115–116	225–226	250	80–95	50
	116–121	226–231	250	95–5	50
	121–129	231–239	250–200	5	35

MaxQuant (see Section 5). For a SILAC standard to generate accurate quantification more than 90% of the proteins contained in the "proteinGroups.txt," file output of MaxQuant should be within a fourfold ratio between the SILAC standard and the sample (ratio H/L), and the $\log_2$ ratios H/L should follow a normal distribution (see Section 5.2.1).

In cases where it is not possible to find suitable SILAC standard, an alternative is the use of the SILAC mouse (Zanivan et al., 2013), which allows accurate quantifications though rather expensive, or label-free quantification (Luber et al., 2010).

3. MS SAMPLE PREPARATION

3.1. Protein digestion

We use two approaches to digest proteins into peptides: in solution, to check incorporation of the SILAC amino acids in the standard and the suitability of the standard and in filter-aided sample preparation (FASP) (Wisniewski, Zougman, Nagaraj, & Mann, 2009), to measure the proteome and phosphoproteome of tissue samples.

3.1.1 In solution

Materials required

DTT
Iodoacetamide (IAA)
Endoproteinase Lys-C
Ammonium bicarbonate (ABC)
Trypsin (MS grade)
Trifluoroacetic acid (TFA)
Solid Phase Extraction Disk, Empore C_{18} (3M # 2215). Prepare C_{18} StageTip with two membranes in a 200 μl pipette tip.
Methanol
Acetonitrile (ACN)
Acetic acid (AA)

Solutions and buffers

DTT stock solution: 100 m*M* in distilled water (MS quality, stored at −20 °C)
IAA stock solution: 550 m*M* in distilled water (MS quality, stored at −20 °C)
Endoproteinase Lys-C stock solution: 1 μg/μl in 50 m*M* ABC (stored at −20 °C)
ABC stock solution: 50 m*M* ABC in distilled water (stored at RT)
Trypsin stock solution: 1 μg/μl in 50 m*M* AA (stored at −80 °C)
Buffer B: 80% ACN, 0.5% AA in distilled water
Buffer A*: 2% ACN, 0.1% TFA in distilled water
Buffer A: 0.5% AA in distilled water.
Urea buffer (see Section 2.2)

This method is simple and allows digestion of small amounts of lysate in urea buffer.

1. Aliquot 10 μg of protein.
2. Add DTT to a final concentration of 1 m*M*. Incubate 1 h at RT.
3. Add IAA (to alkylate reduced cysteines) to a final concentration of 5.5 m*M*, incubate 45 min, RT, protected from light.
4. Add 0.2 μg of endoproteinase Lys-C, incubate 3 h, RT.
5. Dilute the urea buffer to 2 *M* with 50 m*M* ABC (to maintain trypsin activity).
6. Add 0.1 μg of trypsin, incubate overnight, RT.
7. Add TFA to a final concentration of 1% (v/v) to stop digestion.
8. Assemble C_{18} StageTips (Rappsilber, Mann, & Ishihama, 2007) and activate as follows:

   a. Add 50 μl methanol. Spin 1.5 min, 600 × *g*.
   b. Add 50 μl buffer B. Spin.
   c. Add 50 μl buffer A*. Spin.

9. Add digested peptides and centrifuge at 200 × *g*, 10 min.
10. Add 50 μl buffer A. Spin.
11. Elute with 20 μl buffer B, twice.
12. Speed vacuum down to 5 μl to evaporate the ACN.
13. Add 5 μl of buffer A*.
14. Use 2 μl for nLC–MS (2 h gradient, Table 12.1).

If samples are not to be run immediately, eluted peptides (step 11) can be stored at −80 °C and concentrated before adding buffer A* prior to MS analysis.

3.1.2 Filter-aided sample preparation

Materials required

Small FASP: Centrifugal filter unit with 30 kDa cutoff membrane, 500 μl (Sartorius # VN01H22)
Big FASP: Centrifugal filter unit with 30 kDa cutoff membrane, 15 ml (Millipore # UFC903024)
Urea
Tris base, pH 8.5
IAA
ACN

Solutions and buffers

Urea-FASP buffer: 8 *M* urea, 100 m*M* Tris base, pH 8.5
IAA–urea buffer: 0.05 *M* IAA in urea-FASP buffer
SDS lysis buffer (see Section 2.1)
ABC stock solution (see Section 3.1.1)
Endoproteinase Lys-C stock solution (see Section 3.1.1)
Trypsin stock solution (see Section 3.1.1)

With this method, denatured proteins are retained and digested on a 30 kDa filter. This generates pure tryptic peptides, while removing undigested proteins and nucleic acids. Volumes below are given for the small FASP (used for proteome) and those in parentheses are for the big FASP (used for phosphoproteome).

1. Mix the SDS lysates, sample and SILAC standard (suggested ratio 1:1; 100 μg each for proteome and 2.5–10 mg each for phosphoproteome), in the filter unit.

2. Only for small FASP: mix combined lysate with 200 μl urea-FASP buffer.
3. Spin 15 min 14,000 × *g* (30 min, 3000 × *g*) or until lysate has passed.
4. Pass 200 μl (5 ml) of urea-FASP buffer through the filter, 15 min, 14,000 × *g* (25 min, 3000 × *g*). Discard the flow through (FT).
5. Add 100 μl (2.5 ml) of IAA–urea buffer. Incubate 20 min, RT, and centrifuge 10 min, 14,000 × *g* (30 min, 2500 × *g*).
6. Pass through 100 μl (2 ml) urea-FASP buffer three times. Remove the FT.
7. Pass through 100 μl (1 ml) ABC solution three times. Leave the FT to retain humidity.
8. Add 40 μl (2 ml) ABC solution and endoproteinase Lys-C at a concentration of 1 μg for every 100 μg protein. Incubate 3 h, RT.
9. Add 1 μg of trypsin for every 100 μg protein. Incubate overnight, 37 °C in humid chamber.
10. Centrifuge the filters 10 min, 14,000 × *g* (20 min, 2500 × *g*) into new collection tubes and collect the FT.
11. Elute more peptides with 40 μl (2 ml) of 10% ACN in distilled water.
12. Quantify the peptides using absorbance at 280 nm.

The resulting high purity peptides can be stored at −80 °C until preparation for MS analysis. Peptides can be loaded onto C_{18} StageTip, eluted, and run directly at the MS, but for more in-depth proteomic analysis, we further fractionate at the peptide level.

3.2. Peptide fractionation

3.2.1 Proteome analysis

Materials required

Solid Phase Extraction Disk, Empore Anion (3M # 2252). Prepare anion exchange tip with six membranes into a 200 μl pipette tip (Strong anion exchange (SAX) StageTip)

Solid Phase Extraction Disk, Empore C_{18} (3M # 2215). Prepare StageTip with three membranes.

AA

Phoshoric acid

Boric acid

Sodium hydroxide (NaOH)

NaCl

Methanol

TFA

ACN

Solutions and buffers

Stock solution of Britton–Robinson Universal Buffer: prepare 5 × stock solution containing 0.1 *M* AA, 0.1 *M* phosphoric acid, 0.1 *M* boric acid and pH with NaOH to 11, 8, 6, 5, 4, and 3 (Stored at RT)

Britton–Robinson Universal Buffer, pH 3+: dilute stock solution to 20 m*M* and add 250 m*M* NaCl

C_{18} StageTip elution solution: 60% ACN

Buffer A* (See Section 3.1.1)

For in-depth analysis of the proteome, we reduce the complexity of the peptide mixture by fractionation. SAX is a robust method to separate tryptic peptides according to their net charge (Wisniewski, Zougman, & Mann, 2009). The following protocol generates six fractions starting from FASP-generated peptides.

1. Prepare a SAX StageTip and make up the 20 m*M* Britton–Robinson Universal buffer (pH: 11, 8, 6, 5, 4, and 3+)
2. Activate the SAX StageTip with 100 μl methanol. Spin 1.5 min, 2400 × *g*.
3. Equilibrate with 100 μl NaOH (1 *M*). Spin.
4. Equilibrate with 100 μl pH 11 buffer. Spin.
5. Add a further 100 μl pH 11 buffer and leave to prevent membrane drying while preparing six C_{18} StageTips (each with three C_{18} membranes).
6. Activate the C_{18} StageTips (see Section 3.1.1, with the exception that the last washes are performed with water and not buffer A*).
7. Centrifuge the SAX StageTip.
8. Cut the top of the SAX StageTip and insert into the top of the first C_{18} StageTip, place this through the lid of a 2 ml tube bored with a small hole, as in Fig. 12.1.
9. Load 50 μg of FASP–peptides into the anion exchanger. Spin 1.5 min, 2400–3000 × *g*.
10. Add 100 μl of pH 11 buffer. Spin.
11. Transfer the SAX StageTip into the next C_{18} StageTip.
12. Add to the SAX StageTip 100 μl of the next buffer with lower pH. Spin.
13. Repeat steps 11–12 until you elute all fractions onto the six C_{18} StageTips.
14. Wash C_{18} StageTips with 50 μl 0.1% TFA.
15. Elute peptides from each C_{18} StageTip with 20 μl of 60% ACN, twice.

16. Speed vacuum samples down to 3 μl.
17. Add equal volume of buffer A*.
18. Use 3 μl for nLC–MS/MS (4 h gradient, Table 12.1). Each fraction is run as separate samples at the MS.

If samples are not to be run immediately, eluted peptides (step 15) can be stored at −80 °C and concentrated before adding buffer A* prior to MS analysis.

3.2.2 Phosphoproteome analysis

Materials required

Resource S column (strong cation exchange or SCX column, GE Healthcare # 17-1178-01)
Titanium dioxide (TiO_2) beads (GL sciences # 5020–75010)
Monopotassium phosphate (KH_2PO_4)
Potassium chloride (KCl)
2,5-dihydroxybenzoic acid (DHB)
TFA
ACN
Solid Phase Extraction Disk, Empore C_8 (3M # 2214). Prepare C_8 StageTip with two membranes.
C_{18} StageTip with two membranes (see Section 3.1.1)

Solutions and buffers

SCX chromatography buffer A: 5 m*M* KH_2PO_4, 30% ACN and adjust to pH 2.7 with TFA
SCX chromatography buffer B: SCX buffer A and 350 m*M* KCl
DHB stock solution: 80% ACN, 0.1% TFA, and 0.2 *M* DHB (freshly prepared)
TiO_2 suspension: 5 mg of TiO_2 beads in 10 μl of DHB solution
TiO_2 wash solution I: 30% ACN and 3% TFA
TiO_2 wash solution II: 80% ACN and 0.3% TFA
TiO_2 elution solution: 15% ammonium hydroxide and 40% ACN
C_{18} StageTip elution solution: 60% ACN
Buffer A (see Section 3.1.1)
Buffer A* (see Section 3.1.1)

SCX chromatography separates tryptic peptides according to their charge. While singly phosphorylated ones bind to the SCX column and can be eluted with increasing salt concentration, multiply phosphorylated peptides are not retained. We describe here a robust method to fractionate 5–20 mg of FASP-digested peptides on a SCX column, where we also retain the FT.

We then use TiO_2 beads, to enrich for phosphorylated peptides, while keeping the background generated by nonphosphorylated peptides at a minimum using DHB.

1. Acidify tryptic peptides eluted from big FASP (see Section 3.1.2) to pH 2.6 with TFA (just before running the SCX to avoid peptide precipitation).
2. Add ACN to 30% final concentration.
3. Centrifuge sample, 5 min, 3500 × *g*.
4. Separate peptides using a SCX column. See Table 12.2 for gradient and additional details. Collect the FT and pool the 16 fractions as follows: first 8 fractions (richer in phosphopeptides) into 4 and last 8 into 2.
5. Add in each fraction 10 μl of TiO_2 suspension.
6. Incubate 60 min in rotor wheel.
7. Centrifuge 2 min, 800 × *g*
   a. For the fractions: remove supernatant and keep the pellet.
   b. For the FT: transfer the supernatant into fresh tube and repeat steps 5–7 three more times. Every time keep the pellet.
8. Resuspend pellet in 150 μl of TiO_2 wash solution I and transfer into C_8 StageTip.
9. Centrifuge C_8 StageTip, 2 min, 600 × *g*.
10. Add 150 μl TiO_2 wash solution I and repeat step 9.

Table 12.2 Strong cation exchange chromatography gradient and parameters

	Time (min)	% B		
Column equilibration	5	0		
Sample loading	9	0	**Buffer A**	5 m*M* KH_2PO_4/30% acetonitrile adjust pH 2.7 with TFA
Gradient	0–30	0–30	**Buffer B**	Buffer A plus 350 m*M* KCl
	30–32	30	**Flow rate (ml/min)**	1
	32–33	30–100	**Absorbance (nm)**	215 and 280
	33–38	100	**Sample**	Diluted in buffer A to 9 ml
	38–43	0	**Fractions (ml)**	2

11. Add 100 μl of TiO_2 wash solution II.
12. Centrifuge C_8 StageTip, 2 min, 600 × *g*.
13. Repeat steps 11–12 twice.
14. Elute peptides with 20 μl of TiO_2 elution solution, three times.
15. Speed vacuum down to 5 μl.
16. Activate C_{18} StageTip (see Section 3.1.1).
17. Add 200 μl buffer A to each sample.
18. Load sample on C_{18} StageTip and centrifuge, 10 min, 200 × *g*.
19. Wash with 150 μl buffer A and centrifuge, 2 min, 600 × *g*.
20. Elute peptides with 20 μl of 60% ACN, twice.
21. Speed vacuum down to 3 μl.
22. Add 3 μl of buffer A^*.
23. Use 3 μl for nLC–MS/MS (2 h gradient, Table 12.1).

If samples are not to be run immediately, eluted peptides (step 20) can be stored at −80 °C and concentrated before adding buffer A^* prior to MS analysis.

4. INSTRUMENTS

The samples generated with SAX and SCX-TiO_2 protocols contain tryptic peptides that span a broad range of abundances. To measure them with high accuracy and depth, we combine the high resolving power of the reverse phase chromatography with the speed and ultra-high-resolution achieved in the Orbitrap Elite (Michalski et al., 2012). We describe below a peptide separation method based on reverse phase chromatography run on the nLC at submicroliter/min flow. This is combined on-line via a nanoelectrospray ion source with a hybrid linear trap quadrupole (LTQ) Orbitrap Elite mass spectrometer operated in the high collision energy dissociation (HCD) mode to fragment peptides.

4.1. High-resolution chromatography and electrospray ionization

The chromatography and electrospray are composed by:

1. nLC (EASY-nLC II, Thermo Scientific).
2. 20 cm capillary column (ID 75 μm, 8 μm tip diameter, FS360-75-8N-5C20, PicoTip emitter, New Objective), home packed with a slurry of 10 mg Empore-C_{18} beads in 400 μl of methanol (Reprosil-Pur Basic C_{18}, 1.9 μm, Dr. Maisch GmbH) under 80 bar of helium using a nanobaume capillary packing unit (Western fluids). The column is compacted

pumping 200 nl/min of buffer B for 30 min or until stable pressure of ~145 bar. Eventually, 18 cm of the column is filled with packing material. The expected pressure in normal chromatography conditions (200 nl/min, 10% buffer B, 35 °C) is 170–180 bar.

3. Nanoelectrospray source (Nanospray Flex Ion Source, Thermo Scientific) modified in house with
   a. Column oven for reproducible chromatography and reduced column pressure (PRSO-V1, Sonation GmbH)
   b. Ion reduction device (ABIRD, ARC Sciences)
   c. Liquid junction cross 1/32 with integrated platinum electrode to connect transfer line, waste line, and capillary column (ES257, Thermo Scientific).

Figure 12.3 shows how the different parts are assembled, where the column tip is located at ~7 mm offset from the heated capillary, ~30 ° from the horizontal plane.

When the tryptic peptides elute from the capillary column (see Table 12.1 for detailed gradients), they are ionized and fly into the mass spectrometer. During this process, the spray stability and ionization efficiency are critical to obtain maximum ion intensity and smooth peak shape, which

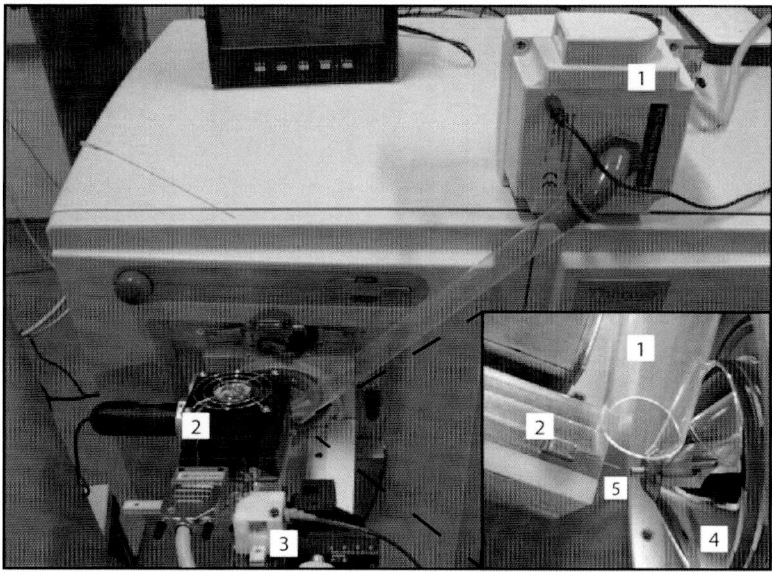

Figure 12.3 *Nanoelectrospray source setup.* 1, Ion reduction device; 2, column oven; 3, liquid junction cross; 4, heating capillary; 5, capillary column.

further result in accurate peptide quantification. To achieve that we use a spray and current voltage of 2.1 kV and ~0.5 μA, respectively, and we require the capillary column not spit over the peptide elution gradient.

4.2. High-resolution mass analyzer

Here, we present the standard settings for the hybrid mass spectrometer LTQ Velos Pro-Orbitrap Elite (Thermo Scientific).

Once ionized, the ions fly through the heated capillary (200 °C) and ion optics (use setup according to the fabric recommendations) into the mass analyzers. The full-scan MS spectra is acquired in the mass range 300–1650 *m/z* with a resolution of 120,000 at *m/z* 400 Th and target value of 1,000,000 charges in the Orbitrap. The top 10 most intense ions are sequentially isolated for fragmentation using HCD and detected in the Orbitrap at the MS^n target value 40,000 ions and recorded at a resolution of 15,000 at *m/z* 400 Th. The ions (single-charged states are excluded) are isolated within a mass window of 2 (*m/z*) and included in the exclusion list to avoid repeated fragmentation of the same ion (see Table 12.3). Data are acquired using the Xcalibur software which generates .RAW files that we analyze with the MaxQuant computational package (Cox & Mann, 2008).

5. MS DATA ANALYSIS

For MS data analysis, we use the MaxQuant package because it has been developed specifically for high-resolution MS data analysis and is freely downloadable (http://www.maxquant.org). This includes the MaxQuant software, which uses the MS .RAW files to identify and accurately quantify peptides and proteins, the Viewer, which allows MS data visualization (not discussed in this chapter) and Perseus, which carries out basic manipulations of the data and performs downstream bioinformatic and statistical analysis of the MaxQuant output tables.

5.1. MaxQuant

MaxQuant runs mostly with default parameters. Below is a general workflow where we highlight parameters that can be applied specifically for the analysis proposed in this chapter. For basic tutorial see Cox et al. (2009).

Table 12.3 Parameters for MS method

	Full-scan MS	MS/MS-dependent scan HCD top 10
AGC target	1×10^6	4×10^4
Microscans	1	1
Maximum injection time (ms)	100	150
Resolution	12×10^4	15×10^3
Minimum signal threshold (counts)		4×10^4
Normalized collision energy (%)		30
Default charge state		2
Isolation width (m/z)		3
Reject charge states		1
Reject mass width (ppm)		Low and high: 10
Exclusion list size		500
Dynamic exclusion duration (s)		60
Exclusion mass width (ppm)		Low and high: 5

1. Load the .RAW files.
2. Generate the experimental design template, where you include the fraction numbers and experiment names.
3. Define a multiplicity of 2 and select Arg10 and Lys8 as heavy labels.
4. Add additional variable modifications:
   a. For the phosphoproteome: Phospho(STY)
   b. For the labeling check: Pro6.
5. Load the FASTA file containing the protein sequences (download from www.uniprot.org). This is used by the MaxQuant search engine Andromeda (Cox et al., 2011) to identify peptides.
6. For the proteome and phosphoproteome, check the "re-quantify" option to obtain a SILAC ratio for peptides where a SILAC pair has not been found; for example, proteins are very low abundant only in the SILAC standard or the sample. For the SILAC incorporation check, do not check this option otherwise there is the risk to underestimate the incorporation.

5.2. Perseus

We provide below some basic steps to generate a list of identified and accurately quantified proteins and phosphorylation sites, starting from .txt tab-delimited tables generated in the MaxQuant output. We further provide some suggestions for statistical tests. All common tasks fall under the Processing and Analysis menus.

5.2.1 Proteome

1. Load "proteinGroups.txt" file and place the columns into the appropriate categories, with the normalized H/L ratios as expression columns.
2. Remove common contaminants, the "reverse" false positives and only identified by site (modified version only) by filtering by category.
3. Check the removed entries and replace proteins of interest.
4. Remove proteins without any unique peptides: this table contains the unambiguously identified proteins.
5. Invert ratio H/L (SILAC standard/sample) for easier sample comparison.
6. Transform by taking $\log_{10}$ of intensities and $\log_2$ of the ratio L/H.
7. Add annotation columns from Uniprot database such as GO annotations.
8. Check for a normal distribution using the histogram and basic statistics and that the replicates are reproducible using column correlation functions. When checking if a SILAC standard is appropriate for the sample of interest, a bimodal distribution of $\log_2$ ratio H/L is an indication that a population of proteins in the sample is not well represented by the SILAC standard.

The resulting dataset can be analyzed in many ways. For direct sample comparison while eliminating the internal standard, you need the ratio of ratios; however, this increases the quantification and statistical error. For analysis without the ratio of ratios, go to step 12.

Ratio of ratios

9. Create groups of the replicates and take the median of ≥ 2 valid values per group: these are your reproducibly quantified proteins.
10. Subtract one ratio from another (divide if not logged):

$$\log_2(\text{ratio } L/H \text{ sample A}) - \log_2(\text{ratio } L/H \text{ sample B}) \\ = \log_2(\text{Sample A}/\text{Sample B})$$

11. Define your significance cutoff (e.g., twofold) or use the "Significance B" (Cox & Mann, 2008). Export outliers. Go to step 14.

SILAC standard ratios

12. Create groups of the replicates and filter by ≥ 2 valid values per group: these are your reproducibly quantified proteins.
13. A *t*-test can be performed between two samples or an ANOVA for multiple samples. Export proteins passing the test.

All analyses

14. Hierarchical clustering can identify groups of regulated proteins.
15. The Fisher exact test provides a *p*-value, corrected for multiple testing hypothesis, associated with whether an annotation category, for example, GO term, is enriched in a portion of your data, for example, within significantly regulated proteins, rather than the expected proportion in relation to the rest of the proteome.

5.2.2 Phosphoproteome

When Phospho(STY) option is used as variable modification during the MaxQuant search, a table called "Phospho(STY)sites.txt" is generated, where each entry contains detailed information for each single serine, threonine, and tyrosine that have been identified to be phosphorylated. The analysis of phospho-data is similar to that described except for the following additional steps:

1. Load "Phospho(STY)sites.txt" and use all sample ratios as expression values (ratio *H*/*L* columns for each site are indicated as _1, _2, or _3, which indicate that the quantification refers to the site when identified in peptides singly, doubly, or multiply phosphorylated, respectively). Include the score difference and localization probability as numerical values.
2. After initial filtering (steps 2–3 of Section 5.2.1), filter by localization probability (remove if ≤ 0.75) and score difference (remove if ≤ 5): the phosphorylation sites that pass this filtering are the class I phosphorylation sites (sites with high phospho-localization probability for a single amino acid).
3. Expand the site table. This option separates the ratio *H*/*L* _1, _2, and _3 into separated entries.
4. Add modification information such as known sites, linear motifs, and kinase–substrate relations (based on publicly available databases, this feature indicates the predicted kinase responsible for the phosphorylation of the indicated site, based on its surrounding amino acid sequence). Regulated phosphorylation sites can be identified as in steps 11 and 13 of Section 5.2.1. Fisher test can be used as in step 15 using the modification information as categories.

At any stage of the analysis, data can be exported for further bioinformatic analysis using other software packages. Examples are Ingenuity (www.ingenuity.com) and KEGG (http://www.genome.jp/kegg/), which are useful to visualize proteins in the context of metabolic pathways, or STRING (http://string-db.org/) and Cytoscape (http://www.cytoscape.org/) to identify and visualize functional connection based, among others, on known protein–protein interaction. For a deeper investigation of the metabolism, computational methods can be used where proteomic data are integrated with genome-scale metabolic model. This approach has already been successfully applied to predict organ specific and tumor cells metabolism (Shlomi, Cabili, Herrgård, Palsson, & Ruppin, 2008; Jerby et al., 2012).

6. ADDITIONAL CONSIDERATIONS

To exploit the full potential of the above described protocol for in depth, high-accuracy proteomic analysis, we provide here some additional tips: (i) for any of the chromatography described, prepare the buffers using MS–LC/MS grade reagents, methanol/ethanol cleaned bottles, and high quality plastic; (ii) use freshly prepared buffers; (iii) do periodical maintenance and calibration of the nLC and mass spectrometer. This must fulfill the factory requirements; (iv) after the maintenance, run standard quality controls to evaluate the optimal performances of the instrument, including smooth and sharp peaks (width of approximately 6 s for 50 fmol of BSA tryptic peptides) for the nLC, and high speed and high success of peptide identification for the MS.

7. CONCLUSIONS

The abundance and PTMs of metabolic enzymes can be used to infer the metabolic state of the cell and we describe in this chapter a SILAC-based MS proteomics approach, which allows accurate and unbiased comparative analysis of the proteome and phosphoproteome of tissues isolated from mice. In the context of cancer, this approach can be applied to gain deeper understanding of metabolic alteration in cancer by measuring proteomic differences: (i) in tumor at different stages, (ii) between primary tumor and metastasis, (iii) between tumors carrying different genetic alterations (including oncogene expressions or knockout for a specific gene), and (vi) between tumors treated or not with anticancer drugs or chemo- or radiotherapy. Additionally, the spike-in SILAC approach is very versatile and can be used to measure a wide variety of samples other than mouse tissues. These include primary cells, mouse xenografts/orthografts, and human tissues. Together with appropriate

Table 12.4 Comparison of SILAC and label-free approaches

	SILAC spike-in	Label-free
Sample preparation	Internal standard has to be fully labeled with SILAC amino acids before experiment	No advanced preparation is needed prior to experiment
	Internal standard preparation increases cost due to the use of special reagents (SILAC amino acids, medium, and dialyzed FBS)	No additional expenses
	SILAC-labeled samples are combined at early stage. This allows complex sample manipulation and excludes quantification errors prior to MS analysis	Robust and reproducible protocols must be used because increased sample manipulation may introduce quantification errors
MS and data analysis	Simultaneous MS analysis of two proteomes (heavy SILAC spike-in and light sample proteome) increases complexity and may reduce the depth of the analysis	Single proteome analysis reduces complexity and gives advantage to in-depth analysis
	Quantification compares peptide pairs in the same MS run, therefore, enables for high accuracy	Quantification compares peptides in different MS runs, therefore, data analysis requires more sophisticated algorithms and more replicates are preferred to quantify peptides accurately

adaptations of the protocols proposed here, other PTMs can be analyzed such as acetylations and succinylations, which have been shown to be key regulators of metabolic enzymes.

At present, SILAC is the most widely used technique in quantitative proteomics because it allows high accurate peptide and protein quantification. Nevertheless, approaches that do not involve any labeling of the sample (referred to as label-free) are becoming more attractive for the future. Indeed, despite the fact that label-free approaches so far provide less accurate quantifications, they are simpler as they do not involve major manipulation of the sample prior to MS analysis. Furthermore, non-labeled samples have reduced complexity compared to labeled ones where two proteomes are mixed together. This gives the possibility to analyze label-free proteomes at a greater depth, while reducing the time of the analysis at the MS (a comparison between SILAC spike-in and label-free quantification methods is provided in Table 12.4). For this reason, great effort is now being devoted to the development of MS and chromatographic setup, and sophisticated algorithms to generate a robust workflow for accurate quantitative proteomics using label-free technologies.

In conclusion, we believe that the protocol described in this chapter, and future advances in MS proteomics, is a powerful approach to gain deeper insights on the metabolic alterations that accompany tumor development and progression.

REFERENCES

Bantscheff, M., Lemeer, S., Savitski, M. M., & Kuster, B. (2012). Quantitative mass spectrometry in proteomics: Critical review update from 2007 to the present. *Analytical and Bioanalytical Chemistry*, *404*, 939–965.

Chaneton, B., & Gottlieb, E. (2012). Rocking cell metabolism: Revised functions of the key glycolytic regulator PKM2 in cancer. *Trends in Biochemical Sciences*, *37*, 309–316.

Chaneton, B., Hillmann, P., Zheng, L., Martin, A. C., Maddocks, O. D., Chokkathukalam, A., et al. (2012). Serine is a natural ligand and allosteric activator of pyruvate kinase M2. *Nature*, *491*, 458–462.

Cox, J., & Mann, M. (2008). MaxQuant enables high peptide identification rates, individualized p.p.b.-range mass accuracies and proteome-wide protein quantification. *Nature Biotechnology*, *26*, 1367–1372.

Cox, J., Matic, I., Hilger, M., Nagaraj, N., Selbach, M., Olsen, J. V., et al. (2009). A practical guide to the MaxQuant computational platform for SILAC-based quantitative proteomics. *Nature Protocols*, *4*, 698–705.

Cox, J., Neuhauser, N., Michalski, A., Scheltema, R. A., Olsen, J. V., & Mann, M. (2011). Andromeda: A peptide search engine integrated into the MaxQuant environment. *Journal of Proteome Research*, *10*, 1794–1805.

Dang, C. V., Le, A., & Gao, P. (2009). MYC-induced cancer cell energy metabolism and therapeutic opportunities. *Clinical Cancer Research*, *15*, 6479–6483.

Deeb, S. J., D'Souza, R. C., Cox, J., Schmidt-Supprian, M., & Mann, M. (2012). Super-SILAC allows classification of diffuse large B-cell lymphoma subtypes by their protein expression profiles. *Molecular & Cellular Proteomics*, *11*, 77–89.

Firth, J. D., Ebert, B. L., & Ratcliffe, P. J. (1995). Hypoxic regulation of lactate dehydrogenase A. Interaction between hypoxia-inducible factor 1 and cAMP response elements. *The Journal of Biological Chemistry*, *270*, 21021–21027.

Geiger, T., Cox, J., Ostasiewicz, P., Wisniewski, J. R., & Mann, M. (2010). Super-SILAC mix for quantitative proteomics of human tumor tissue. *Nature Methods*, *7*, 383–385.

Geiger, T., Madden, S. F., Gallagher, W. M., Cox, J., & Mann, M. (2012). Proteomic portrait of human breast cancer progression identifies novel prognostic markers. *Cancer Research*, *72*, 2428–2439.

Geiger, T., Wisniewski, J. R., Cox, J., Zanivan, S., Kruger, M., Ishihama, Y., et al. (2011). Use of stable isotope labeling by amino acids in cell culture as a spike-in standard in quantitative proteomics. *Nature Protocols*, *6*, 147–157.

Hanahan, D., & Weinberg, R. A. (2011). Hallmarks of cancer: The next generation. *Cell*, *144*, 646–674.

Hitosugi, T., Kang, S., Vander Heiden, M. G., Chung, T. W., Elf, S., Lythgoe, K., et al. (2009). Tyrosine phosphorylation inhibits PKM2 to promote the Warburg effect and tumor growth. *Science Signaling*, *2*, ra73.

Jerby, L., Wolf, L., Denkert, C., Stein, G. Y., Hilvo, M., Oresic, M., et al. (2012). Metabolic associations of reduced proliferation and oxidative stress in advanced breast cancer. *Cancer Research*, *72*, 5712–5720.

Kim, J. W., Tchernyshyov, I., Semenza, G. L., & Dang, C. V. (2006). HIF-1-mediated expression of pyruvate dehydrogenase kinase: A metabolic switch required for cellular adaptation to hypoxia. *Cell Metabolism*, *3*, 177–185.

Lamond, A. I., Uhlen, M., Horning, S., Makarov, A., Robinson, C. V., Serrano, L., et al. (2012). Advancing cell biology through proteomics in space and time (PROSPECTS). *Molecular & Cellular Proteomics*, *11*, O112.017731-1–O112.017731-12.

Luber, C. A., Cox, J., Lauterbach, H., Fancke, B., Selbach, M., Tschopp, J., et al. (2010). Quantitative proteomics reveals subset-specific viral recognition in dendritic cells. *Immunity*, *32*, 279–289.

Marcotte, R., & Muller, W. J. (2008). Signal transduction in transgenic mouse models of human breast cancer—Implications for human breast cancer. *Journal of Mammary Gland Biology and Neoplasia*, *13*, 323–335.

Marin-Valencia, I., Yang, C., Mashimo, T., Cho, S., Baek, H., Yang, X. L., et al. (2012). Analysis of tumor metabolism reveals mitochondrial glucose oxidation in genetically diverse human glioblastomas in the mouse brain in vivo. *Cell Metabolism*, *15*, 827–837.

Michalski, A., Damoc, E, Lange, O., Denisov, E., Nolting, D., Muller, M., et al. (2012). Ultra high resolution linear ion trap Orbitrap mass spectrometer (Orbitrap Elite) facilitates top down LC MS/MS and versatile peptide fragmentation modes. *Molecular & Cellular Proteomics*, *11*(3), O111 013698-1–O111 013698-11.

Monetti, M., Nagaraj, N., Sharma, K., & Mann, M. (2011). Large-scale phosphosite quantification in tissues by a spike-in SILAC method. *Nature Methods*, *8*, 655–658.

Ong, S. E., Blagoev, B., Kratchmarova, I., Kristensen, D. B., Steen, H., Pandey, A., et al. (2002). Stable isotope labeling by amino acids in cell culture, SILAC, as a simple and accurate approach to expression proteomics. *Molecular & Cellular Proteomics*, *1*, 376–386.

Rappsilber, J., Mann, M., & Ishihama, Y. (2007). Protocol for micro-purification, enrichment, pre-fractionation and storage of peptides for proteomics using StageTips. *Nature Protocols*, *2*, 1896–1906.

Rayavarapu, S., Coley, W., Cakir, E., Jahnke, V., Takeda, S., Aoki, Y., et al. (2013). Identification of disease specific pathways using in vivo SILAC proteomics in dystrophin deficient mdx mouse. *Molecular & Cellular Proteomics*, *12*, 1061–1073.

Semenza, G. L. (2010). Defining the role of hypoxia-inducible factor 1 in cancer biology and therapeutics. *Oncogene*, *29*, 625–634.

Shlomi, T., Cabili, M. N., Herrgård, M. J., Palsson, B. Ø., & Ruppin, E. (2008). Network-based prediction of human tissue-specific metabolism. *Nature Biotechnology*, *26*, 1003–1010.

Wang, Q., Zhang, Y., Yang, C., Xiong, H., Lin, Y., Yao, J., et al. (2010). Acetylation of metabolic enzymes coordinates carbon source utilization and metabolic flux. *Science*, *327*, 1004–1007.

Wisniewski, J. R., Zougman, A., & Mann, M. (2009). Combination of FASP and StageTip-based fractionation allows in-depth analysis of the hippocampal membrane proteome. *Journal of Proteome Research*, *8*, 5674–5678.

Wisniewski, J. R., Zougman, A., Nagaraj, N., & Mann, M. (2009). Universal sample preparation method for proteome analysis. *Nature Methods*, *6*, 359–362.

Yeung, S. J., Pan, J., & Lee, M. H. (2008). Roles of p53, MYC and HIF-1 in regulating glycolysis—The seventh hallmark of cancer. *Cellular and Molecular Life Sciences*, *65*, 3981–3999.

Zanivan, S., Meves, A., Behrendt, K., Schoof, E. M., Neilson, L. J., Cox, J., et al. (2013). In vivo SILAC-based proteomics reveals phosphoproteome changes during mouse skin carcinogenesis. *Cell Reports*, *3*, 552–566.

CHAPTER THIRTEEN

Metabolomic Profiling of Neoplastic Lesions in Mice

Xiaojie Lu[*], **Li-Juan Ji**[†], **Jin-Lian Chen**[*,1]
[*]Department of Gastroenterology, Shanghai East Hospital, Tongji University, School of Medicine, Shanghai, China
[†]Department of Rehabilitation, The Second People's Hospital of Huai'an, Huai'an, China
[1]Corresponding author: e-mail address: wqq_021002@163.com

Contents

Abstract

Most cancers develop upon the accumulation of genetic alterations that provoke and sustain the transformed phenotype. Several metabolomic approaches now allow for the global assessment of intermediate metabolites, generating profound insights into the metabolic rewiring associated with malignant transformation. The metabolomic profiling of neoplastic lesions growing in mice, irrespective of their origin, can provide invaluable information on the mechanisms underlying oncogenesis, tumor progression, and response to therapy. Moreover, the metabolomic profiling of tumors growing in mice may result in the identification of novel diagnostic or prognostic biomarkers, which is of great clinical significance. Several methods can be applied to the metabolomic profiling of neoplastic lesions in mice, including mass spectrometry-based techniques (e.g., gas chromatography-, capillary electrophoresis-, or liquid chromatography-coupled mass spectrometry) as well as nuclear magnetic resonance. Here, we compare and discuss the advantages and disadvantages of all these techniques to provide a concise and reliable guide for readers interested in this active area of investigation.

Methods in Enzymology, Volume 543
ISSN 0076-6879
http://dx.doi.org/10.1016/B978-0-12-801329-8.00013-1

1. INTRODUCTION

Humans have been longing for conquering malignant tumor for many years. Unfortunately, up to now, the result is disappointing. The reasons for this are multiple, one of which is the difficulty in revealing the mysterious and complex mechanisms underlying tumorigenesis and progression. Cancer is an acquired genetic disease. Most cancers develop due to accumulated genetic alterations, which overrun the mechanisms that are normally effective in controlling cell cycle and apoptosis, in maintaining cellular homeostasis. Finally, these alterations result in the conversion toward malignant phenotype, which has many distinct characteristics, one of which is its metabolic mode. For example, instead of the more energy-effective tricarboxylic acid cycle pathway, malignant cells tend to metabolize glucose into lactate through glycolytic pathway. The characteristic metabolic modes of malignant cells open a door for cancer research.

The past decade has witnessed great strides in *omic* high-throughput techniques, including genomics, transcriptomics, epigenomics, proteomics, and subsequently, metabolomics. The Greek suffix "ome" refers to collection, and the term omics means the study of collections of DNA, RNA, proteins, and various other molecules, as well as the investigation of biological processes and functions as a whole. Metabolome is the downstream of the transcriptome, and along with genomics, transcriptomics, and proteomics, it is a powerful weapon in revealing the mystery veil of tumor. Metabolites are the end products of the cellular regulatory processes, and metabolomics is a set of methodology that quantitatively measures metabolites in response to environmental stimulus and/or genetic alterations in a dynamic manner, which gives way to full-scale analysis of cellular and tissue metabolism, providing more comprehensive understanding of tumor biology than ever before.

In the field of neoplastic lesions study, mice are the most frequently used animal model. Mice models provide a convenient and feasible tool for the deep investigation into the mechanisms underlying tumor initiation and progression, which are still obscure and not easy to study in humans. Some may argue that many oncological studies can also be conducted in cell lines instead of animal models. Indeed, *in vitro* studies are more convenient and less time-consuming than *in vivo* ones, and, in some cases, as fruitful as *in vivo* studies. But, as we all know, in many cases, the mechanisms and pathways through which normal cells convert toward malignant cells in response to environmental stimulus and genetic alterations have to be assessed in the intact context of a living body (Kwon & Berns, 2013). In this respect, animal models have undoubted advantages over *in vitro* cell culture studies. And among various animal models, mouse model is obviously one of the most frequently used one around the world.

In this review, several commonly used metabolomic methods in profiling neoplastic lesions in mice will be discussed. Generally speaking, there are two major methods in metabolomic analysis of neoplastic lesion, one called mass spectrometry (MS)-based techniques and the other called nuclear magnetic resonance. These two methods have their advantages and disadvantages, respectively. Nevertheless, they can both profile the metabolites in the context of neoplastic lesion in mice model. They are useful for cancer research and management because they make possible the synchronous monitoring of many molecules and functional profiling of cellular pathways. By simultaneously monitoring various molecules in the context of tumor, many potential tumor markers might be discovered. As we know, the prognosis of cancers depends largely on the stage at which they are detected. Although human beings have gained dramatic progress in medical image techniques in recent years, these image techniques are either too expensive as routine physical examinations (such as magnetic resonance imaging), especially in developing areas, or not sensitive enough (such as ultrasound) to detect early carcinomas. Moreover, from the perspective of tumorigenesis, tumors which can be detected by images are already in the late stage of its evolution. So, there is an urgent need of sensitive and specific tumor markers in the clinic to diagnose and monitor tumor. As for this respect, metabolomic profiling holds promise in detecting useful biomarkers. So, it does not surprise us to find that in recent years, there are large amounts of researches intending to find valuable tumor markers, both in animal models and in human bodies.

2. METABOLIC PROFILING AND CANCER BIOLOGY

In the 1920s, Otto Warburg observed a phenomenon that with sufficient glucose, cancer tissues generate great amounts of lactate regardless of whether oxygen is present or absent, which is sharply different from normal tissues in which lactate is produced mainly upon absence of oxygen. This phenomenon is thereafter referred to as the Warburg effect. Subsequently, other metabolic alterations of cancer cells and tissue were found and a new word was invented, Cancer Metabolome, which refers to the complete set of metabolic perturbations that characterize cancer cell, tissue, or organism. After malignant transformation, cancer cells gain several distinguishing features that facilitate their proliferation and invasion. As a result, the metabolic profiling of glucose, protein, and lipid in cancer cells is significantly altered.

As for glucose metabolism, cancer cells have a high rate of aerobic glycolysis in general for which at least three enlightening hypotheses have been proposed. Initially, Warburg hypothesized that cancer cells' reliance on glycolytic metabolism is due to mitochondrial defects that result in impaired aerobic respiration; however, subsequent studies proved that mitochondrial function is not compromised in the majority of cancers. Another hypothesis proposes that glycolytic metabolism is due to hypoxia pressure of tumor cells. As we know, cancer cells must survive periods of hypoxia during tumorigenesis. Moreover, it had been found that pathways that regulate angiogenesis can also promote aerobic glycolysis. The third hypothesis holds that preference of cancer cells to aerobic glycolysis is due to reprogramming of metabolism to enable rapid cell division. In the setting of cancers, cells metabolize glucose via glycolysis–citrate–lipogenesis pathway. Citrate has two alternative routes in the mitochondria, one is Krebs cycle for ATP production and the other is for lipid synthesis. And there is a shift from ATP production to lipogenesis in cancer cells to produce biomass necessary for cell division.

Apart from glucose metabolism, amino acid metabolism is altered in cancers too. The rapid proliferation of tumor cells depends on increased energy consumption and needs high flux of amino acids. When short of glucose, some amino acids can be utilized as energy source, such as glutamine, which is a energy source for cells of rapid division and whose end product alanine can be considered as a potential tumor marker. Besides glutamine and alanine, a number of other amino acids have been studied in various cancers, such as glycine and taurine to name a few.

Theoretically, cancer metabolome research can guide cancer diagnostics and lead to novel chemotherapies, and detailed metabolomic descriptions of cancer can help optimal management of patients. However, in actual, few therapies based on cancer metabolome have been applied in the clinic to date. Therefore, cancer metabolome is still a focus area in cancer research that holds great promise.

3. TECHNIQUES APPLIED IN METABOLOMIC PROFILING OF NEOPLASTIC LESIONS IN MICE

Generally, there are two main categories of metabolomic analysis methods, namely, targeted and nontargeted. The former refers to quantitate a group of previously known metabolites that either have similar molecular structures or belong to a specific metabolic pathway, such as glycolysis cycle or tricarboxylic acid cycle. It relies on a set of preestablished methodologies. While nontargeted analysis is global in scope and has the aim of simultaneously measuring as many metabolites (both known and unknown) as possible from biological samples without bias (Patti, Yanes, & Siuzdak, 2012). It can explore potential tumor markers and identify keys and hubs in tumorigenesis by comparing data that differentiate significantly between various sample groups such as cancer samples versus paracancerous tissues. Different from targeted methods that might center on single pathway with a few analytes, untargeted methods might involve thousands of metabolites in multiple pathways and biological processes.

Metabolomic analysis comprises three steps: tailored sample pretreatment, metabolites extraction, and proper analysis methods. Sample pretreatment and metabolites extraction prior to analysis are sharply different between targeted and nontargeted metabolomic analysis methods. In nontargeted metabolomic research, in order to reduce metabolites loss and maintain metabolome fidelity, sample pretreatment processes are preferred to be as few as possible and organic solvents such as acetonitrile and methanol are usually used to extract metabolites to the largest extent. On the contrary, in targeted metabolomic researches, in order to extract and enrich the targeted metabolites effectively, extensive pretreatment processes are employed, such as protein precipitation and liquid–liquid extraction.

3.1. MS-based techniques

MS-based techniques, such as gas chromatography-MS (GC-MS), capillary electrophoresis-MS (CE-MS), and liquid chromatography-MS (LC-MS), represent the most commonly used techniques in metabolomic study as yet. To analyze samples, MS measures the mass-to-charge ratio of charged particles, which classically consists of three steps: (1) sample molecules are transformed by an ion source from solid/liquid phase into gas phase, (2) individual mass-to-charge ratios of the sample are separated by an analyzer, and (3) the number of ions at each mass-to-charge ratios is recorded by a detector.

MS-based techniques can be used in analyzing a variety of samples, such as body fluid (plasma, serum, urine, etc.), feces, and tissues (such as liver and kidney), and are widely used in the metabolomic research of neoplastic lesions in various mice models of tumor.

3.1.1 Gas chromatography-MS

GC-MS, which is initially applied in botanic studies, is now more and more frequently used in metabolomic studies in tumors, thanks to the rapid progress in analytic techniques.

In GC-MS, GC was applied to separate metabolites after which MS coupled with online database searching was applied to identify the metabolites separated by GC. GC has several advantages, such as high separation efficiency, high sensitivity, high selectivity, easy to use, and low expense. However, as its name infers, gas chromatography is suitable for separation of volatile compounds, which means that in separation of nonvolatile or low-volatile compound, preprocessing of derivatization is needed to enhance volatility. As a result, traditional GC-MS is incapable of analyzing macromolecules as well as nonvolatile compounds which are thermolabile. Fortunately, this shortcoming has been partially overcome in recent years by the rapid progress of techniques. For instance, two-dimensional gas chromatography–time of flight mass spectrometry (GC × GC-TOF-MS), which has higher sensitivity, broader applicability, and fewer preprocessing procedures compared with traditional GC-MS, has been applied in the area of metabolomic analysis of neoplastic lesions. In targeted metabolomic research, if the metabolites of interest are fatty acids, then GC-MS is a great choice.

In one of our previous studies, to investigate the metabolomic profiling of gastric cancer, male mice with severe combined immune deficiency were used as animal model. The model was made by orthotopic implantation of histologically intact tissue of human gastric carcinoma. This process is composed of several steps. First, human gastric cancer cell line SGC-7901 was maintained by passage in subcutaneous tissue of nude mice. Second, tumors were removed aseptically from nude mice and the necrotic tissues were cut off. The adjacent healthy tissues were also harvested and cut into pieces as controls. And finally, after anesthetization, orthotopic implantation of tumors or healthy tissues was conducted. After a certain period, the animals were sacrificed and samples were collected. After pretreatment, the samples were subjected to GC-MS analysis, which successfully identified dozens of differentially expressed metabolites in mice model with orthotopic implantation of gastric tumor tissues (Chen et al., 2010). These metabolites deserve further investigation for their potentiality as tumor markers used for early diagnosis of gastric cancer.

3.1.2 Capillary electrophoresis-MS

The migration of charged particles under the force of an electric field had been discovered more than one century before. Since then, efforts had been made to separate charged particles under the force of a voltage, and accordingly, the term *electrophoresis* was created to describe the phenomenon. However, human's endeavor in applying CE was full of frustrations because of the contradictory between electrophoretic effect and heat effect. Simply speaking, when the voltage increases, the electric current flow in the buffer was also amplified, generating more heat and degrading separation efficacy correspondingly. To address this problem, electrophoresis with a tube as a separation channel is invented and received much attention and interest, which eventually evolved into capillary electrophoresis and is now frequently combined with MS as an indispensable branch of modern separation science.

The fundamental principle of CE-MS is that it separates and analyzes different components according to their different migrating velocities under electrical field. CE-MS is capable of analyzing various samples such as blood serum, blood plasma, urine, cerebrospinal fluid, and tissue samples.

In the study of neoplastic lesions, CE-MS has following merits: separating compounds rapidly with high-resolution power, requiring limited sample pretreatment, tiny amount of samples, and small volume of separation electrolytes. These merits render CE-MS a very promising technique for high-throughput metabolomics. Moreover, in targeted research, CE/MS could be excellent tool for the amino acids. However, its concentration sensitivity is still unsatisfactory and therefore needs to be improved. In this respect, moving reaction boundary (MRB) showed the potential of online sample enrichment (Chen, Fan, & Lu, 2014). MRB was created between the background electrolyte (with H^+) at the anodic side and the alkaline sample at the cathodic side of the capillary. The created MRB can be used for stacking analytes in CE, which is capable of condensating analytes in highly saline matrix (Zhu et al., 2009).

3.1.3 Liquid chromatography-MS

LC-MS refers to a technique which efficiently combines liquid chromatography (for separation) and MS (for analysis). Traditional liquid chromatography is a method characteristic of low column efficacy and time-consuming. During the late 1960s, methods and principles of gas chromatography were introduced into liquid chromatography methodology and as a result, high-performance liquid chromatography (HPLC), also called high pressure liquid chromatography, came into view. This represents the first conversion in LC. The second conversion is that in recent years, a shift from HPLC to Ultra-performance liquid chromatography (UPLC) has taken place. Compared with

HPLC, UPLC can increase resolution sensitivity and peak capacity significantly, and, additionally, UPLC needs smaller sample volumes.

Because of its advantages such as less time-consuming and high sensitivity, LC-MS has been more and more frequently used in the study of neoplastic metabolomics. In a recent study by Loftus et al. (2012), human colorectal cancer cell lines were implanted into subcutaneous tissue of athymic nude male mice. After a certain period, the animals were sacrificed and samples were collected, which were subjected to reversed-phase gradient LC-MS to carry out metabolomic analysis. Subsequently, data gained by MS were identified by online database search. The result of this study on mice model is consistent with nuclear magnetic resonance (NMR)-based analysis of human colorectal tissue, and this model in mice holds the promise to be a practical and valuable model in studying tumor behavior and microenvironment. However, LC-MS also has disadvantages, such as high cost and difficulty in operating.

3.1.4 Matrix-assisted laser/desorption ionization MS

Fifteen years before, matrix-assisted laser/desorption ionization MS (MALDI-MS) was first reported to be applied for molecular imaging of protein distribution in tissue samples (Caprioli, Farmer, & Gile, 1997). The emergency of MALDI-MS is a milestone in tumor metabolomic research because that as a technique of imaging mass spectrometry (IMS), MALDI-MS provides nontargeted and label-free molecular mapping of tumor tissue samples with no loss of histological information. In MALDI, a solid or liquid matrix containing a highly UV-absorbing substance is added to a metal plate and sample molecules are spotted on that plate, subsequently, a laser desorbs and ionizes the sample molecules. The general workflow and principles of MALDI mass spectrometry imaging are complex and out of the range of this chapter; readers who are interested in this can see relevant reference (Neubert & Walch, 2013).

MALDI-MS is a useful tool in various tumor metabolomic researches such as prediction of therapy response and prognosis, molecular tissue classification, and intratumor heterogeneity analysis. Balluff and his coworkers applied histology-based MALDI-IMS in a total of 181 intestinal-type primary resected gastric cancer tissues from two independent patient cohorts and identified seven proteins as potential novel biomarkers in intestinal-type gastric cancer that was associated with an unfavorable overall survival independent of major clinical covariates. Of the seven proteins, three (CRIP1, HNP-1, and S100-A6) were validated immunohistochemically on tissue microarrays as prognostic significance (Balluff et al., 2011). Reyzer has reported the application of MALDI-MS in direct analysis of protein profiles in virus/HER2 transgenic mouse mammary tumor samples after treatment with the erbB receptor inhibitors OSI-774 and Herceptin, and the result showed that MALDI-MS can be used to predict therapeutic response and to test novel therapies at least in this mice tumor model (Reyzer et al., 2004). In a mouse fibrosarcoma model following treatment with the tubulin-binding tumor vascular disrupting agent, combretastatin A-4-phosphate (CA-4-P), Cole and his team applied MALDI-MS to observe the spatial distribution of peptides and to elucidate any pharmacological responses and potential biomarkers. In this study, the researchers reported a clear increase of peptide signals associated with hemoglobin after treatment with CA-4-P on fibrosarcoma mouse models (Cole et al., 2011).

3.1.5 Other MS-based techniques

Apart from techniques been discussed above, there are other MS-based techniques used in metabolomic research in mice model of tumor, such as inductively coupled plasma mass spectrometry (ICP-MS) and Fourier-transform mass spectrometry (FT-MS).

3.1.5.1 Inductively coupled plasma mass spectrometry

Dr. Alan Gray of Applied Research Laboratories in Luton, UK, conducted much of the early research that ultimately led to the commercial development of ICP-MS. In the early 1980s, ICP-MS began to be commercially available. Since then, many developments have been made in areas such as sample introduction, ion transmission, interference removal, and dynamic range.

ICP is a high-temperature ion source whose temperatures approximate 5500 °C that no material withstands (Houk & Praphairaksit, 2001). So, ICP is an effective versatile atomizer and element ionizer. This high temperature results in fragmentation of sample molecules into atomic constituents that serve as detectable surrogates to molecules such as proteins, carbohydrates, and nucleic acids. Additionally, most elements in the periodic table can be accurately detected by ICP-MS, making possible its application in the quantification of almost all kinds of molecules.

Applications of ICP-MS in tumor study are diverse, such as the identification of trace metabolites in urine, blood, and serum. ICP-MS provides quick multiple metabolite analysis even if the metabolites are rather low in concentrations. Apart from monitoring trace metabolites in body fluids and tissue samples, ICP-MS is also capable of isotope ratio measurement in which isotopes are used as label to mark metabolites. And when the labeled metabolites travel through the body, they can be detected and analyzed by ICP-MS. The strengths of ICP-MS include wide dynamic linear range, high sensitivity, high resolution, and rapid analysis process. However, its requirements for reagents and environment are relatively high. ICP-MS techniques grow quickly. Over the next few years, ICP-MS will continue to grow as demands for more sensitive measurement with higher productivity continue to increase. Up to now, there have been several interesting researches of oncologic metabolomic profiling in mice models (Crayton, Elias, Al Zaki, Cheng, & Tsourkas, 2012; Kamaly et al., 2010).

3.1.5.2 Fourier-transform mass spectrometry

FT-MS is a technique that combines the ion trap and Fourier-transform ion cyclotron resonance (FTICR) into a single instrument. The strength of FT-MS is its sensitivity and productivity, except for its high cost. In detail, it provides excellent accuracy with sub-ppm errors, along with ultra-high mass resolving power. What is more, it is capable of identifying analysts' structures, allowing access to ion elemental formulas. With FT-MS, there are two kinds of metabolic profiling procedures: one is direct introduction into the mass spectrometer and the other is relying on LC/CE prior to MS detection. Most of the FT/MS-based metabolomic processes relying on direct introduction of samples have been performed with FTICR (Junot, Madalinski, Tabet, & Ezan, 2010).

Koulman and his colleagues reported that using an extractive instrument called Exactive orbitrap mass spectrometer (Thermo Scientific, Hemel Hempstead, UK) operated at 50,000 resolution power, FT/MS is capable of performing both targeted and untargeted analyses within a single LC/MS acquisition without compromising analytical quality (Koulman et al., 2009). Raw materials were processed by a software called MZ mine to automatically feature the detected metabolites. Moreover, the linearity of metabolite concentration-response curves gained by FT mass spectrometers is satisfactory and sufficient for subsequent quantification.

One challenge FT/MS faces now is how to effectively process thousands of ions and to convert them into metabolite names. A Web server called MassTRIX may offer some help in this respect. Researchers just need to input a peak list gained from mass spectrometer and then the raw masses will be matched against data in the KEGG database.

3.2. Nuclear magnetic resonance

NMR is an analytic method in which strong magnetic fields and radio frequency pulses are applied to the nucleus of atoms, which evokes them to be excited and to arrive at a relatively high energy level. Under a certain energy quanta, the nucleus will resonate, that is, switch between low energy state and high energy state. And, thereafter, radiation is emitted from the atoms and detected by equipments. NMR is a nondestructive analytical method that exploits the magnetic properties of certain nuclei to determine the structure of molecules.

As a major analysis technique, NMR has been applied in many fields of metabolomics, such as identifying genetic differences, monitoring physiological effects, assessing drug safety, and diagnosing disease. In tumor study in mice model, NMR can be applied in *in vivo* analysis as well as body fluid analysis. Compared with MS, NMR is a noninvasive, high-throughput method that is less dependent on operators' experiences and prudence. NMR is capable of analyzing the dynamics of metabolites in living systems, and the results obtained by NMR are more repeatable and objective compared with MS method. Additionally, there is no need of complex sample preprocessing, and the analyzing cost per sample is relatively low. The shortcomings of NMR include its sensitivity, resolution, sample volumes required, and high expenses of equipment cost and maintenance. In NMR, sensitivity depends on the atoms studied (^{1}H, ^{31}P, ^{15}N, ^{13}C, etc.) and/or the introduced isotopes. As a result, NMR has a good sensitivity and specificity only if samples contain these atoms. In addition, the probes used in NMR are large in volume and correspondingly the sample volumes subject to analysis need to be large. Moroz analyzed the metabolic content of urine from NIH III nude mice via NMR before and after inoculation with human glioblastoma multiforme (GBM) cancer cells, and the results demonstrate that metabolomics may be used as a screening tool for GBM cells grown in xenograft models in mice (Moroz, Turner, Slupsky, Fallone, & Syme, 2011).

In order to improve sensitivity, several new techniques have been developed, such as high-resolution NMR, multidimensional NMR, liquid chromatography-NMR, and magic-angle spinning (MAS) NMR. The latter developed in the 1990s is capable of enhancing specificity by eliminating line broadening in analyzing biosamples. In metabolomic research, high-resolution MAS (HR-MAS) magnetic resonance spectroscopy (MRS) can analyze intact tissue qualitatively and quantitatively with minimal sample preparation. MAS-NMR has been successfully introduced to the metabolomic studies in mice model to analyze samples such as mice liver, kidney, and testicles. Because of its high degree of reproducibility and noninvasive nature, specimens and samples analyzed by HR-MAS-MRS can be subsequently evaluated by histopathology, gene microarray, or next-generation sequencing, which facilitates direct comparisons among spectral, morphological, and functional characteristics of the samples to a great extent. In other words, HR-MAS analysis makes possible the combination of metabolomics with genomics or proteomics, which is of great help in studying cancer biology and in clinical aspects of oncology, such as cancer diagnosis, prognosis, and treatment (Griffin & Shockcor, 2004). As for different type of metabolites, HR-MAS-MRS is excellent for glucose and amino acids analysis, but only mobile lipids (refers to lipids that are not membrane-associated) are detectable in MR.

With the aid of HR-MAS-NMR spectroscopy, Backshall and his colleagues utilized metabolic profiling to characterize small bowel and colon tissue of early gastrointestinal tumorigenesis in the ApcMin/+ mouse model, and their result showed that there is a metabolic phenotype that associates with so-called field cancerization, indicating potential biomarkers for disease progression monitoring, chemoprevention evaluation, and outcome and prognosis prediction in this model (Backshall et al., 2009).

4. CONCLUSION

Metabolomics is the systematic study of small-molecular-weight substances that are the final products of genes. It is a valuable tool in tumor research in mice models, which can help to illustrate the underlying mechanism of tumor initiation and progression, and, what is more, can help to find practical tumor markers for tumor diagnosis and surveillance. The two widely used metabolomic methods are MS-based methods and NMR, each having unique merits and drawbacks and should be chosen according to research goal, equipment availability, and personal skills. With the rapid progress in techniques and instruments, it is reasonable to foresee that in the near future, metabolomic profiling of neoplastic in mice will play a more and more important role in oncological research.

REFERENCES

Backshall, A., Alferez, D., Teichert, F., Wilson, I. D., Wilkinson, R. W., Goodlad, R. A., et al. (2009). Detection of metabolic alterations in non-tumor gastrointestinal tissue of the ApcMin/+ mouse by ^{1}H MAS NMR spectroscopy. *Journal of Proteome Research*, *8*, 1423–1430.

Balluff, B., Rauser, S., Meding, S., Elsner, M., Schöne, C., Feuchtinger, A., et al. (2011). MALDI imaging identifies prognostic seven-protein signature of novel tissue markers in intestinal-type gastric cancer. *American Journal of Pathology, 179*(6), 2720–2729.

Caprioli, R. M., Farmer, T. B., & Gile, J. (1997). Molecular imaging of biological samples: Localization of peptides and proteins using MALDI-TOF MS. *Analytical Chemistry, 69*(23), 4751–4760.

Chen, J.-L., Fan, J., & Lu, X.-j. (2014). CE-MS based on moving reaction boundary method for urinary metabolomic analysis of gastric cancer patients. *Electrophoresis, 35*(7), 1032–1039.

Chen, J.-L., Tang, H.-Q., Hu, J.-D., Fan, J., Hong, J., & Gu, J.-Z. (2010). Metabolomics of gastric cancer metastasis detected by gas chromatography and mass spectrometry. *World Journal of Gastroenterology, 16*(46), 5874–5880.

Cole, L. M., Djidja, M. C., Bluff, J., Claude, E., Carolan, V. A., Paley, M., et al. (2011). Investigation of protein induction in tumour vascular targeted strategies by MALDI MSI. *Methods, 54*(4), 442–453.

Crayton, S. H., Elias, D. R., Al Zaki, A., Cheng, Z., & Tsourkas, A. (2012). ICP-MS analysis of lanthanide-doped nanoparticles as a non-radiative, multiplex approach to quantify biodistribution and blood clearance. *Biomaterials, 33*(5), 1509–1519.

Griffin, J. L., & Shockcor, J. P. (2004). Metabolic profiles of cancer cells. *Nature Reviews. Cancer, 4*(7), 551–561.

Houk, R. S., & Praphairaksit, N. (2001). Dissociation of polyatomic ions in the inductively coupled plasma. *Spectrochimica Acta, Part B: Atomic Spectroscopy, 56*, 1069.

Junot, C., Madalinski, G., Tabet, J.-C., & Ezan, E. (2010). Fourier transform mass spectrometry for metabolome analysis. *Analyst, 135*, 2203–2219.

Kamaly, N., Pugh, J. A., Kalber, T. L., Bunch, J., Miller, A. D., McLeod, C. W., et al. (2010). Imaging of gadolinium spatial distribution in tumor tissue by laser ablation inductively coupled plasma mass spectrometry. *Molecular Imaging and Biology, 12*(4), 361–366.

Koulman, A., Woffendin, G., Narayana, V. K., Welchman, H., Crone, C., & Volmer, D. A. (2009). High-resolution extracted ion chromatography, a new tool for metabolomics and lipidomics using a second-generation orbitrap mass spectrometer. *Rapid Communications in Mass Spectrometry, 23*(10), 1411–1418.

Kwon, M.-C., & Berns, A. (2013). Mouse models for lung cancer. *Molecular Oncology, 7*, 156–177.

Loftus, N. J., Lai, L., Wilkinson, R. W., Odedra, R., Wilson, I. D., & Barnes, A. J. (2012). Global metabolite profiling of human colorectal cancer xenografts in mice using HPLC-MS/MS. *Journal of Proteome Research, 12*(6), 2980–2986.

Moroz, J., Turner, J., Slupsky, C., Fallone, G., & Syme, A. (2011). Tumour xenograft detection through quantitative analysis of the metabolic profile of urine in mice. *Physics in Medicine and Biology, 56*, 535–556.

Neubert, P., & Walch, A. (2013). Current frontiers in clinical research application of MALDI imaging mass spectrometry. *Proteomics, 10*(3), 259–273.

Patti, G. J., Yanes, O., & Siuzdak, G. (2012). Metabolomics: The apogee of the omics trilogy. *Nature Reviews. Molecular Cell Biology, 13*, 263–269.

Reyzer, M. L., Caldwell, R. L., Dugger, T. C., Forbes, J. T., Ritter, C. A., Guix, M., et al. (2004). Early changes in protein expression detected by mass spectrometry predict tumor response to molecular therapeutics. *Cancer Research, 64*(24), 9093–9100.

Zhu, W., Zhang, W., Fan, L.-Y., Shao, J., Li, S., Chen, J.-L., et al. (2009). Study on mechanism of stacking of zwitterion in highly saline biologic sample by transient moving reaction boundary created by formic buffer and conjugate base in capillary electrophoresis. *Talanta, 78*, 1194–1200.

CHAPTER FOURTEEN

Metabolomic Profiling of Tumor-Bearing Mice

Hiromi I. Wettersten*, Sheila Ganti†, Robert H. Weiss*,‡,1

*Division of Nephrology, Department of Internal Medicine, University of California, Davis, California, USA
†Department of Comparative Medicine, University of Washington, Seattle, Washington, USA
‡Medical Service, Sacramento VA Medical Center, Sacramento, California, USA
[1]Corresponding author: e-mail address: rhweiss@ucdavis.edu

Contents

Abstract

Metabolomics is one of the newcomers among the “omics” techniques, perhaps also constituting the most relevant for the study of pathophysiological conditions. Metabolomics may indeed yield not only disease-specific biomarkers but also profound insights into the etiology and progression of a variety of human disorders. Various metabolomic approaches are currently available to study oncogenesis and tumor progression *in vivo*, in murine tumor models. Many of these models rely on the xenograft of human cancer cells into immunocompromised mice. Understanding how the metabolism of these cells evolves *in vivo* is critical to evaluate the actual pertinence of xenograft models to human pathology. Here, we discuss various tumor xenograft models and methods for their metabolomic profiling to provide a short guide to investigators interested in this field of research.

Methods in Enzymology, Volume 543
ISSN 0076-6879
http://dx.doi.org/10.1016/B978-0-12-801329-8.00014-3

1. INTRODUCTION

Metabolomics, even as a relative newcomer to the omics disciplines, has already proved itself to be one of the "established omes" (Baker, 2013). This technique has now been utilized in tissue and most biofluids for both biomarker discovery and to dissect out the physiology and pathophysiology of human disease and thereby identify potential therapeutic targets. The superior clinical utility of metabolomics compared to the other established omics lies in the closer proximity of the metabolome to the ultimate phenotype as well as the smaller number of metabolites to be analyzed compared to, for example, proteins and genes (Fig. 14.1).

Despite its abundant theoretical potential, the actual success of metabolomics-directed biomarker discovery has yet to be realized in the clinic, and this may be the result of the nontargeted approach routinely used in existing published studies for metabolite identification in biofluids. A major disadvantage of metabolomics analysis utilizing samples from human subjects is the risk of potential confounders due to the obvious heterogeneity of the (human) subjects evaluated. This disadvantage can be minimized by utilizing syngeneic mouse models, because such models that have

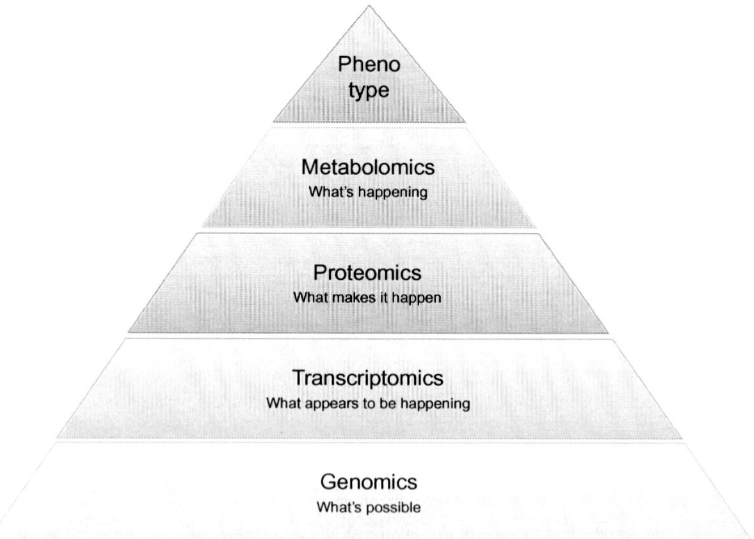

Figure 14.1 *Metabolites are most closely related to phenotype.* The various established omics which lead to metabolomics, which is most closely related to phenotype. Note that there exist substantially fewer metabolites than genes, proteins, and transcripts, a fact which considerably simplifies data acquisition and analysis.

far fewer genetic differences between nontumor bearing and xenograft groups. Furthermore, such animal models allow researchers to trace altered metabolites through all "matrices" (urine, blood, and tissue), an impractical process in human subjects, thereby decreasing the time spent validating some of the spurious metabolites that may end up showing little promise. In this review, we discuss the key issues relating to the implementation of a mouse metabolomics study, and we provide recommendations about how to interpret the resulting data.

2. MOUSE MODELS

For many years, the mouse has been the favorite animal model in oncology due to the convenience of using a small, rapidly breeding animal which bears relatively close resemblance to humans (Sausville & Burger, 2006). The first step of metabolomic profiling in a tumor-bearing mouse is to actually develop the model. In this section, we summarize the various mouse types available for xenografting, as well as the xenograft models already available for primary and metastatic tumor analyses.

Of the many animal models available to study human cancer, one of the most widely used is the human tumor xenograft, which has distinct advantages over other models due to its higher disease fidelity in the use of actual human tissue. In this model, human tumor cells or tissues are subcutaneously grafted into organs (including skin) or intravenously into immune-deficient mice, allowing researchers to test the response of human tumor cells to anticancer agents. In this section, we discuss (1) those mouse strains currently available to establish tumor xenografts; (2) various xenograft models which have already been developed to study different stages of cancer development; and (3) the most recently established mouse model, the humanized tumorgraft, which attempts to overcome many of the disadvantages of earlier models.

2.1. Mouse strains

The establishment of immunocompromised mouse strains has made it possible for researchers to implant human tumor cells into mice without the risk of cell-mediated immunological tissue rejection. However, for the same reason such models are not useful for studying immune-mediated responses to malignancy. Here, we compare the major commonly used immunecompromised mouse strains.

2.1.1 nu/nu (Nude mouse)

Nude mice were the first immunecompromised mouse strain to be generated and have been used for various cancer metabolomics studies including colorectal and kidney cancer (Ganti, Taylor, Kim, et al., 2012; Loftus et al., 2013). These animals show abnormal hair growth and defective development of the thymic epithelium, thus homozygous nude mice lack T cells (Pantelouris, 1973). The total number of circulating lymphocytes is 5 to $6\times$ less in nude mice than in normal animals, and majority of these cells are B cells (Pantelouris, 1968). With the development of nude mice, thymectomy became dispensable for xenografting.

2.1.2 SCID

While nude mice lack only T cells, mice homozygous for the severe combined immune deficiency (SCID) mutation lack both T and B cells and are thus characterized by an absence of both innate and adaptive immune response (Bosma, Custer, & Bosma, 1983). Most homozygous mice also have undetectable IgM, IgG, or IgA (Bosma et al., 1983). While the defects in innate and adaptive immunity in SCID mice provide an ideal *in vivo* environment not only for xenografting, but also for reconstitution with human hematopoietic cells, the disadvantage of the SCID mouse is that the lifespan of these mice is approximately 8.5 months due to their high incidence of thymic lymphomas (Shultz et al., 1995). Recently, Fan et al. performed a metabolomics study in lung cancer utilizing stable isotope with this mouse model (Fan, Lane, Higashi, & Yan, 2011).

2.1.3 NSG *(Shultz et al., 2005)*

The nonobese diabetic (NOD)-scid gamma (NSG) mice have a background of SCID mice and IL-2 receptor gamma chain deficiency. These mice are severely immunodeficient due to the lack of mature B, T, and NK cells; in addition, they lack cytokine signaling such as IL-2, IL-4, IL-7, IL-9, and IL-15. NSG mice live longer than some other immune-deficient mice because unlike SCID mice, they do not develop thymic lymphomas possibly due to the lack of cytokine signaling which is essential for lymphoma development. This model has been used for a metabolomics study for human immunodeficiency virus (Epstein et al., 2013).

2.1.4 Recombination-activating gene (Rag) *(Mombaerts et al., 1992; Shinkai et al., 1992)*

Another mouse model which lacks mature T and B cells (similar to SCID mice) possesses RAG-1 or -2 mutations; these proteins are necessary for V(D)J recombination, a process required for normal B and T cell development. Loss of RAG-1 or -2 causes arrested development of these cells at an early stage in mice homozygous for these mutations. The thymus of these homozygous mice contains significantly fewer cells than heterozygous or wild-type mice and is enriched with immature $CD4^{-}CD8^{-}$ double negative thymocytes expressing the IL-2 receptor.

2.1.5 NRG (Pearson et al., 2008)

NRG mice, which are NOD mice with a RAG-1 as well as an IL-2 receptor common gamma chain mutation, were developed principally for human lymphohematopoietic engraftment. While NSG mice are highly sensitive to irradiation, NRG mice better tolerate irradiation allowing higher levels of human cord blood stem cell engraftment following irradiation conditioning than NSG mice. Thus, NRG mice can prove useful for cell or tissue transplantation studies, especially when radioresistance is required.

2.2. Xenograft models (Table 14.1)

There are several anatomic locations at which tumor cells or tissues can be injected in order to develop xenografts in mice. Each model has advantages and disadvantages as described below.

2.2.1 Subcutaneous models

Subcutaneous xenograft models are often used to determine responsiveness to anticancer treatments by measuring changes in primary tumor volume after treatment. In these models, small tumor tissues or tumor cells are injected subcutaneously by a minor procedure (Morton & Houghton,

Table 14.1 Advantages and disadvantages of tumor xenograft models

	Advantages	Disadvantages
Subcutaneous model	• Minor surgery • Easy to measure tumor volume by caliper	• Artificial microenvironment
Orthotopic model	• Natural microenvironment • Natural metastasis route • Able to assess entire steps for metastasis	• Major surgery • Difficult to measure tumor volume • Relatively long time required for metastasis development
Experimental metastasis model	• Relatively short time required for metastasis development • Controlled number of cells injected for metastasis • Able to specify metastasis sites	• Artificial metastasis route • Evaluation limited for extravasation to colonization

2007). This method of transplantation produces a tumor whose volume is simple to measure without sacrificing the animals with the use of calipers. However, it can be argued that external calipers are relatively inaccurate and encumbered with a significant and size-dependent bias, and that microCT is a more accurate method of tumor measurement (Jensen, Jorgensen, Binderup, & Kjaer, 2008). Thus, while subcutaneous models make the process of tumor injection and volume measurement simple and minimally invasive, such models (that do not represent appropriate sites for human tumors) have not been shown to be as predictive as orthotopic models (i.e., site-specific, *vide infra*) when used to test responses to anticancer drugs (Killion, Radinsky, & Fidler, 1998).

2.2.2 Orthotopic models (spontaneous metastasis models)

In orthotopic models, tumor tissues or cells are injected into the geographical sites of their originating tissues. The principal advantage of this technique over the subcutaneous model is that it more accurately reproduces the tumor microenvironment that allows the emergence of subpopulations of tumor cells (Killion et al., 1998). Indeed, orthotopic models allow for observations of all the metastatic processes, including primary tumor formation, localized invasion, intravasation, extravasation, formation of micrometastasis, and colonization (formation of macrometastasis) (Fig. 14.2). Another advantage of orthotopic mouse models over subcutaneous models is that the former allows researchers to collect both the primary tumor mass arising from original organ as well as metastatic foci; each can be used for separate further analyses including metabolomic profiling to glean data on metastatic mechanisms. The major disadvantage of orthotropic models compared to the subcutaneous models is that they require the animal to undergo major surgery to access the tumor tissue. Additionally, the orthotopic models require imaging including positron emission tomography to measure tumor volume if this information is needed prior to sacrifice (Ray, Tsien, & Gambhir, 2007).

2.2.3 Experimental metastasis models

Experimental metastasis models assess the ability of tumor cells to arrest, extravasate, and grow in a particular organ following intravenous injection (Fig. 14.2). In these models, tumor cells are either injected directly into the venous circulation or into highly vascular organs such as the spleen. The site of injection may vary location of distant metastases throughout the body (reviewed in Khanna & Hunter, 2005). For instance, when the cells are

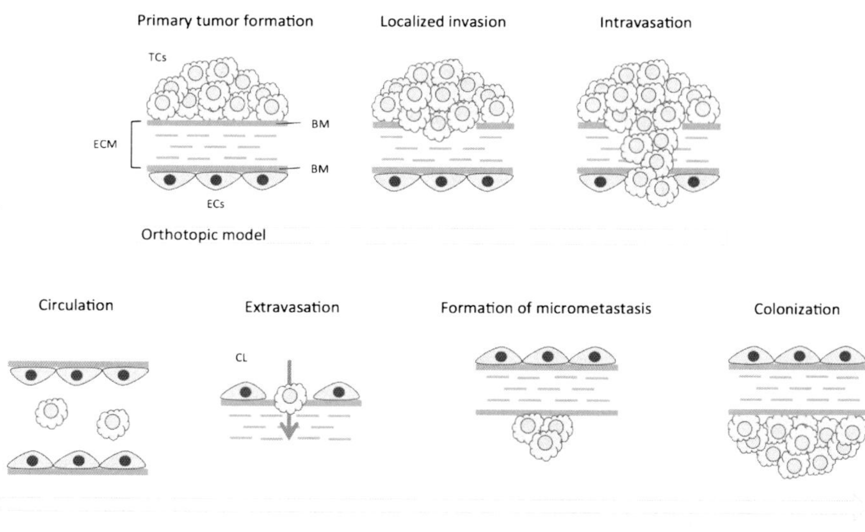

Figure 14.2 *Orthotopic models recapitulate the entire process of metastasis, while other experimental metastasis models cover only post intravasation.* Metastatic progression consists of primary tumor formation, localized invasion, intravasation, circulation, extravasation, formation of micrometastasis, and colonization. The tumor cells injected orthotopically go through all the steps above, while the tumor cells go through only post intravasation steps in experimental metastasis models. Tumor cell, TC; basement membrane, BM; extracellular matrix, ECM; EC, endothelial cell; CL, capillary lumen. (See the color plate.)

injected into the lateral tail vein, the cells generally metastasize to the lung, while intrasplenic or portal vein injections yield liver metastases.

The advantages of experimental metastasis models are mainly disadvantages of spontaneous metastasis models and vice versa. In experimental metastasis models, it is relatively easy to control the number of cells delivered and to target the site of metastasis. However, the evaluation of circulation, extravasation, formation of micrometastases, and colonization in this model is limited.

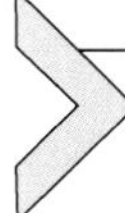

3. SAMPLE PREPARATION AND MASS SPECTROMETRY ANALYSIS (FIG. 14.3)

3.1. Sample collection

After the development of xenografts or allografts in the animal of choice, metabolomics analysis is begun with the collection of samples for MS analysis. Such samples can be any kind of biofluid or tissue, which is a function of

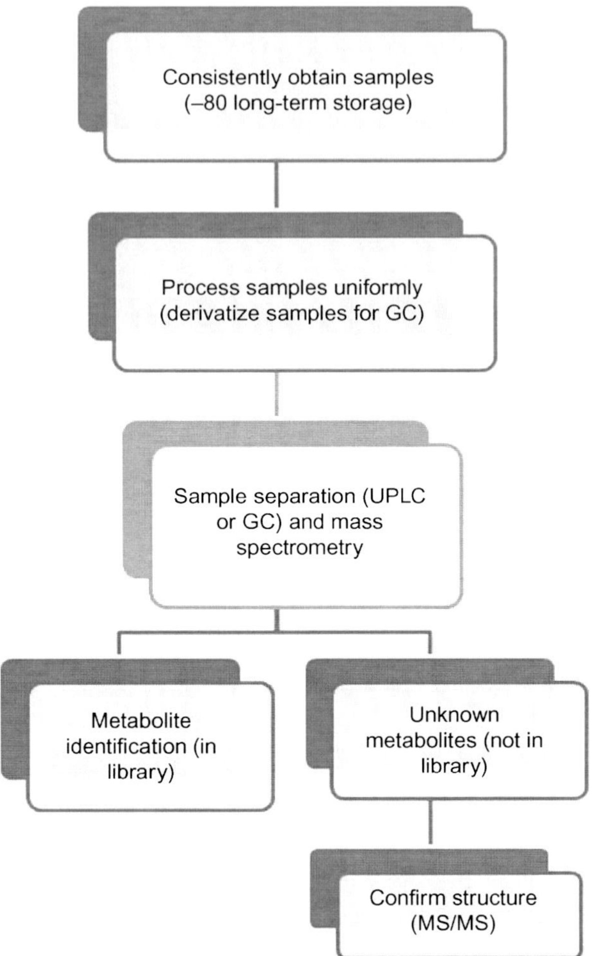

Figure 14.3 *Typical workflow for metabolomics analysis.* In a typical mouse metabolomics experiment, serum, tissue, and/or urine are obtained using consistent collection procedures, are uniformly processed, and then metabolites are separated either by LC or GC. Following mass spectrometry, metabolites are identified based on comparison to a known library. "Unknown" metabolites can be identified by further high-resolution MS/MS or NMR analysis. (See the color plate.)

the specific objectives of the study. The method and consistency of collection and storage are of paramount importance, as handling of samples in systematically different ways can introduce an apparent abundant signal into one batch but not the other, which results in biased outcome (Ang et al., 2012; Kim et al., 2009). However, day-to-day, mealtime and biological variability of the metabolome is relatively small compared to such technical variability, supporting the utility and reproducibility of metabolomics analysis (Kim et al., 2014).In this section, we discuss sample collection methods have been shown to lead to reproducibility of analysis and hence to metabolite stability.

3.1.1 Biofluids

Due to its easy and noninvasive procurement, urine is a common utilized biofluid in metabolomics analyses. In the case of mice, when the study size is small, forced urination by the animals using compression of the suprapubic region is a typical method for collecting fresh samples in mice which is generally representative of clean-catch human urine samples. While there may be temporary stress caused by the procedure, the effect on the urinary metabolome should be minimal since the urine has already been excreted into the bladder by the time of collection. A single mouse will yield between 10 and 250 μl in a single deposit without any intervention or up to 2 ml when collected over a 24-h period (Kurien, Everds, & Scofield, 2004). When collecting large urine volumes, the urine can be pooled by housing animals in metabolic cages for several hours. One study showed that pooling time and storage at ambient temperatures affects stability of metabolites; this study suggested keeping the urine storage tubes in metabolic cages under iced conditions (Bando et al., 2010).

Blood (plasma or serum) is also commonly sampled for metabolomics analyses. The same study which evaluated urine sampling found that sampling site (i.e., jugular vein and abdominal aorta) and anticoagulant (i.e., EDTA and heparin) negatively affect the stability of metabolites; thus, it is recommended to use the same sites for blood collection among the animals and EDTA as an anticoagulant (Bando et al., 2010). Storing samples at −80 °C is recommended for maximum stability of metabolites for most biofluids (Kim et al., 2009).

3.1.2 Tissue

Tissue sampling for metabolomics is complicated by the fact that an organ is an amalgamation of multiple and diverse cell types and structures. This is perhaps most readily illustrated by the kidney, which contains both cortex and medulla with each region containing its own specialized cell types. Furthermore, cancer tissues (including kidney cancer) have been well known to possess heterogeneity (Gerlinger et al., 2012; Ljungberg, Mehle, Stenling, & Roos, 1996). Thus obtaining multiple sections of each tissue is a practical method to cover such heterogeneity. To obtain useful global metabolic profiles, tissues should be collected on dry ice to minimize further metabolism which might occur after collection, and the samples should either be processed immediately upon collection or be snap-frozen on liquid nitrogen and stored at −80 °C in order to avoid metabolite degradation (Want et al., 2013). If the sample is large, freezing and storage of smaller pieces or aliquots is optimal due to more rapid freezing of these samples.

3.2. Platform setup

3.2.1 UPLC for acidic and basic species

Ultra-high-performance liquid chromatography (UPLC) coupled to mass spectrometry (MS) is an excellent separation/detection method to identify amino acids, nucleic acids, lipids, and a range of other larger organic molecules (Ganti & Weiss, 2011). Depending on the mobile phase and column chosen, metabolites can be separated by size, polarity, and/or charge. Combining an acidic or basic mobile phase with a C18 reverse phase column allows for enhanced separation by analyte polarity as well as charge. Very polar metabolites will not be retained on the column and will elute with the void volume (Dettmer, Aronov, & Hammock, 2007). An acidic mobile phase is typically a mixture of 0.1% formic acid/water (solvent A) and 0.1% formic acid/methanol or acetonitrile (solvent B) run as a gradient (Evans, DeHaven, Barrett, Mitchell, & Milgram, 2009; Liu et al., 2013). The basic mobile phase is made up of 6.5 m*M* ammonium bicarbonate/water pH 8.0 (solvent A) and 6.5 m*M* ammonium bicarbonate/95% methanol/5% water (solvent B). This separation scheme can be used to analyze the composition of biological samples such as serum, tissue, and urine. For tissue, the typical separation time is between 15 and 30 min per sample with 2–3 min of reequilibration between each sample (Want et al., 2013). Metabolites are identified based on their retention time, mass to charge (m/z) ratio, and fragmentation pattern, which can vary based on the mass spectrometer used. Therefore, the metabolite libraries used for this identification are not as well developed for UPLC–MS compared to GC–MS (Yoshida et al., 2012).

3.2.2 GC for volatile metabolites

Gas chromatography coupled to mass spectrometry (GC–MS) is frequently used in parallel with UPLC–MS in order to increase the likelihood of covering the full spectrum of metabolites. While UPLC–MS is better suited for a nontargeted approach since it can identify a larger number of metabolites, the metabolite libraries for GC–MS are more complete such that a higher percent of detected metabolites can be identified (Ryan, Robards, Prenzler, & Kendall, 2011). The typical run time for a GC–MS experiment is about 25 min/sample (Dunn et al., 2011). Since GC–MS selects for thermally stable, volatile metabolites, samples need to be derivatized to increase the number of metabolites that fall into this category—a major drawback to using GC–MS (see below for more details) (Evans et al., 2009). Such metabolites include steroids/lipids, oils/fossil fuels, plant metabolites, and intermediates of drug metabolism, (Yoshida et al., 2012). This separation technique involves the use an inert gas (helium, argon, or nitrogen) as a mobile phase. In order to introduce the sample into the mobile phase, it must be heated steadily where it interacts with the stationary phase in the column. Separation is achieved by affinity chromatography with the stationary phase. Like UPLC–MS, metabolites are identified based the retention time, m/z, and fragmentation pattern. Since GC–MS has been in existence longer than LC–MS, the libraries in the former technique are more extensive and are available based on the derivatization method utilized (Yoshida et al., 2012). GC–MS is a powerful approach that offers high sensitivity and lower cost (than LC–MS) but requires additional steps for sample preparation that can introduce external sources of error. If the derivatization process is suboptimal, variation between runs can increase substantially, especially where quantification of metabolite abundance is desired (Theodoridis, Gika, Want, & Wilson, 2012).

3.2.3 MS platform

When samples enter the mass spectrometer (MS) from the column, they must first be ionized which requires an electron source. Electron impact ionization at −70 eV is the most commonly used with GC and electrospray ionization for UPLC (Kind & Fiehn, 2010). Several different MS platforms can be used to obtain m/z ratio and fragmentation pattern independent of which separation technique is utilized.

Time-of-flight (TOF) MS relies on accelerating ions using an electric field through space and measuring the time it takes for an ion to reach a detector. If the ions all have the same charge, then the TOF only depends on the ion's mass. In this way, the m/z can be determined. The MS is run in positive mode to detect positively charged ions (collected with the acidic mobile phase) and in negative mode for negatively charged species (from the basic mobile phase) (Ganti, Taylor, Abu Aboud, et al., 2012). Peak resolution can be improved by coupling TOF-MS to a quadrupole mass spectrometer (qTOFMS). In this setup, four parallel rods create a radio frequency that selects ions with a specific m/z to pass through to the detector. By altering the radio frequency, a large range of ions with varying m/z can be analyzed from a single sample. Alternatively, a linear ion trap quadrupole mass spectrometer can be used. After the sample is ionized, this mass analyzer physically traps ions using a radio frequency in a two-dimensional field allowing only ions with a specific m/z pass through to the detector, similar to a quadrupole (Douglas, Frank, & Mao, 2005).

3.3. Sample preparation for tissue, urine, and blood

3.3.1 Sample extraction

Metabolite concentrations can vary between tissues and between urine samples. Several measurements and manipulations, primarily relating to issues of normalization of metabolite concentrations to biofluid dilutions, must be taken prior to running these samples. Tissue metabolites can be normalized based on tissue weight (Ganti, Taylor, Abu Aboud, et al., 2012). Urine, however, should not be normalized by volume due to the wide range of solute concentration as a function of fluid intake, time of day, etc.; thus, urinary spot creatinine concentration or osmolality for normalization purposes should be determined prior to analyzing the sample. The peak area calculated for all metabolites for a given sample should be divided by that samples weight in the case of tissue, or by creatinine or osmolality in the case of urine. Either creatinine or osmolarity are acceptable methods for normalizing urinary content. However, in diseases which result in compromised kidney function, urinary creatinine should not be used since secretion of this metabolite will be variably impaired (Hellerstein et al., 1988). In those situations where kidney function is not a factor, normalizing by creatinine is the preferred method. Blood (serum or plasma) need not be normalized in this manner as it is relatively invariant with respect to solute concentration (Zuppi et al., 1997).

Independent of the separation method, the first step of sample preparation is to precipitate out proteins to prevent column clogging. This can be done by mixing the sample with an isopropanol:acetonitrile:water mixture (v/v/v) at 3:2:2 or an 8:2 methanol:acetonitrile (v/v) mixture and subsequently centrifuging and collecting the supernatant (Strathmann, Laha, & Hoofnagle, 2011; Taylor et al., 2010). The samples are then evaporated and reconstituted with a solution containing internal standards in an appropriate solvent (the same as the mobile phase for LC (Evans et al., 2009) or chloroform for GC (Taylor et al., 2010)). These standards, also sometimes called retention markers, are metabolites with known structures and well-documented retention times which assist in identification and quantitation of "unknown" peaks. They also may be modified or labeled with stable isotopes (Trupp et al., 2012) to distinguish them from the endogenous metabolites; if quantitation is needed, these isotopic standards can be injected at known concentrations to create a calibration curve for the endogenously occurring metabolite. Retention markers act as a sort of "grid" to account for drifting retention times normally occurring, while running a larger sample set and with column age (Ganti & Weiss, 2011) and offers a quality control step.

3.3.2 Derivatization for GC–MS

One additional step must be undertaken when preparing the GC samples. As relatively few endogenously occurring metabolites are volatile, a derivatization step is generally necessary to increase the number of detectable metabolites. After the sample is evaporated, a methylsilyl-containing compound (either BSTFA or MSTFA) is added to the sample. The highly nonpolar silicon-based functional group replaces the hydrogen in –NH, –OH, –COOH, and –SH bonds, as compounds containing these functional groups are more likely to form intermolecular hydrogen bonds (Halket et al., 2005). Derivatizing the samples increases the thermal stability of such hydrogen-containing metabolites and allows them to be separated and detected by GC–MS.

3.4. Metabolite identification

In this section, we discuss the various methods available for metabolite identification (Table 14.2).

3.4.1 LC- and GC–MS Libraries

Metabolite identification for both LC– and GC–MS relies on the determination of the m/z ratio and the mass spectral fragmentation pattern. These factors are compared against a library of compounds in order to assign a metabolite identity. The number of metabolites that can be appropriately identified is entirely dependent on the completeness of the library, and the same library cannot be used for both LC– and GC–MS. Frequently used GC–MS libraries/databases include NIST, Wiley, and the Fiehn Metabolomics Library, while LC–MS libraries/databases include MetAlign, MzMine, and XCMS. The identity of a metabolite should always be confirmed by comparison to an authentic standard.

3.4.2 Determining structures for unknowns

A large-scale, nontargeted metabolomics analysis will lead to the identification of a large number of "unknown" metabolites—discrete authentic compounds with no immediately assigned chemical identity. Frequently, these unknowns comprise nearly two-thirds of all the metabolites isolated from a single sample (Ganti, Taylor, Abu Aboud, et al., 2012; Kim et al., 2011;

Table 14.2 Software and database used for metabolite identification and data analysis (alphabetical order)

Software/Database	Source	Summary
BioSpider	http://www.biospider.ca	Database to search Websites for drugs and metabolites
The Human Metabolome Database (HMDB)	http://www.hmdb.ca	Database for searching detailed information about human metabolites
Kyoto Encyclopedia of Genes and Genomes (KEGG)	http://www.genome.jp/kegg/	Database for systems, genomic, chemical, and health information including metabolites and proteins
Lipid Metabolites and Pathways Strategy (LIPID MAPS)	http://www.lipidmaps.org	System to identify and quantify lipids
MassBank	http://www.massbank.jp	Mass spectral database for organic compounds
Mathematica package for differential analysis of metabolite profiles (MathDAMP)	http://mathdamp.iab.keio.ac.jp	Software used for sample analysis. Compensate for retention time deviation
MetAlign	http://www.wageningenur.nl/nl/Expertises-Dienstverlening/Onderzoeksinstituten/rikilt/show/MetAlign.htm	Software for preprocessing and comparison of LC– and GC–MS data
METLIN	http://metlin.scripps.edu/index.php	Repository of metabolite information and tandem mass spectrometry data
Mzmine	http://mzmine.sourceforge.net	Software for mass spectrometry data processing (mainly LC–MS)
Pubchem	http://pubchem.ncbi.nlm.nih.gov	Database for chemical molecules and their activities (bioassay, compound, and substance)
XCMS	http://metlin.scripps.edu/xcms/	Software for preprocessing LC–MS data

Taylor et al., 2010). While some studies utilize a "fast-lane" approach in which metabolite identity is not required (Kind, Tolstikov, Fiehn, & Weiss, 2007), in most cases of translational research such identification will be necessary. Under those circumstances, several approaches can be undertaken to assign an identity to these unknowns. For example, tandem MS (MS/MS or MS^2) can be used to obtain more detailed structural information with which to reconstruct the metabolite. MS^2 is an incredibly useful tool to add to existing libraries and is suitable for metabolites separated by either LC or GC (Dunn et al., 2011). This MS^2 fragmentation pattern is compared to patterns in existing libraries/databases (see Table 14.2). A second method to identify unknowns (which is usually used in conjunction with tandem MS) is simply to run a pure compound and compare the fragmentation pattern to the metabolite of interest. This approach is appropriate when some information about the metabolite structure is already known. A detailed example of how to identify unknowns is provided by Schultz et al. (Zhu et al., 2013).

3.4.3 Methods for controlling quality

In order for any metabolomics-based study to have clinical translational potential, there exists the requirement that the data be reproducible. A pooled sample can be used to control for quality of data. The pooled sample consists of small aliquots of each biological sample to be analyzed; one pooled sample for each tissue/biofluid is required (Dunn et al., 2011). This pooled sample is then injected at a regular interval (once every 5–10 samples), and each injection is compared. The spectral data for each replicate should be analyzed to ensure little variation (within 30%) in signal intensity of the peaks. Researchers should also confirm that the retention time does not shift more than 2% between these QC samples (Theodoridis et al., 2012).

4. STATISTICS

After identifying the metabolites, the data need to be processed appropriately to adjust the measured intensity levels and statistical analyses need to be performed for group comparisons. While an exhaustive discussion of statistics as applied to metabolomics is beyond the scope of this chapter, the readers are referred to available reviews (Weiss & Kim, 2012). In this section, we briefly summarize the data analysis processes; software packages available for metabolomics data analyses were summarized in Table 14.2.

4.1. Preprocessing

The initial step in data processing in a metabolomics study is normalization of the data in order to remove biases resulting from varying sample recovery during extraction or differences between sample ionization in the MS that can confound biological interpretation of the data (reviewed in Katajamaa & Oresic, 2007 and Weiss & Kim, 2012). In the next step, data pretreatment methods can be used to correct for those aspects which confound biological interpretation of the metabolomics data sets. These methods are intended to improve the veracity of the biological interpretation of the data. There are various methods for pretreatment including centering, scaling, and transformations (van den Berg, Hoefsloot, Westerhuis, Smilde, & van der Werf, 2006). The choice for a data pretreatment method depends on the biological information to be obtained and on the data analysis method chosen. These methods often fail when a metabolite is not identified in all samples either because it is missing from the sample or its abundance is below the instrument's limit of detection. In these instances, the metabolomic dataset may suffer from "missing values," that is, quantitative "zero" values for the metabolite. When present, these missing values can be imputed with a representative quantity such as a value of half the lowest detected peak or the lowest detected peak; this avoids difficulties with transformation by preventing division by zero.

4.2. Univariate analysis and multivariate analysis

The next step in the statistical analysis is to attempt to identify those potential biomarkers with both biological and statistical significance. This process can be done by univariate or multivariate analysis (reviewed in Weiss & Kim, 2012) and will be discussed below.

4.2.1 Univariate analysis

Univariate analysis is used to identify those individual metabolites which, either singly or multiplexed, are capable of differentiating between biological groups, such as separating tumor bearing mice from nontumor bearing (control) mice. Statistical procedures used for this analysis include a *t*-test, ANOVA, Mann–Whitney U test, Wilcoxon signed-rank test, and logistic regression. These tests are used to individually or globally screen the measured metabolites for an association with a disease.

4.2.2 Multivariate analysis

A single metabolite biomarker is generally insufficient to differentiate between groups. For this reason, a multivariate analysis, which identifies sets of metabolites (e.g., patterns or clusters) in the data, can result in a higher likelihood of group separation. Statistical methods for this analysis include unsupervised methods, such as principle component analysis (PCA) or cluster analysis, and supervised methods, such as latent Dirichlet allocation (LDA), partial least squares (PLS), PLS Discriminant Analysis (PLS-DA), artificial neural network (ANN), and machine learning methods. These methods provide an overview of a large dataset that is useful for identifying patterns and clusters in the data and expressing the data to visually highlight similarities and differences. Unsupervised methods may reduce potential bias since the classes are unlabeled (Kim et al., 2009).

Regardless of one's choice of method for statistical analysis, it is necessary to subsequently validate the identified potential biomarkers and therapeutic targets by examining them in new and separate sample sets (for biomarkers), and *in vitro* and/or *in vivo* experiments evaluating the identified pathways or molecules (for therapeutic targets). These validation methods will be discussed in the next section.

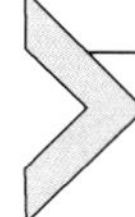

5. BIOMARKER VALIDATION

5.1. Confirming findings in a separate cohort

Once the metabolite(s) that distinguish disease from control groups are identified, these results must be validated in a second cohort, the so-called test set or validation set to ensure accuracy and ultimate clinical translatability of the experiment. Samples from the second cohort are prepared and analyzed using identical methods as the first but are generally blinded in some fashion to the investigators. Significantly altered metabolites from the first cohort, often called the "training set," are used to develop a model to appropriately distinguish disease from control (Kobayashi et al., 2013). In this particular study with human samples, the training set included 42 pancreatic cancer samples and 41 healthy controls, while the validation set included 41 cancer and 40 healthy samples. The amount of variability in the first data set should be taken into consideration. For example, less genetic/dietary/environmental variability is expected in a mouse study as compared to a human study, so fewer samples can be used in each set. The model generated by the training set may not include all of the significantly altered metabolites but could contain a subset of them. The accuracy of separation in each model can be using receiver operating characteristic (ROC) curves and measuring the area under the curve (AUC). This model is then applied to the blinded test set; ROC curves can be generated for the test set and the AUC compared to the training set (Nishiumi et al., 2012) for purposes of validation.

5.2. Determining biological significance

An extremely important, and unfortunately often overlooked, aspect of metabolomics analysis is to always keep one's eye and mind on the biological significance of the data! Not all significantly altered metabolites will be biologically relevant. Many serum, urinary, and tissue metabolites are derived from the diet, gut flora, or result from drug metabolism which, depending on the study design (i.e., gastrointestinal vs. renal disorders) can make these metabolites irrelevant or even confounders to the dataset. It is important to note that some investigators utilize these metabolites in their metabolomics analyses—in particular those studying gastrointestinal diseases (Martin, Collino, Rezzi, & Kochhar, 2012). Experiments to alter metabolite levels should be undertaken to determine their biological relevance with either *in vitro* or *in vivo* studies.

In vitro studies may include experiments where metabolites of interest are added to the culture medium in which cancer cells are growing (Ganti, Taylor, Kim, et al., 2012) or pharmacologically inhibiting/activating a pathway to cause metabolite accumulation/depletion (Ganti, Taylor, Abu Aboud, et al., 2012). To determine biological relevance, a readout of activity is required, such as apoptosis, inflammation, or proliferation.

In vivo studies can be used in much the same way. A mouse xenograft model was developed (Ganti, Taylor, Abu Aboud, et al., 2012) to validate findings from a human urine metabolomics study (Kim et al., 2011). The mouse xenograft study allowed simultaneous matrix (urine, tissue, and serum) metabolomics to determine if the metabolites identified in the human study originated from the tumor itself or if they were derived from a systemic response due to the presence of the tumor. Others have used xenograft models to examine changes in the metabolome as a result of disease progression. One study compared serum biomarkers of mice with alcohol liver disease to mice with xenograft hepatocellular carcinoma (Li et al., 2011). The *in vitro* studies lead to a first pass at validation by providing evidence of perturbed molecular signaling, while the *in vivo* studies demonstrate the overall response to a cancerous mass.

6. SUMMARY

As one of the relative newcomers to the omics disciplines, the field of metabolomics is now beginning to show great potential in the discovery of clinically useful biomarkers and targets for therapy in human disease. Given the obvious heterogeneity of the human population, the use of mouse models to identify metabolic alterations due to disease in multiple matrices prior to patient evaluation has shown considerable promise. However, it is of paramount importance to keep in mind the biological significance of the data obtained throughout the course of the study. In this chapter, we have discussed the basic techniques for using such animal models in metabolomics experiments. Furthermore, we provide guidelines and make recommendations for carrying out a successful metabolomics study which will lead to effective clinical translation and thus earlier detection and treatment of human disease.

REFERENCES

Ang, J. E., Revell, V., Mann, A., Mantele, S., Otway, D. T., Johnston, J. D., et al. (2012). Identification of human plasma metabolites exhibiting time-of-day variation using an untargeted liquid chromatography-mass spectrometry metabolomic approach. *Chronobiology International*, *29*(7), 868–881. http://dx.doi.org/10.3109/07420528.2012.699122.

Baker, M. (2013). Big biology: The 'omes puzzle. *Nature*, *494*(7438), 416–419. http://dx.doi.org/10.1038/494416a.

Bando, K., Kawahara, R., Kunimatsu, T., Sakai, J., Kimura, J., Funabashi, H., et al. (2010). Influences of biofluid sample collection and handling procedures on GC-MS based metabolomic studies. *Journal of Bioscience and Bioengineering*, *110*(4), 491–499. http://dx.doi.org/10.1016/j.jbiosc.2010.04.010.

Bosma, G. C., Custer, R. P., & Bosma, M. J. (1983). A severe combined immunodeficiency mutation in the mouse. *Nature*, *301*(5900), 527–530.

Dettmer, K., Aronov, P. A., & Hammock, B. D. (2007). Mass spectrometry-based metabolomics. *Mass Spectrometry Reviews*, *26*(1), 51–78. http://dx.doi.org/10.1002/mas.20108.

Douglas, D. J., Frank, A. J., & Mao, D. (2005). Linear ion traps in mass spectrometry. *Mass Spectrometry Reviews*, *24*, 1–29.

Dunn, W. B., Broadhurst, D., Begley, P., Zelena, E., Francis-McIntyre, S., Anderson, N., et al. (2011). Procedures for large-scale metabolic profiling of serum and plasma using gas chromatography and liquid chromatography coupled to mass spectrometry. *Nature Protocols*, *6*(7), 1060–1083. http://dx.doi.org/10.1038/nprot.2011.335.

Epstein, A. A., Narayanasamy, P., Dash, P. K., High, R., Bathena, S. P., Gorantla, S., et al. (2013). Combinatorial assessments of brain tissue metabolomics and histopathology in rodent models of human immunodeficiency virus infection. *Journal of Neuroimmune Pharmacology*, *8*(5), 1224–1238. http://dx.doi.org/10.1007/s11481-013-9461-9.

Evans, A. M., DeHaven, C. D., Barrett, T., Mitchell, M., & Milgram, E. (2009). Integrated, nontargeted ultrahigh performance liquid chromatography/electrospray ionization tandem mass spectrometry platform for the identification and relative quantification of the small-molecule complement of biological systems. *Analytical Chemistry*, *81*, 6656–6667.

Fan, T. W., Lane, A. N., Higashi, R. M., & Yan, J. (2011). Stable isotope resolved metabolomics of lung cancer in a SCID mouse model. *Metabolomics*, 7(2), 257–269. http://dx.doi.org/10.1007/s11306-010-0249-0.

Ganti, S., Taylor, S. L., Abu Aboud, O., Yang, J., Evans, C., Osier, M. V., et al. (2012). Kidney tumor biomarkers revealed by simultaneous multiple matrix metabolomics analysis. *Cancer Research*, *72*, 3471–3479.

Ganti, S., Taylor, S. L., Kim, K., Hoppel, C. L., Guo, L., Yang, J., et al. (2012). Urinary acylcarninities are altered in human kidney cancer. *International Journal of Cancer*, *130*, 2791–2800.

Ganti, S., & Weiss, R. H. (2011). Urine metabolomics for kidney cancer detection and biomarker discovery. *Urological Oncology: Seminars and Original Investigations*, *29*, 551–557.

Gerlinger, M., Rowan, A. J., Horswell, S., Larkin, J., Endesfelder, D., Gronroos, E., et al. (2012). Intratumor heterogeneity and branched evolution revealed by multiregion sequencing. *New England Journal of Medicine*, *366*(10), 883–892. http://dx.doi.org/10.1056/NEJMoa1113205.

Halket, J. M., Waterman, D., Przyborowska, A. M., Patel, R. K. P., Fraser, P. D., & Bramley, P. M. (2005). Chemical derivataztion and mass spectral libraries in metabolic profiling by GC/MS and LC/MS. *Journal of Experimental Botany*, *56*(410), 219–243.

Hellerstein, S., Hunter, J. L., & Warady, B. A. (1988). Creatinine excretion rates for evaluation of kidney function in children. *Pediatric Nephrology*, *2*(4), 419–424.

Jensen, M. M., Jorgensen, J. T., Binderup, T., & Kjaer, A. (2008). Tumor volume in subcutaneous mouse xenografts measured by microCT is more accurate and reproducible than determined by 18F-FDG-microPET or external caliper. *BMC Medical Imaging*, *8*, 16. http://dx.doi.org/10.1186/1471-2342-8-16.

Katajamaa, M., & Oresic, M. (2007). Data processing for mass spectrometry-based metabolomics. *Journal of Chromatography A*, *1158*(1–2), 318–328. http://dx.doi.org/10.1016/j.chroma.2007.04.021.

Khanna, C., & Hunter, K. (2005). Modeling metastasis in vivo. *Carcinogenesis*, *26*(3), 513–523. http://dx.doi.org/10.1093/carcin/bgh261.

Killion, J. J., Radinsky, R., & Fidler, I. J. (1998). Orthotopic models are necessary to predict therapy of transplantable tumors in mice. *Cancer and Metastasis Reviews*, *17*(3), 279–284.

Kim, K., Aronov, P., Zakharkin, S. O., Anderson, D., Perroud, B., Thompson, I. M., et al. (2009). Urine metabolomics analysis for kidney cancer detection and biomarker discovery. *Molecular & Cellular Proteomics*, *8*(3), 558–570. http://dx.doi.org/10.1074/mcp.M800165-MCP200.

Kim, K., Mall, C., Taylor, S. L., Hitchcock, S., Zhang, C., Wettersten, H. I., et al. (2014). Mealtime, Temporal, and Daily Variability of the Human Urinary and Plasma Metabolomes in a Tightly Controlled Environment. *PloS One*, *9*(1), e86223.

Kim, K., Taylor, S. L., Ganti, S., Guo, L., Osier, M. V., & Weiss, R. H. (2011). Urine metabolomic analysis identifies potential biomarkers and pathogenic pathways in kidney cancer. *OMICS*, *15*(5), 293–303.

Kind, T., & Fiehn, O. (2010). Advances in structure elucidation of small molecules using mass spectrometry. *Bioanalytical Reviews*, *2*, 23–60.

Kind, T., Tolstikov, V., Fiehn, O., & Weiss, R. H. (2007). A comprehensive urinary metabolomic approach for identifying kidney cancerr. *Analytical Biochemistry*, *363*(2), 185–195. http://dx.doi.org/10.1016/j.ab.2007.01.028.

Kobayashi, T., Nishiumi, S., Ikeda, A., Yoshie, T., Sakai, A., Matsubara, A., et al. (2013). A novel serum-based diagnostic approach to pancreatic cancer. *Cancer Epidemiology, Biomarkers & Prevention*, *22*(4), 571–579.

Kurien, B. T., Everds, N. E., & Scofield, R. H. (2004). Experimental animal urine collection: A review. *Laboratory Animals*, *38*(4), 333–361.

Li, S., Liu, H., Jin, Y., Lin, S., Cai, Z., & Jiang, Y. (2011). Metabolomics study of alcohol-induced liver injury and hepatocellular carcinoma xenografts in mice. *Journal of Chromatography B, Analytical Technologies in the Biomedical and Life Sciences*, *879*, 2369–2375.

Liu, L., Wang, M., Yang, X., Bin, M., Na, L., Niu, Y., et al. (2013). Fasting serum lipid and dehydroepiandosterone sulfate as important metabolites for detecting isolated postchallenge diabetes: Serum metabolomics via ultra-high-performance liquid chromatography/mass spectrometry. *Clinical Chemistry*, *59*(9), 1338–1348.

Ljungberg, B., Mehle, C., Stenling, R., & Roos, G. (1996). Heterogeneity in renal cell carcinoma and its impact no prognosis—A flow cytometric study. *British Journal of Cancer*, *74*(1), 123–127.

Loftus, N. J., Lai, L., Wilkinson, R. W., Odedra, R., Wilson, I. D., & Barnes, A. J. (2013). Global metabolite profiling of human colorectal cancer xenografts in mice using HPLC-MS/MS. *Journal of Proteome Research*, *12*(6), 2980–2986. http://dx.doi.org/10.1021/pr400260h.

Martin, F.-P., Collino, S., Rezzi, S., & Kochhar, S. (2012). Metabolomic applications to decipher gut microbial metabolic influence in health and disease. *Frontiers in Physiology*, *3*, 113.

Mombaerts, P., Iacomini, J., Johnson, R. S., Herrup, K., Tonegawa, S., & Papaioannou, V. E. (1992). RAG-1-deficient mice have no mature B and T lymphocytes. *Cell*, *68*(5), 869–877.

Morton, C. L., & Houghton, P. J. (2007). Establishment of human tumor xenografts in immunodeficient mice. *Nature Protocols*, *2*(2), 247–250. http://dx.doi.org/10.1038/nprot.2007.25.

Nishiumi, S., Kobayashi, T., Ikeda, A., Yoshie, T., Kibi, M., Izumi, Y., et al. (2012). A novel serum metabolomics-based diagnostic approach for colorectal cancer. *PLoS One*, 7(7), e40459.

Pantelouris, E. M. (1968). Absence of thymus in a mouse mutant. *Nature*, *217*(5126), 370–371.

Pantelouris, E. M. (1973). Athymic development in the mouse. *Differentiation*, *1*(6), 437–450.

Pearson, T., Shultz, L. D., Miller, D., King, M., Laning, J., Fodor, W., et al. (2008). Non-obese diabetic-recombination activating gene-1 (NOD-Rag1 null) interleukin (IL)-2 receptor common gamma chain (IL2r gamma null) null mice: A radioresistant model for human lymphohaematopoietic engraftment. *Clinical and Experimental Immunology*, *154*(2), 270–284. http://dx.doi.org/10.1111/j.1365-2249.2008.03753.x.

Ray, P., Tsien, R., & Gambhir, S. S. (2007). Construction and validation of improved triple fusion reporter gene vectors for molecular imaging of living subjects. *Cancer Research*, *67*(7), 3085–3093. http://dx.doi.org/10.1158/0008-5472.CAN-06-2402.

Ryan, D., Robards, K., Prenzler, P. D., & Kendall, M. (2011). Recent and potential developments in the analysis of urine: A review. *Analytica Chimica Acta*, *684*, 17–29.

Sausville, E. A., & Burger, A. M. (2006). Contributions of human tumor xenografts to anticancer drug development. *Cancer Research*, *66*(7), 3351–3354. http://dx.doi.org/10.1158/0008-5472.CAN-05-3627, discussion 3354.

Shinkai, Y., Rathbun, G., Lam, K. P., Oltz, E. M., Stewart, V., Mendelsohn, M., et al. (1992). RAG-2-deficient mice lack mature lymphocytes owing to inability to initiate V(D)J rearrangement. *Cell*, *68*(5), 855–867.

Shultz, L. D., Lyons, B. L., Burzenski, L. M., Gott, B., Chen, X., Chaleff, S., et al. (2005). Human lymphoid and myeloid cell development in NOD/LtSz-scid IL2R gamma null mice engrafted with mobilized human hemopoietic stem cells. *Journal of Immunology*, *174*(10), 6477–6489.

Shultz, L. D., Schweitzer, P. A., Christianson, S. W., Gott, B., Schweitzer, I. B., Tennent, B., et al. (1995). Multiple defects in innate and adaptive immunologic function in NOD/LtSz-scid mice. *Journal of Immunology*, *154*(1), 180–191.

Strathmann, F. G., Laha, T. J., & Hoofnagle, A. N. (2011). Quantification of 1α,25 dihydroxy vitamin D by immunoextraction and liquid chromatography-tandem mass spectrometry. *Clinical Chemistry*, *57*(9), 1279–1285.

Taylor, S. L., Ganti, S., Bukanov, N. O., Chapman, A., Fiehn, O., Osier, M. V., et al. (2010). A metabolomics approach using juvenile cystic mice to identify urinary biomarkers and altered pathways in polycystic kidney disease. *American Journal of Physiology Renal Physiology*, *298*, F909–F922.

Theodoridis, G. A., Gika, H. G., Want, E. J., & Wilson, I. D. (2012). Liquid chromatography-mass spectrometry based global metabolite profiling: A review. *Analytica Chimica Acta*, *711*, 7–16. http://dx.doi.org/10.1016/j.aca.2011.09.042.

Trupp, M., Zhu, H., Wikoff, W. R., Baillie, R. A., Zeng, Z.-B., Karp, P. D., et al. (2012). Metabolomics reveals amino acids contribute to variation in response to simvastatin treatment. *PLoS One*, 7(7), e38386.

van den Berg, R. A., Hoefsloot, H. C., Westerhuis, J. A., Smilde, A. K., & van der Werf, M. J. (2006). Centering, scaling, and transformations: Improving the biological information content of metabolomics data. *BMC Genomics*, 7, 142. http://dx.doi.org/10.1186/1471-2164-7-142.

Want, E. J., Masson, P., Michopoulos, F., Wilson, I. D., Theodoridis, G., Plumb, R. S., et al. (2013). Global metabolic profiling of animal and human tissues via UPLC-MS. *Nature Protocols*, *8*(1), 17–32. http://dx.doi.org/10.1038/nprot.2012.135.

Weiss, R. H., & Kim, K. (2012). Metabolomics in the study of kidney diseases. *Nature Reviews Nephrology*, *8*(1), 22–33. http://dx.doi.org/10.1038/nrneph.2011.152.

Yoshida, M., Hatano, N., Nishiumi, Irino Y., Izumi, Y., Takenawa, T., & Azuma, T. (2012). Diagnosis of gastroenterological disease by metabolome analysis using gas chromatography–mass spectrometry. *Journal of Gastroenterology*, *47*, 9–20.

Zhu, Z. J., Schultz, A. W., Wang, J., Johnson, C. H., Yannone, S. M., Patti, G. J., et al. (2013). Liquid chromatography quadrupole time-of-flight mass spectrometry characterization of metabolites guided by the METLIN database. *Nature Protocols*, *8*(3), 451–460. http://dx.doi.org/10.1038/nprot.2013.004.

Zuppi, C., Messana, I., Forni, F., Rossi, C., Pennacchietti, L., Ferrari, F., et al. (1997). ^{1}H NMR spectra of normal urines: Reference ranges of the major metabolites. *Clinica Chimica Acta*, *265*, 85–97.

CHAPTER FIFTEEN

Metabolomic Studies of Patient Material by High-Resolution Magic Angle Spinning Nuclear Magnetic Resonance Spectroscopy

Hector Keun[1]

Department of Surgery and Cancer, Imperial College London, South Kensington, London, United Kingdom
[1]Corresponding author: e-mail address: h.keun@imperial.ac.uk

Contents

Abstract

Magic angle spinning nuclear magnetic resonance (MAS-NMR) spectroscopy offers a convenient means for the rapid determination of metabolic profiles from intact malignant tissues with high resolution. The implementation of MAS-NMR spectroscopy requires minimal sample processing, hence being compatible with complementary histological or biochemical analyses. The metabolites routinely detected in ^{1}H MAS-NMR spectra can simultaneously inform on many of the metabolic alterations that characterize malignant cells, including altered choline metabolism and the so-called Warburg effect. Clinical MAS-NMR profiles have been attributed with diagnostic or prognostic value, correlating to disease subtype, tumor stage/grade, response to chemotherapy, and patient survival. Herein, the scientific rationale behind MAS-NMR and its

Methods in Enzymology, Volume 543
ISSN 0076-6879
http://dx.doi.org/10.1016/B978-0-12-801329-8.00015-5

utility for translational cancer research and patient stratification is summarized. Moreover, a basic protocol for the analysis of tumor samples by MAS-NMR spectroscopy is detailed.

1. MAS-NMR SPECTROSCOPY AS A TOOL FOR METABOLIC PROFILING

Nuclear magnetic resonance (NMR) spectroscopy has been used for several decades to characterize in an untargeted manner the metabolite content of biological samples (a.k.a. metabolic profiling, metabolomics, metabonomics). By reporting in a quantitative manner on all highly abundant molecules present in the sample that contain a nucleus of interest (usually ^{1}H, ubiquitous in biomolecules and the most naturally abundant and sensitively detected isotope), NMR spectra provide a useful compositional profile of up to ~50 simultaneous metabolite measurements without the need for preselecting the molecules to be detected. Typically, NMR in this context is applied as a solution-state technique to samples such as urine or blood plasma (Beckonert et al., 2007), and it is the free rotation of molecules in solution that leads ultimately to the well-resolved spectra from which characteristic metabolite signals can be individually recognized and/or quantified. However, such mobility is restricted in a tissue sample. Magic angle spinning offers a means by which to obtain high-resolution NMR spectra, resembling solution-state spectra, from an intact "semi-solid" biological specimen, for example, a human tissue biopsy (Cheng et al., 1996). In a solid-state sample, spectral resonances are significantly broadened by several factors including dipolar coupling and chemical shift anisotropy. As demonstrated in the 1950s (Andrew & Newing, 1958; Lowe, 1959), by rapidly spinning the sample in a rotor placed at a specific angle to the static magnetic field (54.7°), these effects can be significantly alleviated and spectral quality dramatically improved.

In order to conduct MAS-NMR, a dedicated NMR probe is required, coupled to an additional console which controls sample rotation. Samples are prepared to the appropriate size and then placed into a ceramic rotor with Kel-F inserts to fix the tissue in place (Beckonert et al., 2010). More recent implementations have inserts that can be preloaded and tissue locked within (Jimenez et al., 2013). These inserts can then be transferred from frozen storage to the rotor for spectral acquisition and then returned to storage. Sample sizes are typically 10–50 mg wet weight, but it is possible to acquire spectra on as little as 5–6 mg using standard equipment or at nanoliter volumes using custom probes (Wong et al., 2012).

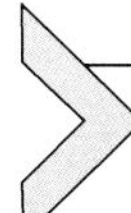

2. MAS-NMR VERSUS SOLUTION-STATE NMR SPECTROSCOPY OF TUMORS

MAS-NMR offers several potential advantages over tissue homogenization and extraction followed by solution-state NMR. The first is sensitivity; this is improved by reducing the active volume measured during the experiment from a couple of 100 μl in standard solution-state probes to a few tens of microliter, improving the probe fill factor (Wong et al., 2012). However, this is mitigated by the fact solution-state sensitivity can be boosted by the use of cryoprobes (Keun et al., 2002) and microliter probes (Griffin et al., 2002) which are also available. A more important advantage of avoiding the extraction process and analyzing intact tissue is that sample preparation is minimal, typically just placing a specimen of an appropriate size into a rotor with a small volume of deuterated saline. This means that a tissue can be prepared and a spectrum obtained with a minimum delay of just 10–20 min compared to many hours for extraction. This makes rapid assessment of tissue possible, potentially even reporting back information, for example, about resection margins to a clinical team during a surgical procedure (Bathen et al., 2013). In principle much of the physical architecture of the tissue specimen can be preserved during MAS-NMR analysis, including the compartmentation of metabolites into, for example, intracellular and extracellular pools or the presence of lipid aggregates. These are destroyed upon extraction and in tumor tissue could contain important information: for example, the presence of increased mobile lipid signals has often been reported in tumors; other important markers such as phosphocholine (PC) and citrate may be more relevant in the intracellular or interstitial fluid pool, respectively. It is also possible in principle to edit the components visible in the spectrum on the basis of relative mobility as for solution-state NMR (Wang et al., 2003). The degree of tissue damage/metabolite decomposition during MAS-NMR analysis is in practice dependent on the tissue analyzed, the rate at which the sample is spun (Opstad, Bell, Griffiths, & Howe, 2008a) and the temperature of analysis (Swanson et al., 2006). Importantly, it has been demonstrated that biochemical analysis by other means is possible post-MAS-NMR, including histology and RNA/DNA measurements. In one study, it was concluded that the results of histopathologic and microarray analysis were statistically similar between pre- and post-MAS prostatic surgical and biopsy specimens, but some loss of quality was observed by biopsy (Santos et al., 2010). The capacity to conduct complementary analyses in parallel on the same specimen offers many opportunities for improved biomarker analysis and for refining analysis of the MAS-NMR data, for example, excluding samples with low tumor content or high levels of necrosis.

There are also potentially disadvantages to MAS-NMR of intact tissue compared to solution-state analysis. The spectral resolution is typically inferior and some metabolites may be in chemical exchange or restricted in mobility such that these are not detected in the intact tissue. For example, lipid resonances are very sensitive to changes in mobility and thus temperature. These effects add to difficulty of quantification of metabolites in MAS-NMR and standard solution-state internal standards such as trimethylsilyl propionate (TSP) are not very reliable. Furthermore, the optimal conditions for acquisition of quantitative NMR data require a delay between acquisitions of approximately 5 × the longest T_1 placing increased demands on experiment time; however, this can be alleviated by shorter pulse angles. More recently, techniques such as ERETIC (The Electronic REference To access *In vivo* Concentrations) (Akoka, Barantin, & Trierweiler, 1999) that involves introducing a synthetic signal into the spectrum has been shown to provide a reliable means of quantification in MAS-NMR (Celda et al., 2009; Martinez-Bisbal et al., 2009). Spectral artifacts such as "spinning side bands" can appear in the spectrum and may be incompletely eliminated at rotation speeds that cause minimal damage to the specimen. The sample remains biochemically active and so the metabolic composition may change during analysis, limiting longer acquisitions, and complicating 2D NMR analysis. This activity is in addition to the release of metabolites from mechanical stress due to spinning (Santos et al., 2010), for example, we and others (Rocha et al., 2010) have observed changes in the relative levels of choline metabolites during prolonged spinning (over several hours) in some samples. These effects favor shorter acquisitions conducted at lower temperatures. Probably, the most significant challenges in MAS-NMR analysis are not limited to the technique but are the generic issues of sample collection and tissue heterogeneity. The amount of tissue required for analysis and the rapid changes in metabolism in tissue while it is unfrozen make it difficult to isolate precisely a well-defined section of tissue prior to analysis. The fraction of tumor, necrotic tissue, fatty deposits, immune cell infiltration, and distance from the tumor all introduce spectral variability that can obscure the underlying signature of malignant tissue and may make the metabolic profile of the particular specimen unrepresentative of the metabolic phenotype of the overall tumor. Furthermore, it is likely that metabolic gradients exist within a tumor due to hypoxia and nutrient deprivation in regions distant from the vasculature (Gatenby & Gillies, 2004) and possibly also due to substrate exchange and recycling (Sonveaux et al., 2008). However, some studies are beginning to take advantage of the sensitivity of MAS-NMR to these phenomena and use the technique to probe explicitly for the metabolic signatures associated with field cancerization and tumor heterogeneity (Jimenez et al., 2013; Yakoub, Keun, Goldin, & Hanna, 2010).

3. MAS-NMR TUMOR PROFILING IN CLINICAL STUDIES

MAS-NMR has been used to examine human tumors from a wide range of sites, most extensively those of the brain (Cheng, Chang, Louis, & Gonzalez, 1998; Gonzalez-Velez et al., 2009; Martinez-Bisbal et al., 2004; Opstad, Bell, Griffiths, & Howe, 2009; Peet et al., 2007; Righi, Roda, et al., 2009; Tzika et al., 2002; Wright et al., 2010), breast (Cheng, Chang, Smith, & Gonzalez, 1998; Sitter et al., 2006; Sitter, Sonnewald, Spraul, Fjosne, & Gribbestad, 2002), prostate (Cheng, Wu, Smith, & Gonzalez, 2001; Kurhanewicz, Swanson, Nelson, & Vigneron, 2002; Swanson et al., 2003, 2006), colon and rectum (Chan et al., 2009; Piotto et al., 2009), esophagus (Yakoub et al., 2010; Yang et al., 2013), stomach (Calabrese et al., 2008; Wong et al., 2012), cervix (Mahon, deSouza, et al., 2004), head and neck (Somashekar et al., 2011), and gall bladder (Bharti et al., 2013). Malignant lymph nodes have also been characterized by MAS-NMR (Cheng et al., 1996; Kumar et al., 2012), and there have been specific studies of metastasis to brain from tumors of different primary origin (Sjobakk et al., 2008, 2013). A number of other studies have used MAS-NMR to study tumors generated in a range of rodent models that will not be discussed here. Across these studies, ^{1}H NMR spectroscopy has consistently been demonstrated to be able to distinguish transformed from normal tissue by detection of a wide range of molecules that report on a multiple metabolic pathways that are perturbed in malignancy (Fig. 15.1). Moreover, it would seem possible to use MAS-NMR profiles—particularly in brain and breast cancers—to aid the classification of tumors into defined histologic/molecular subtypes that may have different prognoses and may require different therapeutic strategies (Andronesi et al., 2008, 2012; Chinnaiyan et al., 2012; Denkert et al., 2012; Millis et al., 1999; Tzika et al., 2007).

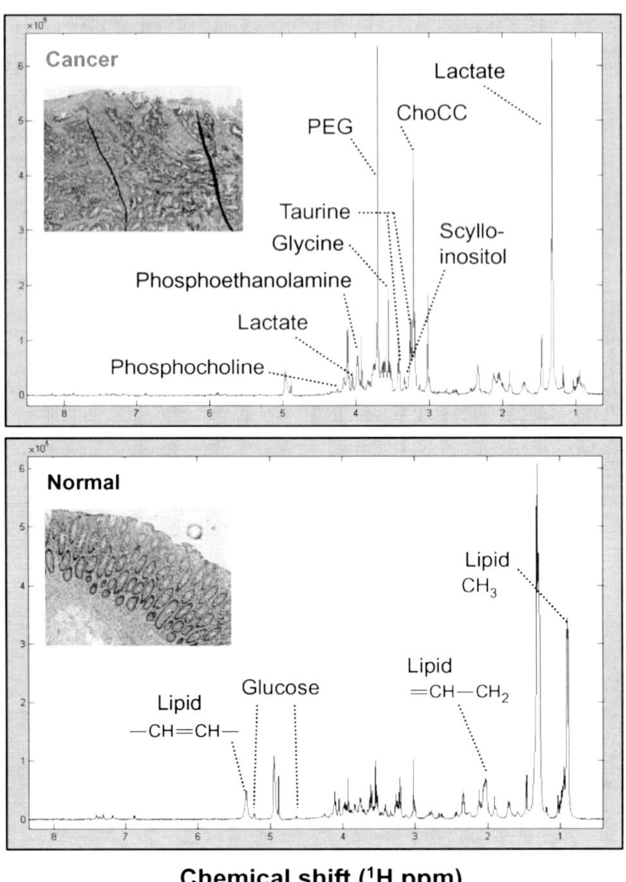

Figure 15.1 *Typical* ^1H *MAS-NMR spectra of malignant and normal colon tissues.* ChoCC, choline-containing compounds. *Reproduced from Chan et al. (2009).* (See the color plate.)

Metabolite levels detected by MAS-NMR spectroscopy have also been demonstrated to correlate either directly to patient outcomes or established prognostic factors including: grade in astrocytomas (Righi, Roda, et al., 2009); survival (Cao, Sitter, et al., 2012; Giskeodegard et al., 2012), grade and lymph node status (Bathen et al., 2007; Giskeodegard et al., 2010), and hormone receptor status (Bathen et al., 2007) in breast cancer; recurrence (Maxeiner et al., 2010), Gleason score (Bertilsson et al., 2012; Cheng et al., 2005b; van Asten et al., 2008), proliferation (Ki67) (Keshari et al., 2011; Stenman et al., 2011), and PSA velocity (Dittrich et al., 2012) in prostate cancer; survival and stage (Jimenez et al., 2013;

Mirnezami et al., 2013) in colon cancer. Finally, it has been shown that MAS-NMR metabolite profiles can differentiate response to drug treatment, for example, neoadjuvant chemotherapy with epirubicin or paclitaxel in breast cancer patients (Cao, Giskeodegard, et al., 2012). Collectively, there is considerable evidence for the value of pursing translational research using MAS-NMR spectroscopy of tumors in the context of diagnosis, prognosis, patient stratification, guiding surgical intervention, and predicting treatment response.

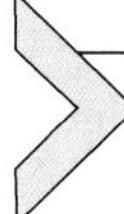

4. METABOLIC PERTURBATIONS IN TUMORS DETECTED BY MAS-NMR SPECTROSCOPY

4.1. Choline metabolism

Perhaps, the most widely reported NMR-detectable phenotypes associated with cancer are alterations in choline metabolism. For several decades, magnetic resonance studies *in vivo* and *ex vivo* have consistently reported increased levels of total choline in tumor tissues, dominated by an increase in PC resonances and often presenting as an increase in the PC:GPC (glycerophosphocholine) ratio. These effects can be broadly recapitulated in cell and animal models as cells are transformed from a nontumorgenic state to a malignant phenotype (Glunde, Jie, & Bhujwalla, 2004; Iorio et al., 2010; Teahan, Bevan, Waxman, & Keun, 2011), although some exceptions have been reported (Teichert et al., 2008). These effects are thought to be largely an indicator/consequence of rapid proliferation and the resulting demand for synthesis of membrane lipids such as phosphatidylcholine. The uptake of choline and activity of choline kinase (ChoK), the enzyme that produces PC, have been shown to be elevated in a number of tumors (Glunde & Bhujwalla, 2007), and ChoK inhibitors have been demonstrated to block tumor cell growth and decrease intracellular PC levels (Al-Saffar et al., 2006). Both the MR choline signal and PET choline tracers can be used image choline metabolism and localize tumors or detect metabolic response to therapy. Recently, other drivers for altered choline have been reported which indicate that changes in choline metabolism may not simply be a passive consequence of increased proliferation. For example, the glycerophosphodiesterase EDI3, which converts GPC to G3P and free choline, is associated with poorer outcomes in gynecological cancers and promotes tumor cell migration and invasion (Stewart et al., 2012). Such phenotypes may be mediated by signaling lipids; ChoK has been reported to activate MAPK and PI3K/AKT signaling possibly via increased production of phosphatidic acid (Yalcin et al., 2010). Current MAS-NMR studies are helping to understand the link between altered choline metabolism and patient outcomes or known prognostic factors such as grade and stage, as well as chemotherapy response, and are continuing to reveal new phenomena associated with this metabolic phenotype such as field effects. Field cancerization refers to the clonal expansion of cells which harbor early, preneoplastic genetic, or epigenetic lesions which predispose the tissue to tumor formation. As indicated by studies in cell lines, altered choline metabolism appears to be an early event in tumorigenesis, detectable by MAS-NMR in models of gastrointestinal cancer prior to the onset of tumors (Backshall et al., 2009) and in histologically normal tissue distal from the primary tumor from patients with colorectal (Jimenez et al., 2013), esophageal (Yakoub et al., 2010), and prostate cancer (Cheng et al., 2005a). Thus, choline metabolite levels detected by MAS-NMR in the tumor and in noninvolved tissue might provide a diagnostic, predictive, or prognostic biomarker that can report on more than just tumor proliferation.

4.2. Amino acids

MAS-NMR has routinely reported on a number of significant differences in tumor amino acid composition. Tumors often possess high rates of glutaminolysis and glutamine or glutamate levels in tissue can associate with disease progression (Yakoub et al., 2010). Several reports describe an association between MAS-NMR-detected high tumor glycine levels or/and poor prognosis in breast cancer (Cao, Giskeodegard, et al., 2012; Sitter et al., 2010). Glycine is also correlated with grade in brain tumors (Davies et al., 2010; Righi, Andronesi, Mintzopoulos, Black, & Tzika, 2010). Glycine has a number of significant roles in central carbon metabolism, acting as a precursor for protein, nucleotides, and glutathione. *De novo* synthesis of glycine from glucose has been shown to be strongly correlated to proliferation rate in cancer cells (Jain et al., 2012) and availability of the precursor extracellular serine may affect the ability of the tumor cell to cope with oxidative stress (Maddocks et al., 2013). Increases in the amino acid taurine are also frequently reported in MAS-NMR spectra of tumors (Chan et al., 2009; Jimenez et al., 2013; Peet et al., 2007; Piotto et al., 2009), but the reason for this event is not yet clear. One study has observed a positive correlation between MAS-NMR-detected taurine levels in astrocytomas and apoptosis (TUNEL staining) independently of necrosis (Opstad et al., 2009).

4.3. Lipids

The lipid-derived resonances visible in MAS-NMR spectra are not straightforward to characterize in terms of specific molecular structures, but rather report on the distributions of lipid aggregates with varying size/mobility/composition in the tissue. Many publications have reported a dramatic increase in mobile lipid pools, possibly cytoplasmic droplets rich in triglycerides and unsaturated fatty acids, within malignant samples (Mahon, Cox, et al., 2004; Opstad, Bell, Griffiths, & Howe, 2008b; Righi, Mucci, et al., 2009; Zietkowski et al., 2010). Lipid signals, not necessarily of the same composition, are also observed to increase during apoptosis (Griffin et al., 2003).

4.4. Glycolysis and lactate production

MAS-NMR can detect lactate and glucose in tumors and other metabolites including pyruvate that can reflect the increase of glycolysis in tumors (the Warburg effect). Elevated lactate levels have been associated with poorer outcomes in patients suggesting a prognostic value (Giskeodegard et al., 2012; Yokota et al., 2007) and may also be sensitive to chemotherapy response (Cao, Giskeodegard, et al., 2012).

4.5. Polyamines and citrate

While most of the above features are common to many solid tumors, some tissue-specific metabolic responses have been identified by MAS-NMR. This is particularly the case in prostate tumors, where the normal prostatic glandular epithelium exhibits a truncated TCA cycle and net citrate production (Costello & Franklin, 2006). The tissue also possesses high levels of polyamines which are NMR detectable. This phenotype is reversed in prostate cancer cells, thus dramatic decreases in the tissue citrate pool and polyamine resonances are clearly observed in NMR spectra of malignant prostate tissue (Bertilsson et al., 2012; Cheng et al., 2005a; Swanson et al., 2003). Together with changes in choline, lactate, and alanine, citrate and polyamine signals form a distinctive signature that differentiates prostate tumors on the basis of stage and grade and patient outcomes (Giskeodegard et al., 2013; Tessem et al., 2008).

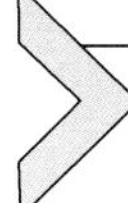

5. PROTOCOL FOR THE ANALYSIS OF TISSUE BY ^{1}H MAS-NMR SPECTROSCOPY

The following protocol is based on Bruker instrumentation for MAS-NMR spectroscopy, for other instrument manufacturers some modifications will be necessary. The rotor kit described here consists of a zirconium rotor, a rotor cap, an insert and a grub screw. Additional tools are required such as holding screw for the insert, a small screwdriver, and a spacer tool for positioning the insert. Alternatively, disposable 30 μl inserts are available for single use with each sample.

5.1. Sample preparation

1. Place fresh or frozen sample in a Petri dish (which is ideally cooled on ice). Cut tissue to an appropriate size for the rotor using a tissue punch or scalpel (typically 10–50 mg wet weight, <50 μl).
2. Add tissue to preweighed (tare weight) eppendorf and weigh to determine mass.
3. Return tissue to Petri dish, rinse for a few seconds with ~10 μl of D_2O or 0.9% (w/v) saline in D_2O before placing into the zirconium rotor. Add further D_2O or D_2O/saline if necessary to fill the expected sample volume avoiding air bubbles. Place the insert into the rotor and position the insert using the spacer tool. Seal the insert with the grub screw. Wipe away any excess fluid.
4. Alternative to step 3, the sample can be prepared in advance into a disposable insert, using D_2O or D_2O/saline to fill the remaining insert space while avoiding air bubbles. Once sealed, this insert can be frozen prior to analysis or placed into the rotor.
5. Place the cap onto the rotor, taking care not to damage the teeth of the cap which are required for accurate rotation. *Note*: removing the cap can be assisted by placing the rotor briefly in liquid nitrogen.
6. Mark half the outside edge of the base of the rotor with permanent black marker to allow the detection of rotation. The total time taken for steps 1–6 should be ~5–10 min.

5.2. NMR spectrometer set up

7. Place the sample into the magnet bore and allow to descend into the MAS probehead with the rotor cap at the top.
8. Ensure that the sample is properly inserted into the probe and at the magic angle, then begin rotation and equilibrate sample to the desired spin speed (up to 5000 Hz at 14 T or 600 MHz ^{1}H) and temperature (1–10 °C, lower temperatures preserve the sample better but reduce the visibility of lipid resonances).

9. "Lock" the spectrometer to the D_2O resonance, tune and match the probe. Check lineshape and adjust shims if necessary. It is usually sufficient to optimize just Z, X, ZX, and X^2–Y^2 shims. *Note*: a good starting point can be obtained by shimming using rotor filled with just a solution of 10 m*M* TSP/D_2O. More details on shimming MAS probes are given in (Piotto, Elbayed, Wieruszeski, & Lippens, 2005; Sodickson & Cory, 1997). Occasionally, it may be necessary to calibrate the magic angle alignment of the probe; this can be achieved with a KBr standard.
10. Calibrate the 90° pulse and presaturation frequency.

5.3. NMR spectral acquisition

11. Typical pulse sequences to be used are a standard 1D presaturation sequence such as "noesy presat" and the 1D Carr–Purcell–Meiboom–Gill (CPMG); this uses a spin echo to edit the spectrum according to T_2 which tends to reduce signals from large molecules such as lipid aggregates. The receiver gain should be checked for each sample/sequence, and the dynamic range of the instrument not exceed. Details of other sequences such as for 2D experiments, for editing on other factors (e.g., diffusion) or for detection of nuclei other than ^{1}H can be found in Beckonert et al. (2010).
12. Typical experimental parameters operating at 600 MHz ^{1}H frequency would be to record 128–256 free induction decays (FIDs) into 32 K complex data points with a spectral width of 20 ppm giving an acquisition time of 1.36 s and using a relaxation delay of 2 s.
13. Noesy-presat parameters: RD–90°–3 μs–90°–T_1 delay–90°–acquire. The T_1 delay is typically ~100 ms—this parameter usually does not impact significantly on the improvement in solvent suppression.
14. CPMG-specific parameters: RD–90°–$(\tau–180–\tau)_n$–acquire. The total T_2 delay is given by $2 \times \tau \times n$, and a series of n values should be evaluated on a test sample to achieve the desired suppression of resonances from high-molecular weight species. The τ delay should be kept fixed and chosen to avoid J-coupling artifacts or excessive heating. At 600 MHz ^{1}H typically $\tau = 200$–400 μs with a total T_2 delay of up to 240 ms.
15. Recorded FIDs are typically then processed with zero filling to 64 K and exponential line broadening of 0.3–1 Hz. Calibration of the frequency axis can be made by the α-glucose anomeric proton resonance (5.23 ppm) or the alanine methyl resonance (1.47 ppm).

REFERENCES

Akoka, S., Barantin, L., & Trierweiler, M. (1999). Concentration measurement by proton NMR using the ERETIC method. *Analytical Chemistry*, *71*(13), 2554–2557.

Al-Saffar, N. M. S., Troy, H., de Molina, A. R., Jackson, L. E., Madhu, B., Griffiths, J. R., et al. (2006). Noninvasive magnetic resonance spectroscopic pharmacodynamic markers of the choline kinase inhibitor MN58b in human carcinoma models. *Cancer Research*, *66*(1), 427–434.

Andrew, E. R., & Newing, R. A. (1958). The narrowing of nuclear magnetic resonance spectra by molecular rotation in solids. *Proceedings of the Physical Society*, *72*, 959–972.

Andronesi, O. C., Blekas, K. D., Mintzopoulos, D., Astrakas, L., Black, P. M., & Tzika, A. A. (2008). Molecular classification of brain tumor biopsies using solid-state magic angle spinning proton magnetic resonance spectroscopy and robust classifiers. *International Journal of Oncology*, *33*(5), 1017–1025.

Andronesi, O. C., Kim, G. S., Gerstner, E., Batchelor, T., Tzika, A. A., Fantin, V. R., et al. (2012). Detection of 2-hydroxyglutarate in IDH-mutated glioma patients by in vivo spectral-editing and 2D correlation magnetic resonance spectroscopy. *Science Translational Medicine*, *4*(116), 116ra114.

Backshall, A., Alferez, D., Teichert, F., Wilson, I. D., Wilkinson, R. W., Goodlad, R. A., et al. (2009). Detection of metabolic alterations in non-tumor gastrointestinal tissue of the Apc(Min/+) mouse by H-1 MAS NMR spectroscopy. *Journal of Proteome Research*, *8*(3), 1423–1430.

Bathen, T. F., Geurts, B., Sitter, B., Fjosne, H. E., Lundgren, S., Buydens, L. M., et al. (2013). Feasibility of MR metabolomics for immediate analysis of resection margins during breast cancer surgery. *PLoS One*, *8*(4), e61578.

Bathen, T. F., Jensen, L. R., Sitter, B., Fjosne, H. E., Halgunset, J., Axelson, D. E., et al. (2007). MR-determined metabolic phenotype of breast cancer in prediction of lymphatic spread, grade, and hormone status. *Breast Cancer Research and Treatment*, *104*(2), 181–189.

Beckonert, O., Coen, M., Keun, H. C., Wang, Y., Ebbels, T. M., Holmes, E., et al. (2010). High-resolution magic-angle-spinning NMR spectroscopy for metabolic profiling of intact tissues. *Nature Protocols*, *5*(6), 1019–1032.

Beckonert, O., Keun, H. C., Ebbels, T. M. D., Bundy, J. G., Holmes, E., Lindon, J. C., et al. (2007). Metabolic profiling, metabolomic and metabonomic procedures for NMR spectroscopy of urine, plasma, serum and tissue extracts. *Nature Protocols*, *2*(11), 2692–2703.

Bertilsson, H., Tessem, M. B., Flatberg, A., Viset, T., Gribbestad, I., Angelsen, A., et al. (2012). Changes in gene transcription underlying the aberrant citrate and choline metabolism in human prostate cancer samples. *Clinical Cancer Research: An Official Journal of the American Association for Cancer Research*, *18*(12), 3261–3269.

Bharti, S. K., Behari, A., Kapoor, V. K., Kumari, N., Krishnani, N., & Roy, R. (2013). Magic angle spinning NMR spectroscopic metabolic profiling of gall bladder tissues for differentiating malignant from benign disease. *Metabolomics*, *9*(1), 101–118.

Calabrese, C., Pisi, A., Di Febo, G., Liguori, G., Filippini, G., Cervellera, M., et al. (2008). Biochemical alterations from normal mucosa to gastric cancer by ex vivo magnetic resonance spectroscopy. *Cancer Epidemiology, Biomarkers & Prevention: A Publication of the American Association for Cancer Research, Cosponsored by the American Society of Preventive Oncology*, *17*(6), 1386–1395.

Cao, M. D., Giskeodegard, G. F., Bathen, T. F., Sitter, B., Bofin, A., Lonning, P. E., et al. (2012). Prognostic value of metabolic response in breast cancer patients receiving neoadjuvant chemotherapy. *BMC Cancer*, *12*, 39.

Cao, M. D., Sitter, B., Bathen, T. F., Bofin, A., Lonning, P. E., Lundgren, S., et al. (2012). Predicting long-term survival and treatment response in breast cancer patients receiving neoadjuvant chemotherapy by MR metabolic profiling. *NMR in Biomedicine, 25*(2), 369–378.

Celda, B., Martinez-Bisbal, M. C., Monleon, D., Assemat, O., Piotto, M., Piquer, J., et al. (2009). Determination of metabolite concentrations in human brain tumour biopsy samples using HR-MAS and ERETIC measurements. *NMR in Biomedicine, 22*(2), 199–206.

Chan, E. C., Koh, P. K., Mal, M., Cheah, P. Y., Eu, K. W., Backshall, A., et al. (2009). Metabolic profiling of human colorectal cancer using high-resolution magic angle spinning nuclear magnetic resonance (HR-MAS NMR) spectroscopy and gas chromatography mass spectrometry (GC/MS). *Journal of Proteome Research, 8*(1), 352–361.

Cheng, L. L., Burns, M. A., Taylor, J. L., He, W., Halpern, E. F., McDougal, W. S., et al. (2005a). Metabolic characterization of human prostate cancer with tissue magnetic resonance spectroscopy. *Cancer Research, 65*(8), 3030–3034.

Cheng, L. L., Chang, I. W., Louis, D. N., & Gonzalez, R. G. (1998). Correlation of high-resolution magic angle spinning proton magnetic resonance spectroscopy with histopathology of intact human brain tumor specimens. *Cancer Research, 58*(9), 1825–1832.

Cheng, L. L., Chang, I. W., Smith, B. L., & Gonzalez, R. G. (1998). Evaluating human breast ductal carcinomas with high-resolution magic-angle spinning proton magnetic resonance spectroscopy. *Journal of Magnetic Resonance, 135*(1), 194–202.

Cheng, L. L., Lean, C. L., Bogdanova, A., Wright, S. C., Jr., Ackerman, J. L., Brady, T. J., et al. (1996). Enhanced resolution of proton NMR spectra of malignant lymph nodes using magic-angle spinning. *Magnetic Resonance in Medicine: Official Journal of the Society of Magnetic Resonance in Medicine/Society of Magnetic Resonance in Medicine, 36*(5), 653–658.

Cheng, L. L., Wu, C., Smith, M. R., & Gonzalez, R. G. (2001). Non-destructive quantitation of spermine in human prostate tissue samples using HRMAS 1H NMR spectroscopy at 9.4T. *FEBS Letters, 494*(1–2), 112–116.

Chinnaiyan, P., Kensicki, E., Bloom, G., Prabhu, A., Sarcar, B., Kahali, S., et al. (2012). The metabolomic signature of malignant glioma reflects accelerated anabolic metabolism. *Cancer Research, 72*(22), 5878–5888.

Costello, L. C., & Franklin, R. B. (2006). The clinical relevance of the metabolism of prostate cancer; zinc and tumor suppression: Connecting the dots. *Molecular Cancer, 5*, 17.

Davies, N. P., Wilson, M., Natarajan, K., Sun, Y., MacPherson, L., Brundler, M. A., et al. (2010). Non-invasive detection of glycine as a biomarker of malignancy in childhood brain tumours using in-vivo 1H MRS at 1.5 tesla confirmed by ex-vivo high-resolution magic-angle spinning NMR. *NMR in Biomedicine, 23*(1), 80–87.

Denkert, C., Bucher, E., Hilvo, M., Salek, R., Oresic, M., Griffin, J., et al. (2012). Metabolomics of human breast cancer: New approaches for tumor typing and biomarker discovery. *Genome Medicine, 4*(4), 37.

Dittrich, R., Kurth, J., Decelle, E. A., Defeo, E. M., Taupitz, M., Wu, S., et al. (2012). Assessing prostate cancer growth with citrate measured by intact tissue proton magnetic resonance spectroscopy. *Prostate Cancer and Prostatic Diseases, 15*(3), 278–282.

Gatenby, R. A., & Gillies, R. J. (2004). Why do cancers have high aerobic glycolysis? *Nature Reviews. Cancer, 4*(11), 891–899.

Giskeodegard, G. F., Bertilsson, H., Selnaes, K. M., Wright, A. J., Bathen, T. F., Viset, T., et al. (2013). Spermine and citrate as metabolic biomarkers for assessing prostate cancer aggressiveness. *PLoS One, 8*(4), e62375.

Giskeodegard, G. F., Grinde, M. T., Sitter, B., Axelson, D. E., Lundgren, S., Fjosne, H. E., et al. (2010). Multivariate modeling and prediction of breast cancer prognostic factors using MR metabolomics. *Journal of Proteome Research, 9*(2), 972–979.

Giskeodegard, G. F., Lundgren, S., Sitter, B., Fjosne, H. E., Postma, G., Buydens, L. M. C., et al. (2012). Lactate and glycine-potential MR biomarkers of prognosis in estrogen receptor-positive breast cancers. *NMR in Biomedicine, 25*(11), 1271–1279.

Glunde, K., & Bhujwalla, Z. M. (2007). Choline kinase alpha in cancer prognosis and treatment. *The Lancet Oncology*, *8*(10), 855–857.

Glunde, K., Jie, C., & Bhujwalla, Z. M. (2004). Molecular causes of the aberrant choline phospholipid metabolism in breast cancer. *Cancer Research*, *64*(12), 4270–4276.

Gonzalez-Velez, H., Mier, M., Julia-Sape, M., Arvanitis, T. N., Garcia-Gomez, J. M., Robles, M., et al. (2009). HealthAgents: Distributed multi-agent brain tumor diagnosis and prognosis. *Applied Intelligence*, *30*(3), 191–202.

Griffin, J. L., Lehtimaki, K. K., Valonen, P. K., Grohn, O. H., Kettunen, M. I., Yla-Herttuala, S., et al. (2003). Assignment of 1H nuclear magnetic resonance visible polyunsaturated fatty acids in BT4C gliomas undergoing ganciclovir-thymidine kinase gene therapy-induced programmed cell death. *Cancer Research*, *63*(12), 3195–3201.

Griffin, J. L., Nicholls, A. W., Keun, H. C., Mortishire-Smith, R. J., Nicholson, J. K., & Kuehn, T. (2002). Metabolic profiling of rodent biological fluids via H-1 NMR spectroscopy using a 1 mm microlitre probe. *Analyst*, *127*(5), 582–584.

Iorio, E., Ricci, A., Bagnoli, M., Pisanu, M. E., Castellano, G., Di Vito, M., et al. (2010). Activation of phosphatidylcholine cycle enzymes in human epithelial ovarian cancer cells. *Cancer Research*, *70*(5), 2126–2135.

Jain, M., Nilsson, R., Sharma, S., Madhusudhan, N., Kitami, T., Souza, A. L., et al. (2012). Metabolite profiling identifies a key role for glycine in rapid cancer cell proliferation. *Science*, *336*(6084), 1040–1044.

Jimenez, B., Mirnezami, R., Kinross, J., Cloarec, O., Keun, H. C., Holmes, E., et al. (2013). 1H HR-MAS NMR spectroscopy of tumor-induced local metabolic "field-effects" enables colorectal cancer staging and prognostication. *Journal of Proteome Research*, *12*(2), 959–968.

Keshari, K. R., Tsachres, H., Iman, R., Delos Santos, L., Tabatabai, Z. L., Shinohara, K., et al. (2011). Correlation of phospholipid metabolites with prostate cancer pathologic grade, proliferative status and surgical stage—Impact of tissue environment. *NMR in Biomedicine*, *24*(6), 691–699.

Keun, H. C., Beckonert, O., Griffin, J. L., Richter, C., Moskau, D., Lindon, J. C., et al. (2002). Cryogenic probe C-13 NMR spectroscopy of urine for metabonomic studies. *Analytical Chemistry*, *74*(17), 4588–4593.

Kumar, S., Kumar, S., Singh, A., Goel, M. M., Roy, R., & Agarwal, G. (2012). Evaluation of axillary nodal metastasis with high resolution magic angle proton magnetic resonance spectroscopy in breast cancer patients—A pilot study. *European Journal of Cancer*, *48*, S73.

Kurhanewicz, J., Swanson, M. G., Nelson, S. J., & Vigneron, D. B. (2002). Combined magnetic resonance imaging and spectroscopic imaging approach to molecular imaging of prostate cancer. *Journal of Magnetic Resonance Imaging*, *16*(4), 451–463.

Lowe, I. J. (1959). Free induction decays of rotating solids. *Physical Review Letters*, *2*, 285–287.

Maddocks, O. D., Berkers, C. R., Mason, S. M., Zheng, L., Blyth, K., Gottlieb, E., et al. (2013). Serine starvation induces stress and p53-dependent metabolic remodelling in cancer cells. *Nature*, *493*(7433), 542–546.

Mahon, M. M., Cox, I. J., Dina, R., Soutter, W. P., McIndoe, G. A., Williams, A. D., et al. (2004). H-1 magnetic resonance spectroscopy of preinvasive and invasive cervical cancer: In vivo-ex vivo profiles and effect of tumor load. *Journal of Magnetic Resonance Imaging*, *19*(3), 356–364.

Mahon, M. M., deSouza, N. M., Dina, R., Soutter, W. P., McIndoe, G. A., Williams, A. D., et al. (2004). Preinvasive and invasive cervical cancer: An ex vivo proton magic angle spinning magnetic resonance spectroscopy study. *NMR in Biomedicine*, *17*(3), 144–153.

Martinez-Bisbal, M. C., Marti-Bonmati, L., Piquer, J., Revert, A., Ferrer, P., Llacer, J. L., et al. (2004). H-1 and C-13 HR-MAS spectroscopy of intact biopsy samples ex vivo and in vivo H-1 MRS study of human high grade gliomas. *NMR in Biomedicine*, *17*(4), 191–205.

Martinez-Bisbal, M. C., Monleon, D., Assemat, O., Piotto, M., Piquer, J., Llacer, J. L., et al. (2009). Determination of metabolite concentrations in human brain tumour biopsy samples using HR-MAS and ERETIC measurements. *NMR in Biomedicine, 22*(2), 199–206.

Maxeiner, A., Adkins, C. B., Zhang, Y. F., Taupitz, M., Halpern, E. F., McDougal, W. S., et al. (2010). Retrospective analysis of prostate cancer recurrence potential with tissue metabolomic profiles. *Prostate, 70*(7), 710–717.

Millis, K., Weybright, P., Campbell, N., Fletcher, J. A., Fletcher, C. D., Cory, D. G., et al. (1999). Classification of human liposarcoma and lipoma using ex vivo proton NMR spectroscopy. *Magnetic Resonance in Medicine: Official Journal of the Society of Magnetic Resonance in Medicine/Society of Magnetic Resonance in Medicine, 41*(2), 257–267.

Mirnezami, R., Jimenez, B., Li, J. V., Kinross, J. M., Veselkov, K., Goldin, R. D., et al. (2013). Rapid diagnosis and staging of colorectal cancer via high-resolution magic angle spinning nuclear magnetic resonance (HR-MAS NMR) spectroscopy of intact tissue biopsies. *Annals of Surgery*.

Opstad, K. S., Bell, B. A., Griffiths, J. R., & Howe, F. A. (2008a). An assessment of the effects of sample ischaemia and spinning time on the metabolic profile of brain tumour biopsy specimens as determined by high-resolution magic angle spinning (1)H NMR. *NMR in Biomedicine, 21*(10), 1138–1147.

Opstad, K. S., Bell, B. A., Griffiths, J. R., & Howe, F. A. (2008b). An investigation of human brain tumour lipids by high-resolution magic angle spinning 1H MRS and histological analysis. *NMR in Biomedicine, 21*(7), 677–685.

Opstad, K. S., Bell, B. A., Griffiths, J. R., & Howe, F. A. (2009). Taurine: A potential marker of apoptosis in gliomas. *British Journal of Cancer, 100*(5), 789–794.

Peet, A. C., McConville, C., Wilson, M., Levine, B. A., Reed, M., Dyer, S. A., et al. (2007). H-1 MRS identifies specific metabolite profiles associated with MYCN-amplified and non-amplified tumour subtypes of neuroblastoma cell lines. *NMR in Biomedicine, 20*(7), 692–700.

Piotto, M., Elbayed, K., Wieruszeski, J. M., & Lippens, G. (2005). Practical aspects of shimming a high resolution magic angle spinning probe. *Journal of Magnetic Resonance, 173*(1), 84–89.

Piotto, M., Moussallieh, F. M., Dillmann, B., Imperiale, A., Neuville, A., Brigand, C., et al. (2009). Metabolic characterization of primary human colorectal cancers using high resolution magic angle spinning (1)H magnetic resonance spectroscopy. *Metabolomics, 5*(3), 292–301.

Righi, V., Andronesi, O. C., Mintzopoulos, D., Black, P. M., & Tzika, A. A. (2010). High-resolution magic angle spinning magnetic resonance spectroscopy detects glycine as a biomarker in brain tumors. *International Journal of Oncology, 36*(2), 301–306.

Righi, V., Mucci, A., Schenetti, L., Bacci, A., Agati, R., Leonardi, M., et al. (2009). Identification of mobile lipids in human cancer tissues by ex vivo diffusion edited HR-MAS MRS. *Oncology Reports, 22*(6), 1493–1496.

Righi, V., Roda, J. M., Paz, J., Mucci, A., Tugnoli, V., Rodriguez-Tarduchy, G., et al. (2009). H-1 HR-MAS and genomic analysis of human tumor biopsies discriminate between high and low grade astrocytomas. *NMR in Biomedicine, 22*(6), 629–637.

Rocha, C. M., Barros, A. S., Gil, A. M., Goodfellow, B. J., Humpfer, E., Spraul, M., et al. (2010). Metabolic profiling of human lung cancer tissue by 1H high resolution magic angle spinning (HRMAS) NMR spectroscopy. *Journal of Proteome Research, 9*(1), 319–332.

Santos, C. F., Kurhanewicz, J., Tabatabai, Z. L., Simko, J. P., Keshari, K. R., Gbegnon, A., et al. (2010). Metabolic, pathologic, and genetic analysis of prostate tissues: Quantitative evaluation of histopathologic and mRNA integrity after HR-MAS spectroscopy. *NMR in Biomedicine, 23*(4), 391–398.

Sitter, B., Bathen, T. F., Singstad, T. E., Fjosne, H. E., Lundgren, S., Halgunset, J., et al. (2010). Quantification of metabolites in breast cancer patients with different clinical prognosis using HR MAS MR spectroscopy. *NMR in Biomedicine, 23*(4), 424–431.

Sitter, B., Lundgren, S., Bathen, T. F., Halgunset, J., Fjosne, H. E., & Gribbestad, I. S. (2006). Comparison of HR MAS MR spectroscopic profiles of breast cancer tissue with clinical parameters. *NMR in Biomedicine, 19*(1), 30–40.

Sitter, B., Sonnewald, U., Spraul, M., Fjosne, H. E., & Gribbestad, I. S. (2002). High-resolution magic angle spinning MRS of breast cancer tissue. *NMR in Biomedicine, 15*(5), 327–337.

Sjobakk, T. E., Johansen, R., Bathen, T. F., Sonnewald, U., Juul, R., Torp, S. H., et al. (2008). Characterization of brain metastases using high-resolution magic angle spinning MRS. *NMR in Biomedicine, 21*(2), 175–185.

Sjobakk, T. E., Vettukattil, R., Gulati, M., Gulati, S., Lundgren, S., Gribbestad, I. S., et al. (2013). Metabolic profiles of brain metastases. *International Journal of Molecular Sciences, 14*(1), 2104–2118.

Sodickson, A., & Cory, D. G. (1997). Shimming a high-resolution MAS probe. *Journal of Magnetic Resonance, 128*(1), 87–91.

Somashekar, B. S., Kamarajan, P., Danciu, T., Kapila, Y. L., Chinnaiyan, A. M., Rajendiran, T. M., et al. (2011). Magic angle spinning NMR-based metabolic profiling of head and neck squamous cell carcinoma tissues. *Journal of Proteome Research, 10*(11), 5232–5241.

Sonveaux, P., Vegran, F., Schroeder, T., Wergin, M. C., Verrax, J., Rabbani, Z. N., et al. (2008). Targeting lactate-fueled respiration selectively kills hypoxic tumor cells in mice. *Journal of Clinical Investigation, 118*(12), 3930–3942.

Stenman, K., Stattin, P., Stenlund, H., Riklund, K., Grobner, G., & Bergh, A. (2011). H HRMAS NMR derived bio-markers related to tumor grade, tumor cell fraction, and cell proliferation in prostate tissue samples. *Biomarker Insights, 6*, 39–47.

Stewart, J. D., Marchan, R., Lesjak, M. S., Lambert, J., Hergenroeder, R., Ellis, J. K., et al. (2012). Choline-releasing glycerophosphodiesterase EDI3 drives tumor cell migration and metastasis. *Proceedings of the National Academy of Sciences of the United States of America, 109*(21), 8155–8160.

Swanson, M. G., Vigneron, D. B., Tabatabai, Z. L., Males, R. G., Schmitt, L., Carroll, P. R., et al. (2003). Proton HR-MAS spectroscopy and quantitative pathologic analysis of MRI/3D-MRSI-targeted postsurgical prostate tissues. *Magnetic Resonance in Medicine, 50*(5), 944–954.

Swanson, M. G., Zektzer, A. S., Tabatabai, Z. L., Simko, J., Jarso, S., Keshari, K. R., et al. (2006). Quantitative analysis of prostate metabolites using H-1 HR-MAS spectroscopy. *Magnetic Resonance in Medicine, 55*(6), 1257–1264.

Teahan, O., Bevan, C. L., Waxman, J., & Keun, H. C. (2011). Metabolic signatures of malignant progression in prostate epithelial cells. *International Journal of Biochemistry & Cell Biology, 43*(7), 1002–1009.

Teichert, F., Verschoyle, R. D., Greaves, P., Edwards, R. E., Teahan, O., Jones, D. J. L., et al. (2008). Metabolic profiling of transgenic adenocarcinoma of mouse prostate (TRAMP) tissue by H-1-NMR analysis: Evidence for unusual phospholipid metabolism. *Prostate, 68*(10), 1035–1047.

Tessem, M. B., Swanson, M. G., Keshari, K. R., Albers, M. J., Joun, D., Tabatabai, Z. L., et al. (2008). Evaluation of lactate and alanine as metabolic biomarkers of prostate cancer using 1H HR-MAS spectroscopy of biopsy tissues. *Magnetic Resonance in Medicine: Official Journal of the Society of Magnetic Resonance in Medicine/Society of Magnetic Resonance in Medicine, 60*(3), 510–516.

Tzika, A. A., Astrakas, L., Cao, H., Mintzopoulos, D., Andronesi, O. C., Mindrinos, M., et al. (2007). Combination of high-resolution magic angle spinning proton magnetic resonance spectroscopy and microscale genomics to type brain tumor biopsies. *International Journal of Molecular Medicine*, *20*(2), 199–208.

Tzika, A. A., Cheng, L. L., Goumnerova, L., Madsen, J. R., Zurakowski, D., Astrakas, L. G., et al. (2002). Biochemical characterization of pediatric brain tumors by using in vivo and ex vivo magnetic resonance spectroscopy. *Journal of Neurosurgery*, *96*(6), 1023–1031.

van Asten, J. J., Cuijpers, V., Hulsbergen-van de Kaa, C., Soede-Huijbregts, C., Witjes, J. A., Verhofstad, A., et al. (2008). High resolution magic angle spinning NMR spectroscopy for metabolic assessment of cancer presence and Gleason score in human prostate needle biopsies. *Magma*, *21*(6), 435–442.

Wang, Y., Bollard, M. E., Keun, H., Antti, H., Beckonert, O., Ebbels, T. M., et al. (2003). Spectral editing and pattern recognition methods applied to high-resolution magic-angle spinning 1H nuclear magnetic resonance spectroscopy of liver tissues. *Analytical Biochemistry*, *323*(1), 26–32.

Wong, A., Jimenez, B., Li, X., Holmes, E., Nicholson, J. K., Lindon, J. C., et al. (2012). Evaluation of high resolution magic-angle coil spinning NMR spectroscopy for metabolic profiling of nanoliter tissue biopsies. *Analytical Chemistry*, *84*(8), 3843–3848.

Wright, A. J., Fellows, G. A., Griffiths, J. R., Wilson, M., Bell, B. A., & Howe, F. A. (2010). Ex-vivo HRMAS of adult brain tumours: Metabolite quantification and assignment of tumour biomarkers. *Molecular Cancer*, *9*, 66.

Yakoub, D., Keun, H. C., Goldin, R., & Hanna, G. B. (2010). Metabolic profiling detects field effects in nondysplastic tissue from esophageal cancer patients. *Cancer Research*, *70*(22), 9129–9136.

Yalcin, A., Clem, B., Makoni, S., Clem, A., Nelson, K., Thornburg, J., et al. (2010). Selective inhibition of choline kinase simultaneously attenuates MAPK and PI3K/AKT signaling. *Oncogene*, *29*(1), 139–149.

Yang, Y., Wang, L., Wang, S., Liang, S., Chen, A., Tang, H., et al. (2013). Study of metabonomic profiles of human esophageal carcinoma by use of high-resolution magic-angle spinning 1H NMR spectroscopy and multivariate data analysis. *Analytical and Bioanalytical Chemistry*, *405*(10), 3381–3389.

Yokota, H., Guo, J. F., Matoba, M., Higashi, K., Tonami, H., & Nagao, Y. (2007). Lactate, choline, and creatine levels measured by vitro H-1-MRS as prognostic parameters in patients with non-small-cell lung cancer. *Journal of Magnetic Resonance Imaging*, *25*(5), 992–999.

Zietkowski, D., Davidson, R. L., Eykyn, T. R., De Silva, S. S., Desouza, N. M., & Payne, G. S. (2010). Detection of cancer in cervical tissue biopsies using mobile lipid resonances measured with diffusion-weighted (1)H magnetic resonance spectroscopy. *NMR in Biomedicine*, *23*(4), 382–390.

CHAPTER SIXTEEN

Analysis of Metabolomic Profiling Data Acquired on GC–MS

Imhoi Koo, Xiaoli Wei, Xiang Zhang[1]
Department of Chemistry, University of Louisville, Louisville, Kentucky, USA
[1]Corresponding author: e-mail address: xiang.zhang@louisville.edu

Contents

Abstract

Gas chromatography–mass spectrometry (GC–MS) is one of the three most popular analytical platforms for metabolomics and is largely employed for the study of oncometabolism. Large volumes of data are usually generated in a GC–MS experiment, and many analytical steps are required to extract biologically relevant information from GC–MS data. These steps include (1) spectrum deconvolution, to convert raw data into a peak list; (2) metabolite identification, to recognize metabolites associated to chromatographic peaks; (3) quantification, to compare the abundance of a specific metabolite in different samples; (4) association network analysis, to reveal correlations among the changes in the abundance of multiple metabolites; and (5) pathway analysis, to understand the biochemical interrelationship between several metabolites that vary in a coordinated or differential manner. Here, we describe in detail the analytical steps that are necessary to interpret a GC–MS dataset.

Methods in Enzymology, Volume 543
ISSN 0076-6879
http://dx.doi.org/10.1016/B978-0-12-801329-8.00016-7

1. INTRODUCTION

Metabolome represents the collection of small compound metabolites in an organism, typically under 1000 Da. All metabolites in a metabolome form a large network of metabolic reactions, where outputs from one enzymatic chemical reaction are inputs to other chemical reactions. For this reason, metabolome is time sensitive, and much more dynamic than proteome and genome. Metabolites, as the products of gene expression and all other biological processes, have a high degree of diversity in their chemical properties. For example, metabolites have a wide range of molecular weights and large variations in concentration, and metabolites can be polar or nonpolar, as well as organic or inorganic molecules. The diverse characteristics of metabolites make the chemical separation and detection challenging technical steps in metabolomics.

Due to the extreme complexity of metabolome, three types of analytical platforms are currently employed in metabolomics: gas chromatography–mass spectrometry (GC–MS), liquid chromatography–mass spectrometry (LC–MS), and nuclear magnetic resonance (NMR). Technological advances now allow collection of enormous quantities of metabolomics data at the organelle, cell, tissue, organ, and organism level. These data not only provide critical information about biomolecular function but also raise new questions concerning the biochemical, spatial, and temporal relationships of metabolites. Effective use of these voluminous molecular data will enable a better understanding of the biological system being investigated.

In this chapter, we introduce the bioinformatics approaches for analyzing metabolomic profiling data acquired on GC–MS systems. During the GC–MS analysis, metabolites are first separated on GC column based on their partition coefficients between mobile phase and stationary phase. The separated metabolites are further subjected to mass spectrometer for detection. Therefore, the GC–MS system measures retention time and mass spectrum of each metabolite. Large volumes of data are usually generated in a GC–MS experiment that contains information on a broad range of compound classes. Many data analysis steps are involved to uncover the biological knowledge from the GC–MS data. We will introduce the current practice in spectrum deconvolution, metabolite identification, quantification, association network analysis, and pathway analysis, for analysis of GC–MS data.

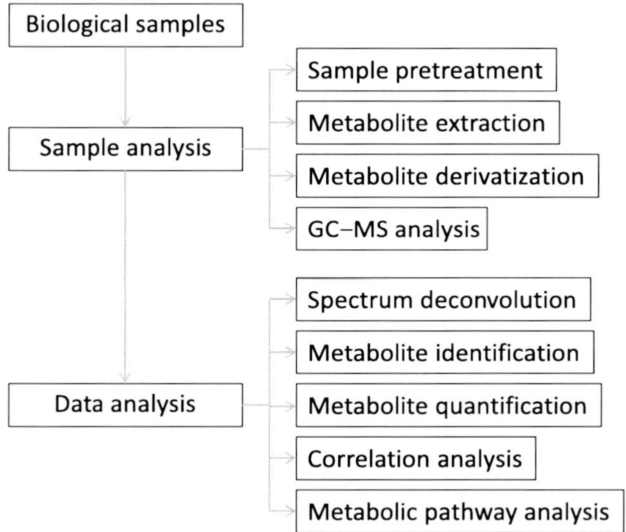

Figure 16.1 Workflow of analyzing GC–MS data in metabolomics.

Figure 16.1 depicts the workflow of GC–MS analysis. During the sample analysis step, the biological samples are usually preprocessed for metabolite extraction, such as homogenization of tissue samples. The extracted metabolites are derivatized to make the metabolites volatile and thermally stable for GC–MS analysis. Many parameters can affect the performance of GC–MS, including column selection, temperature gradient rate, flow rate, sample rate of data acquisition, etc. Therefore, quality control is a critical component in GC–MS-based metabolomics. A sample practice is that a blank and a pooled sample are prepared in parallel together with the real biological samples, respectively. The blank sample is prepared without the addition of metabolite extract, while the pooled sample is prepared by mixing a portion of multiple biological samples. During the GC–MS analysis, the biological samples are analyzed in a random order to avoid systematic bias. Both blank and pooled samples are analyzed on GC–MS after every five biological samples. Experimental data of the blank sample are used as negative control to monitor the background signal intensity and carry-overs during GC–MS analyses, and the data of the pooled sample are used to monitor the sensitivity and stability of the GC–MS system. The statistical clustering and other type of analysis of the data of QC samples can reveal the integrity of the chromatographic method.

2. SPECTRUM DECONVOLUTION

The purpose of spectral deconvolution is to differentiate signals arising from the real analyte as opposed to signals arising from contaminants or instrumental noise. In GC–MS spectrum deconvolution, a typical approach is to first construct selected ion chromatogram (XIC) for each mass-to-charge (*m/z*) value, that is, mass channel, from a GC–MS dataset. After detecting peaks in each XIC, all peaks that exhibit the same GC peak shape over their XICs are clustered together to reconstruct the chromatogram of a metabolite.

A number of algorithms and software tools have been developed for deconvoluting GC–MS data, including AMDIS (Stein & Scott, 1994), ChromaTOF, multivariate curve resolution (Jonsson et al., 2005, 2006), and ADAP (Ni et al., 2012). While these existing computational algorithms and associated software tools have been widely used in metabolomics, further development for accurately deconvoluting the GC–MS data is still needed. A common limitation in the current analysis packages is that the existing algorithms require manually assigned signal-to-noise ratio threshold and/or denoising parameters. These user-defined parameters are usually not optimized in the existing peak detection algorithms, resulting in a certain rate of false-positive and/or false-negative peaks.

3. METABOLITE IDENTIFICATION

Most of the GC–MS instruments are equipped with electron ionization (EI). In an EI source, the electrons are accelerated to 70 eV. The concentrated election beam with such a high energy can directly fragment a metabolite into multiple fragments to form an EI mass spectrum. It is possible that molecular ion of a metabolite may not present in the resulting mass spectrum. Therefore, compound identification in GC–MS is usually achieved by directly comparing a query EI mass spectrum with reference mass spectra in a library via spectrum matching without the use of molecular ion *m/z* value.

Several mass spectral libraries have been created, and various mass spectral similarity measures have been developed, including composite similarity (Stein & Scott, 1994), Fourier and wavelet transform-based composite measures (Koo, Zhang, & Kim, 2011), and mixture semi-partial and partial correlation-based measures (Kim, Koo, Jeong, et al., 2012). Furthermore, some efforts have been also devoted to find the optimal weight factors corresponding to both intensity and m/z value (Kim, Koo, Wei, & Zhang, 2012; Koo, Kim, & Zhang, 2013).

An EI mass spectrum only reveals partial molecular information of a metabolite. It is necessary to incorporate the metabolite separation information to achieve high accuracy of identification. Since retention time in GC depends on experiment conditions, retention index was introduced to reduce such dependency and further used for metabolite identification. Although a group of compounds may have very high degree of mass spectral similarity with one or multiple reference mass spectra, the difference of retention index, that is, normalized retention time, among them may be large enough to differentiate the true compound from others. A few software tools have been developed to incorporate metabolite's mass spectrum and retention index for identification (Zhang et al., 2011, 2012). However, the existing methods use the retention index and mass spectrum in two separate analysis steps, by employing retention index as a filter to remove the potential false-positive identifications generated by mass spectrum matching. The sequential nature of the two-step analysis strategy increases the risk of introducing errors from each independent stage since there is no way to correct the errors caused by the previous step.

Even though combination of mass spectral matching and retention index filtering can improve the identification accuracy, the extreme complexity of a metabolome significantly diminishes the accuracy of metabolite identification. Furthermore, the incompleteness of the existing mass spectral library not only introduces a certain degree of false-positive identifications but also leaves a number of chromatographic peaks without any compound identification. Therefore, the gold standard method for metabolite identification is still the use of authentic standards of the metabolites of interest.

4. METABOLITE QUANTIFICATION

4.1. Cross-sample alignment

Ideally, the retention time and mass spectrum of a metabolite should be the same in different samples if these samples are analyzed on the same GC–MS system under the identical experimental conditions. However, this is not the case due to experimental variation. The purpose of cross-sample alignment is to recognize the metabolite peaks generated by the same metabolite in different samples.

Two types of alignment approaches have been developed: profile alignment and peak alignment. The profile alignment directly uses the entire chromatographic data, that is, the raw instrument data, as the input data. In the peak alignment approach, the raw instrument data are first deconvoluted to peak list, and the peak lists of multiple samples are then used as the input data for alignment. MetPP software employs a two-step peak alignment approach, full alignment and partial alignment, for analysis of GC–MS and GC × GC–MS data (Wei et al., 2013). During the alignment, the metabolite peaks in different samples are aligned based on their similarity of retention time(s) and mass spectrum measured by a mixture score. The full alignment recognizes all potential landmark peaks that are generated by the same metabolite and presented in all samples. After full alignment, the retention time values of remaining peaks in each peak list are corrected based on the retention time difference of the landmark peaks between each peak list and reference peak list. After that, the partial alignment is performed to align all the non-landmark peaks in each peak list to the peaks in reference peak list.

4.2. Normalization

To allow multi-experiment analyses, it is important to first normalize the data to make the samples comparable. Normalization step targets to quantitatively reduce overall peak intensity variations due to experimental errors such as varying amounts of samples loaded onto GC–MS, variations generated during sample preparation, etc. Several normalization methods have been proposed. Cyclic loess (Dudoit, Yang, Callow, & Speed, 2002) and contrast-based (Bolstad, Irizarry, Astrand, & Speed, 2003) normalizations are two extensions of the difference in log expression values versus the average of the log expression values method. Trimmed contrast mean and trimmed contrast median are two scaling-based normalization methods. Group-based quantile normalization first performs quantile normalization for the samples that belong to the same sample group and then employs a trimmed constant mean method to normalize all samples across the sample groups (Wei et al., 2011).

4.3. Statistical significance tests

The purpose of statistical significance analysis is to find metabolites that have significantly different abundance levels between sample groups. Multiple conventional statistical significance test methods can be used, such as two-tailed t-test, two-sample Kolmogorov–Smirnov test, and Wilcoxon rank sum test. To increase the confidence of statistical testing, sample permutation can be applied to each conventional statistical significance test. Permutation test, also known as re-randomization test, is a nonparametric test. It first randomly exchanges sample labels and then performs one of the above-mentioned conventional statistical test methods. Additionally, analysis of variance and Kruskal–Wallis test can be used for multiple groups' test.

Hundreds of metabolites can be detected in a GC–MS experiment. It is necessary to control the false discovery rate by correcting the multiple comparison issue. Multiple methods can be used, including the Bonferroni correction, the Benjamini–Hochberg method (Benjamini & Hochberg, 1995), and the q-value method (Storey, 2002).

4.4. Pattern recognition

Pattern recognition aims to study the differences of the metabolite expression profiles acquired under different physiological conditions. There are two main categories in pattern recognition: supervised and unsupervised learning. The former one uses the labeled training data to train a model or classifier for the regression or classification tasks, while the latter one groups data without prior knowledge.

Many supervised learning algorithms have been developed, such as linear discriminant classifier, decision tree, nearest-neighbor algorithm, artificial neural networks, and support vector machine (SVM). The SVM classifier has been demonstrated as the most popular and effective classifier in various data classification tasks. The main idea of SVM is to transform the original input space to a high-dimensional feature space by using a kernel function and then achieve optimum classification in this new feature space by choosing the hyperplane that maximizing the margin from the transformed features.

The most frequently used pattern recognition method in analysis of GC–MS data is the unsupervised learning. To reduce dimensions of the multivariate, multi-channeled data to a manageable subset, principal component analysis and partial least squares can be applied to improve the efficiency of clustering. Several clustering methods can be used to group the data into meaningful clusters, for example, k-means clustering, agglomerative hierarchical clustering, spectral clustering, fuzzy c-means clustering (Bezdek, 1981), and density-based spatial clustering of applications with noise (Ester, Kriegel, Sander, & Xu, 1996). To measure the clustering performance, the clustering accuracy can be represented as the number of correctly grouped samples divided by the number of all samples.

5. ASSOCIATION NETWORK ANALYSIS

Although the statistical significance test and pattern recognition can provide the regulation information of each metabolite related to a case–control study, the whole picture of why significant metabolites cause the related disease or cellular function remains unclear. An active research topic in metabolomics is to study the metabolite associations at the system level to obtain physiological snapshot of cellular status.

Pearson's correlation coefficient and rank-based correlation coefficients such as Spearman's and Kendall tau correlation coefficients are usually used to calculate correlation network. However, these approaches are unable to distinguish the direct from the indirect relationship among metabolites. Gaussian graphical model (GGM) was proposed to find direct correlations (Whittaker, 1990). GGM represents an association by a partial correlation coefficient to measure conditional dependence or independence between two metabolites controlling other metabolites. The human population cohort study showed that GGM found strong correlation of intracellular pathways in fasting blood serum samples (Krumsiek, Suhre, Illig, Adamski, & Theis, 2011). However, it is common that the number of samples is usually smaller than the number of metabolites in real experimental condition, resulting in several statistical and computational challenges. Multiple methods have been developed to address the sample size problem (Kramer, Schafer, & Boulesteix, 2009; Pihur, Datta, & Datta, 2008). However, finding the appropriate correlation algorithms for the calculation of metabolite association network remains as a challenge in metabolomics.

6. METABOLIC PATHWAY ANALYSIS

The quantitative metabolite analysis was performed on the abundance of each individual metabolite, and the interrelationships between the various metabolites were not considered. It is necessary to incorporate the analytical discovery with the metabolite pathway analysis to further filter and/or enrich the analytical discovery. For instance, Ingenuity Pathway Analysis (IPA) is a metabolic pathway analysis software package. It correlates specifically targeted metabolites with potential metabolic pathways for data analysis that helps researchers to model, analyze, and understand complex biological and chemical systems at the core of life science research. Details of using IPA can be found at http://ingenuity.com.

Compared with LC–MS system, GC–MS can only analyze volatile and thermally stable compounds. In addition, an extra analytical step, chemical derivatization, is always required prior to GC–MS for analysis of biological samples, such as plasma, urine, feces, and tissue. In order to achieve high metabolite coverage, the same biological samples are usually analyzed on both GC–MS and LC–MS systems. The metabolites detected in both analytical platforms can be used to evaluate the platform variation.

ACKNOWLEDGMENTS

This work was supported by NIH grant RO1GM087735 through the National Institute of General Medical Sciences.

REFERENCES

Benjamini, Y., & Hochberg, Y. (1995). Controlling the false discovery rate—A practical and powerful approach to multiple testing. *Journal of the Royal Statistical Society: Series B: Methodological*, *57*, 289–300.

Bezdek, J. C. (1981). *Pattern recognition with fuzzy objective function algorithms*. New York: Plenum Press.

Bolstad, B. M., Irizarry, R. A., Astrand, M., & Speed, T. P. (2003). A comparison of normalization methods for high density oligonucleotide array data based on variance and bias. *Bioinformatics*, *19*, 185–193.

Dudoit, S., Yang, Y. H., Callow, M. J., & Speed, T. P. (2002). Statistical methods for identifying differentially expressed genes in replicated cDNA microarray experiments. *Statistica Sinica*, *12*, 111–139.

Ester, M., Kriegel, H., Sander, J., & Xu, X. (1996). A density-based algorithm for discovering clusters in large spatial databases with noise. In *Proceedings of 2nd international conference on knowledge discovery and data mining* (p. 6).

Jonsson, P., Johansson, A. I., Gullberg, J., Trygg, J. A. J., Grung, B., & Moritz, T. (2005). High-throughput data analysis for detecting and identifying differences between samples in GC/MS-based metabolomic analyses. *Analytical Chemistry*, 77(17), 5635–5642.

Jonsson, P., Johansson, E. S., Wuolikainen, A., Lindberg, J., Schuppe-Koistinen, I., Kusano, M., et al. (2006). Predictive metabolite profiling applying hierarchical multivariate curve resolution to GC-MS data—A potential tool for multi-parametric diagnosis. *Journal of Proteome Research*, *5*(6), 1407–1414.

Kim, S., Koo, I., Jeong, J., Wu, S. W., Shi, X., & Zhang, X. (2012). Compound identification using partial and semipartial correlations for gas chromatography–mass spectrometry data. *Analytical Chemistry*, *84*(15), 6477–6487.

Kim, S., Koo, I., Wei, X. L., & Zhang, X. (2012). A method of finding optimal weight factors for compound identification in gas chromatography–mass spectrometry. *Bioinformatics*, *28*(8), 1158–1163.

Koo, I., Kim, S., & Zhang, X. (2013). Comparative analysis of mass spectral matching-based compound identification in gas chromatography–mass spectrometry. *Journal of Chromatography. A*, *1298*, 132–138.

Koo, I., Zhang, X., & Kim, S. (2011). Wavelet- and Fourier-transform-based spectrum similarity approaches to compound identification in gas chromatography/mass spectrometry. *Analytical Chemistry*, *83*(14), 5631–5638.

Kramer, N., Schafer, J., & Boulesteix, A. L. (2009). Regularized estimation of large-scale gene association networks using graphical Gaussian models. *BMC Bioinformatics*, *10*, 384.

Krumsiek, J., Suhre, K., Illig, T., Adamski, J., & Theis, F. J. (2011). Gaussian graphical modeling reconstructs pathway reactions from high-throughput metabolomics data. *BMC Systems Biology*, *5*, 21.

Ni, Y., Qiu, Y. P., Jiang, W. X., Suttlemyre, K., Su, M. M., Zhang, W. C., et al. (2012). ADAP-GC 2.0: Deconvolution of coeluting metabolites from GC/TOF-MS data for metabolomics studies. *Analytical Chemistry*, *84*(15), 6619–6629.

Pihur, V., Datta, S., & Datta, S. (2008). Reconstruction of genetic association networks from microarray data: A partial least squares approach. *Bioinformatics*, *24*(4), 561–568.

Stein, S. E., & Scott, D. R. (1994). Optimization and testing of mass-spectral library search algorithms for compound identification. *Journal of the American Society for Mass Spectrometry*, *5*(9), 859–866.

Storey, J. D. (2002). A direct approach to false discovery rates. *Journal of the Royal Statistical Society Series B-Statistical Methodology*, *64*, 479–498.

Wei, X. L., Shi, X., Koo, I., Kim, S., Schmidt, R. H., Arteel, G. E., et al. (2013). MetPP: A computational platform for comprehensive two-dimensional gas chromatography time-of-flight mass spectrometry-based metabolomics. *Bioinformatics*, *29*(14), 1786–1792.

Wei, X. L., Sun, W. L., Shi, X., Koo, I., Wang, B., Zhang, J., et al. (2011). MetSign: A computational platform for high-resolution mass spectrometry-based metabolomics. *Analytical Chemistry*, *83*(20), 7668–7675.

Whittaker, J. (1990). *Graphical models in applied multivariate statistics*. New York: Wiley.

Zhang, J., Fang, A. Q., Wang, B., Kim, S. H., Bogdanov, B., Zhou, Z. X., et al. (2011). iMatch: A retention index tool for analysis of gas chromatography–mass spectrometry data. *Journal of Chromatography. A*, *1218*(37), 6522–6530.

Zhang, J., Koo, I., Wang, B., Gao, Q. W., Zheng, C. H., & Zhang, X. (2012). A large scale test dataset to determine optimal retention index threshold based on three mass spectral similarity measures. *Journal of Chromatography. A*, *1251*, 188–193.

AUTHOR INDEX

Note: Page numbers followed by "*f*" indicate figures "*t*" indicate tables and "*np*" indicate footnote.

B

C

D

E

F

H

I

T

U

V

X

Y

Z

SUBJECT INDEX

Note: Page numbers followed by "*f*" indicate figures and "*t*" indicate tables.

I

L

Q

R

S

T

X

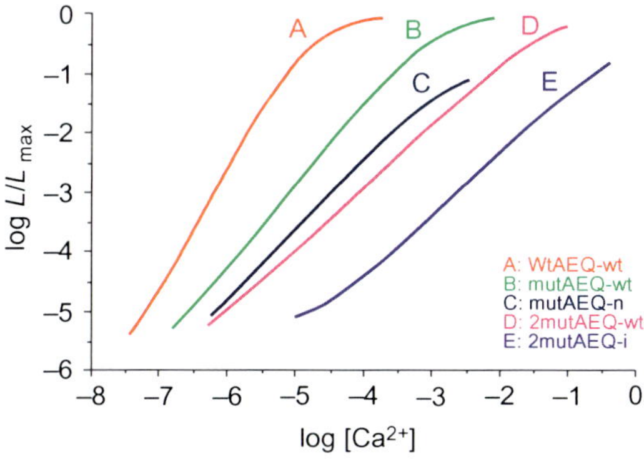

Denis Ottolini *et al.*, Figure 2.1 *Relationship between photon emission and Ca^{2+} concentration of different aequorin variants reconstituted with wt,* n *or* i *coelenterazine*. The curves were obtained by calculating the ratio between the luminescence obtained at a given [Ca^{2+}] (L) and the total remaining luminescent at that moment (L_{max}). L/L_{max} values are represented against the [Ca^{2+}].

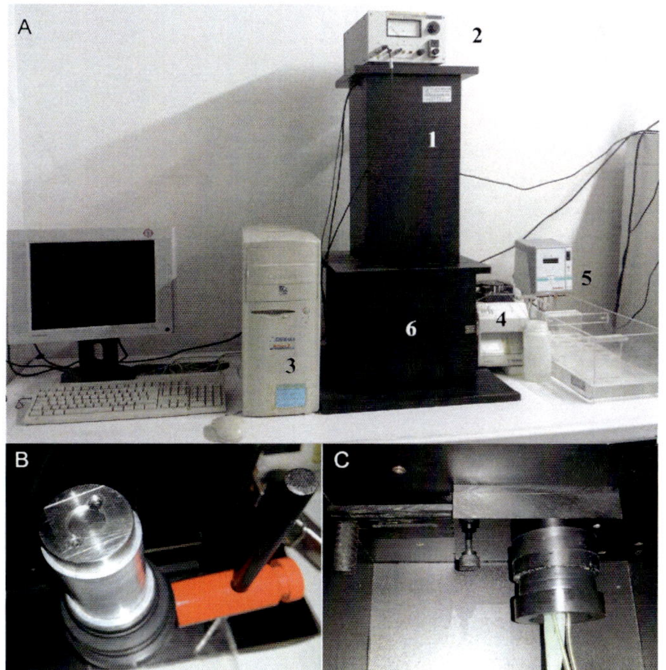

Denis Ottolini *et al.*, Figure 2.2 *The aequorin measuring system*. Panel A shows the photomultiplier connected with the amplifier/discriminator and kept in the dark by inserting it in a black box (1). It is connected to a power supply (2) and to a counting board inserted into a PC (3). A peristaltic perfusion pump (4) is connected with the chamber which accommodates the coverslip with the transfected cells (see it in details in panel B). The water jacket chamber is also connected with a thermostatic bath to keep it at the desired temperature (5). During the experiment, the perfusion chamber with the coverslip is placed in close proximity to the phototube by inserting it in a connected black box (6). Panel C shows in detail the position of the perfusion chamber during the Ca^{2+} measurements.

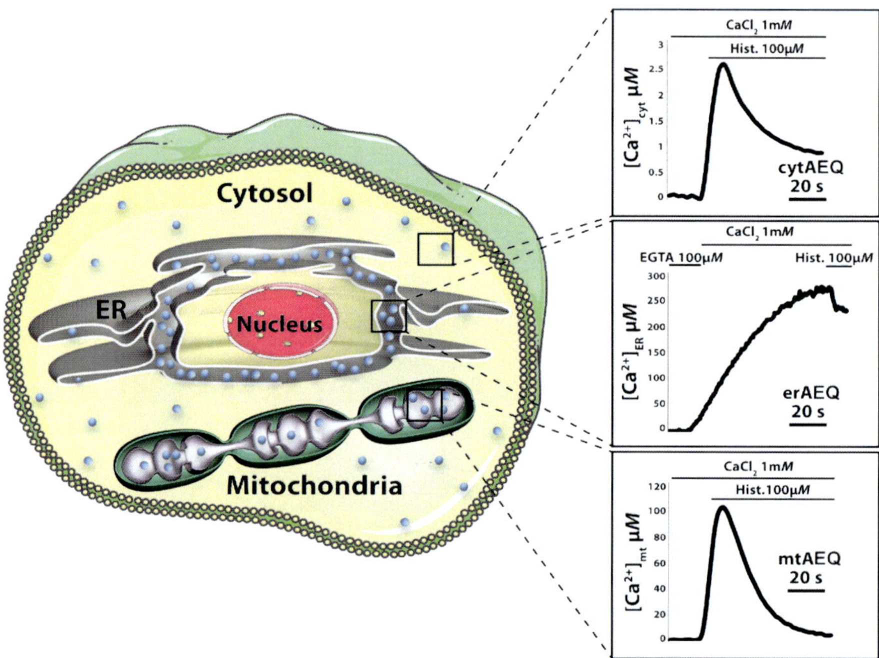

Denis Ottolini *et al.*, Figure 2.3 *Typical Ca^{2+} measurements in eukaryotic cells.* The traces refer to Ca^{2+} concentration transients monitored in HeLa cells expressing aequorin targeted to cytosol (cytAEQ), to endoplasmic reticulum (erAEQ) and to the mitochondrial matrix (mtAEQ). The cells were perfused with the indicated buffer and stimulated with histamine, an $InsP_3$-generating agonist.

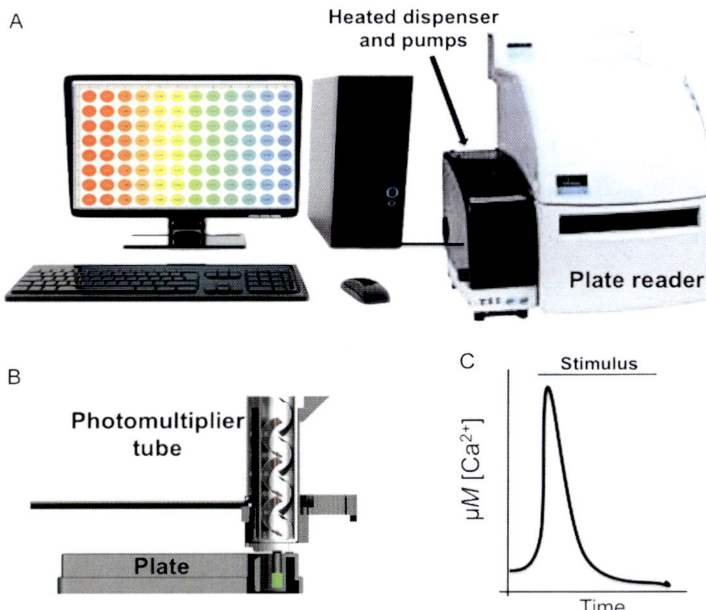

Denis Ottolini *et al.*, Figure 2.4 *Typical multiwell plate system for the automatic acquisition of luminescence*. The EnVision™ multilabel plate reader (A) (Perkin Elmer) is based on a PC-controlled platform that permits a quantitative detection of light-emitting or light-absorbing markers. The photomultiplier (B) is located in close proximity to the well and moves along the plate to collect the light from each single well. The output data can be visualized as luminescence intensity (A) or as calibrated Ca^{2+} transient trace (C).

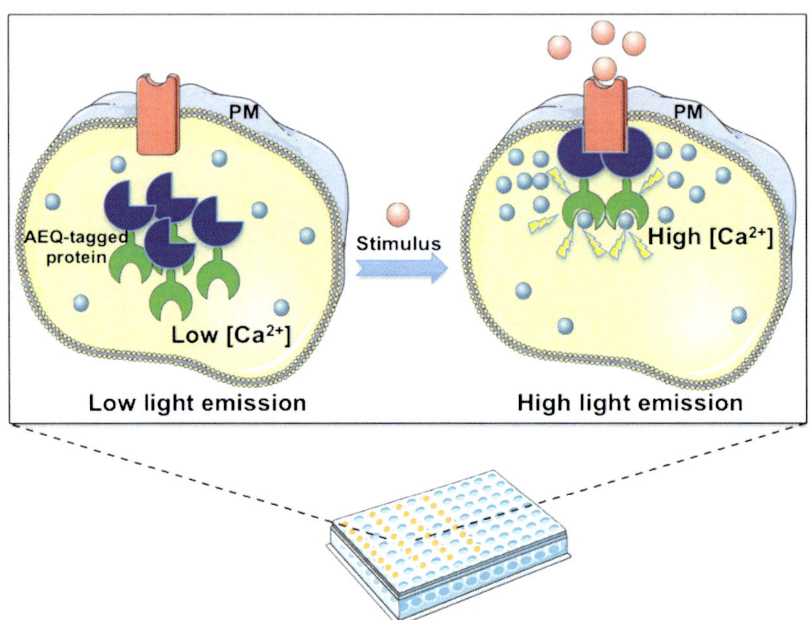

Denis Ottolini *et al.*, Figure 2.5 *Use of aequorin-tagged proteins to study the translocation of specific proteins in subcellular compartments characterized by different Ca^{2+} concentrations.* Upon cell stimulation a aequorin-tagged protein located in the cytosol translocates to the plasma membrane, where Ca^{2+} concentration is at least 1 order of magnitude higher than in the cytosol, giving rise to an easily detectable increase in light emission (yellow flashes). The changes in light emission reflect the translocation of the aequorin-tagged protein.

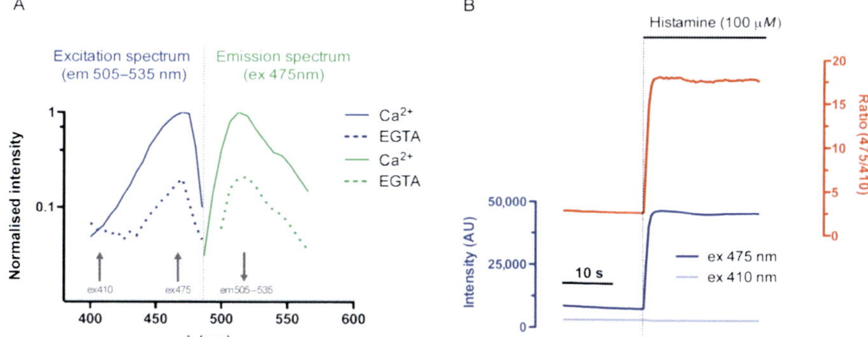

Julia M. Hill *et al.*, Figure 3.1 *Mitochondrial [Ca^{2+}] measured with the 2mtGCaMP6m circularly permuted GFP-based high-affinity Ca^{2+} probe.* (A) Excitation and emission spectra of the probe expressed in HeLa cells, obtained at 475 nm excitation and 505–535 nm emission, respectively. Spectra were measured either in the presence of Ca^{2+} or in intracellular buffer in the presence of EGTA following digitonin permeabilization to determine the Ca^{2+}-free traces, as previously described (Szabadkai et al., 2001). Arrows indicate the excitation (410 nm: isosbestic point; 475 nm) and emission wavelengths used in the experiments. (B) Basal and histamine (100 μ*M*) stimulated mitochondrial Ca^{2+} signals in HeLa cells transiently transfected with 2mtGCaMP6m. The two single wavelengths at 410 and 475 nm excitations as well as the ratio are shown. The experiments are representative of at least three different preparations.

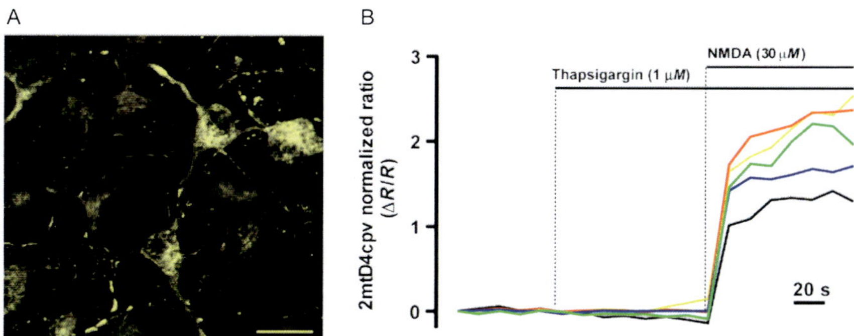

Julia M. Hill *et al.*, Figure 3.2 *Mitochondrial Ca^{2+} dynamics in neurons measured with the 2mtD4cpv cameleon probe.* (A) Micrograph shows the subcellular localization of the probe. Images were acquired in the cpv channel. Scale bar: 20 μm. (B) Neurons in the left panel were exposed to thapsigargin (1 μ*M*) and NMDA (30 μ*M*) as indicated, and the changes in the 535/485 ratio of the 2mtD4cpv probe were measured and normalized to the control period (Δ*R*/*R*). Traces show no increase in the FRET ratio after the thapsigargin-induced slow and low concentration cytosolic Ca^{2+} rise due to the low affinity of the probe. However, mitochondrial Ca^{2+} uptake upon NMDA receptor activation causes a fast, two- to fourfold increase in the ratio, indicating strong Ca^{2+} load.

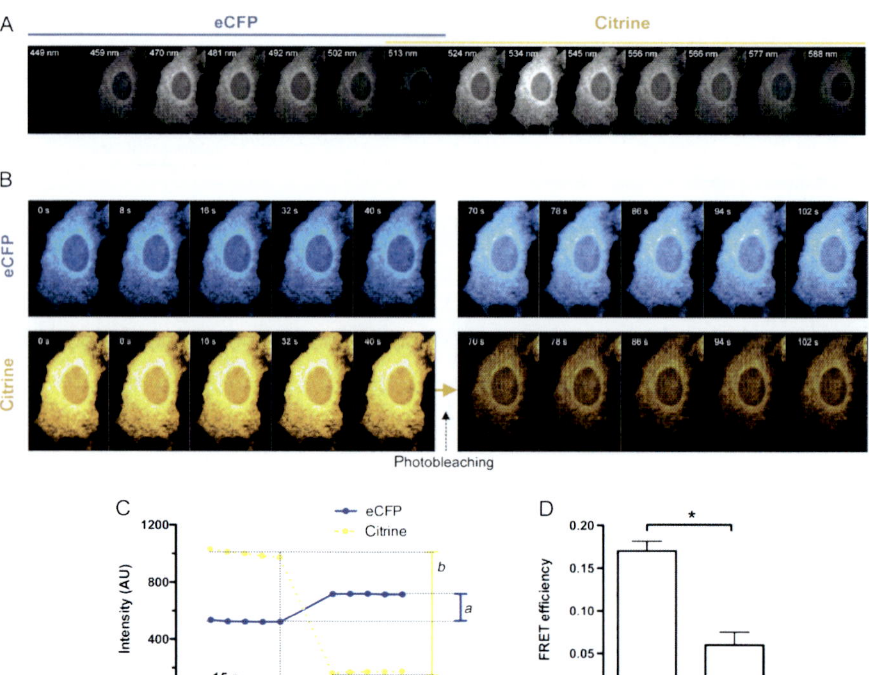

Julia M. Hill *et al.*, Figure 3.3 *Quantification of steady-state [Ca^{2+}] in the ER by acceptor photobleaching of the ERD1 FRET-based Ca^{2+} sensor.* (A) Emission spectra of the eCFP and citrine signals of the FRET probe in the 450–588 nm range following excitation at 405 nm. (B) Emission spectra were acquired during a short time lapse, composed of five image series, followed by photobleaching of the citrine probe (using 488 and 514 nm laser lines) and further five image series. The spectra were used for spectral deconvolution and separation of the specific eCFP and citrine signals using the Zeiss Meta system. (C) Quantification of FRET efficiency using the fluorescent intensities measured in (B). FRET efficiency was defined as the ratio between the increase in CFP and decrease in YFP following bleaching. (D) Steady-state ERD1 FRET efficiency in control, nontreated (NT) and thapsigargin (1 μ*M*, 30 min, in Ca^{2+}-free KRB) treated (Tg) A549 cells. The SERCA inhibitor Tg causes depletion of the ER Ca^{2+} content, reflected by the reduction in the FRET efficiency. Mean ± SEM of three independent experiments are shown. $^{*}p < 0.01$ (*t*-test).

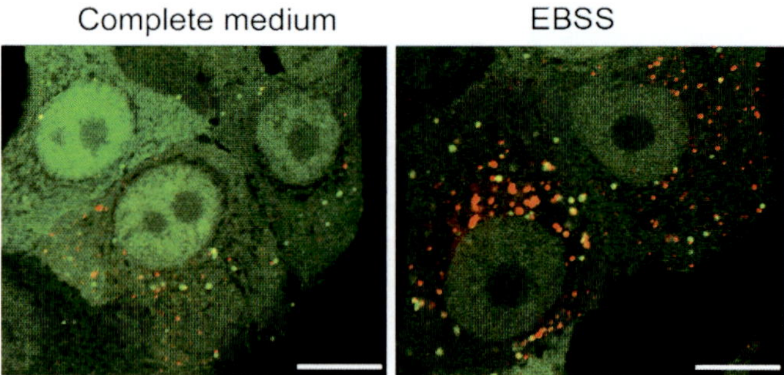

Nicolas Dupont *et al.*, Figure 4.2 *Monitoring the autophagic flux by the tandem mRFP-GFP LC3 probe.* HeLa cells stably expressing mRFP-GFP LC3 were or were not subjected to starvation (EBSS) for 2 hours, fixed, and then examined by confocal microscopy. Bar indicates 10 μm.

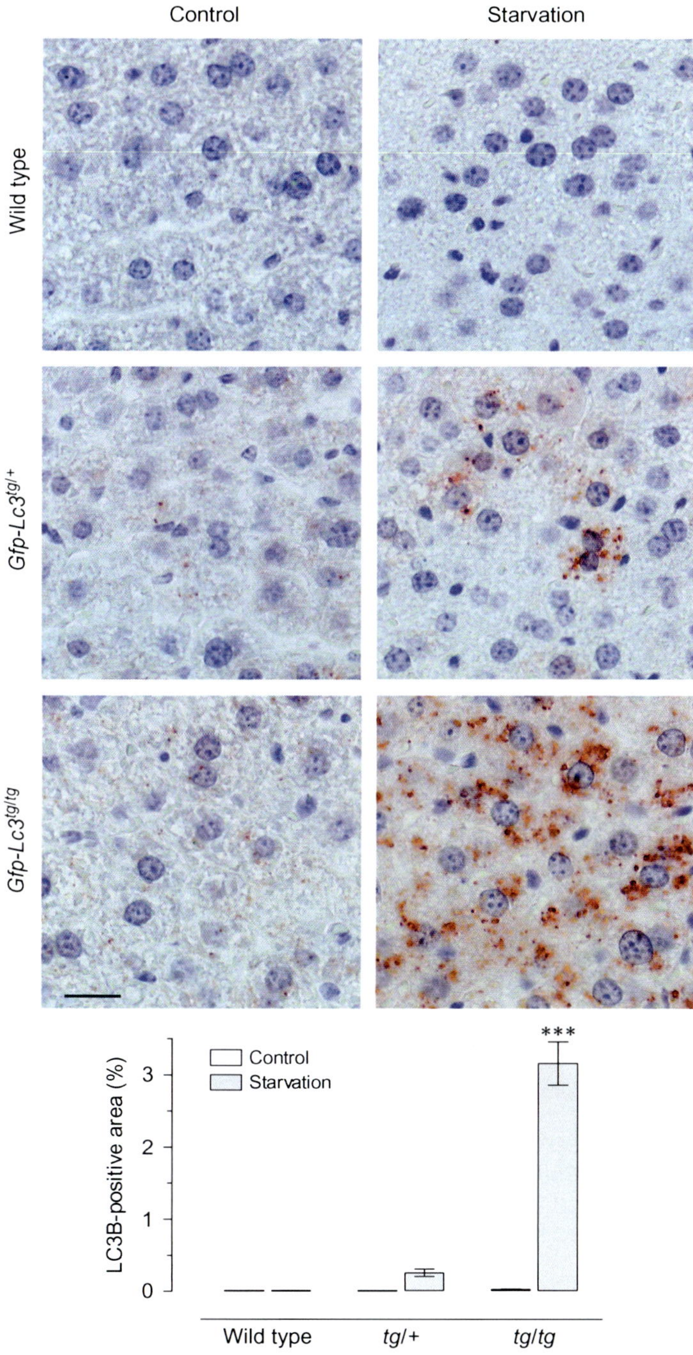

Wim Martinet *et al.*, Figure 5.8 *Immunohistochemical detection of LC3B in GFP-LC3 transgenic mice.* Liver samples were isolated from fed wild-type, *Gfp-Lc3*$^{tg/+}$, or *Gfp-Lc3*$^{tg/tg}$ mice (control) and from mice that underwent starvation for 48 h to induce autophagy. After fixation in neutral buffered formalin, tissues were paraffin-embedded and stained for LC3B using rabbit monoclonal anti-LC3B (clone D11; Cell signaling, 1:100 (wild type); 1:300 (*Gfp-Lc3*$^{tg/+}$) or 1:1000 (*Gfp-Lc3*$^{tg/tg}$)) and Envision+. Scale bar = 20 μm. The LC3B-positive area was quantified. $^{***}P < 0.001$ versus control (two-way ANOVA, followed by Bonferroni's posttest, $n = 10$).

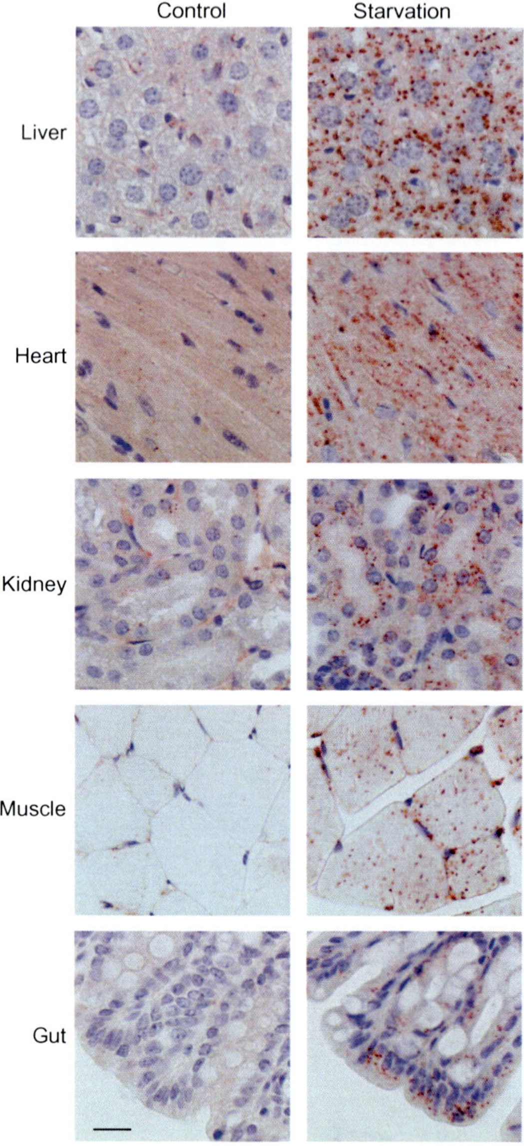

Wim Martinet *et al.*, Figure 5.9 *Immunohistochemical detection of LC3B in different organs of starved Gfp-Lc3$^{tg/tg}$ mice. Gfp-Lc3$^{tg/tg}$* mice were fed regular chow (control) or underwent starvation for 48 h. Tissue samples from different organs were collected and fixed in neutral buffered formalin. Thereafter, tissues were paraffin-embedded and stained for LC3B using mouse monoclonal anti-LC3B (clone 5F10; Nanotools, 1:1000) and Envision+. Scale bar = 20 μm.

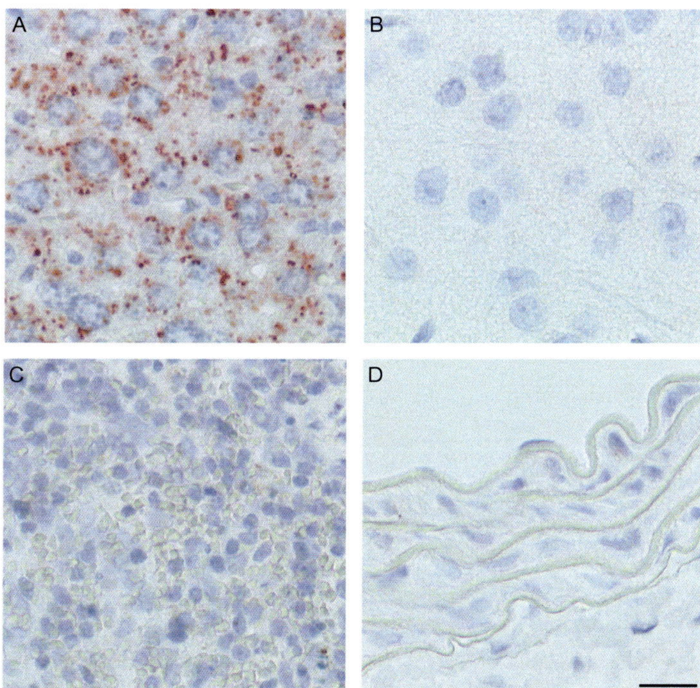

Wim Martinet *et al.*, Figure 5.10 *Starvation-induced autophagy in mice is not uniform but tissue-specific. Gfp-Lc3*$^{tg/tg}$ mice underwent starvation for 48 h. Tissue samples from liver (A), brain (B), spleen (C), and aorta (D) were collected and fixed in neutral buffered formalin. Thereafter, tissues were paraffin-embedded and stained for LC3B using rabbit monoclonal anti-LC3B (clone D11; Cell signaling, 1:1000) and Envision+. Scale bar = 20 μm.

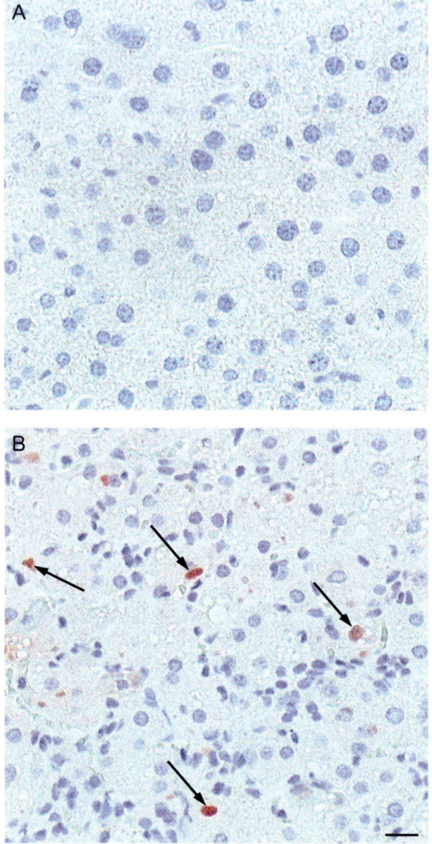

Wim Martinet *et al.*, Figure 5.11 *LC3A isoforms accumulate in autophagy-deficient tissues and may cause false positive staining*. Liver samples were isolated from fed autophagy-competent (*Atg7*$^{+/+}$ *Alb-Cre*$^{+}$) mice (A) or liver-specific autophagy-deficient (*Atg7*$^{F/F}$ *Alb-Cre*$^{+}$) mice (B). After fixation in neutral buffered formalin, tissues were paraffin-embedded and stained for LC3A using rabbit polyclonal anti-LC3A (Abgent, 1:100) in combination with Vectastain ABC. Globular structures of different size (arrows) stain positive for LC3A in liver from *Atg7*$^{F/F}$ *Alb-Cre*$^{+}$ mice. Similar results were obtained with antibodies against LC3B. Scale bar = 20 μm.

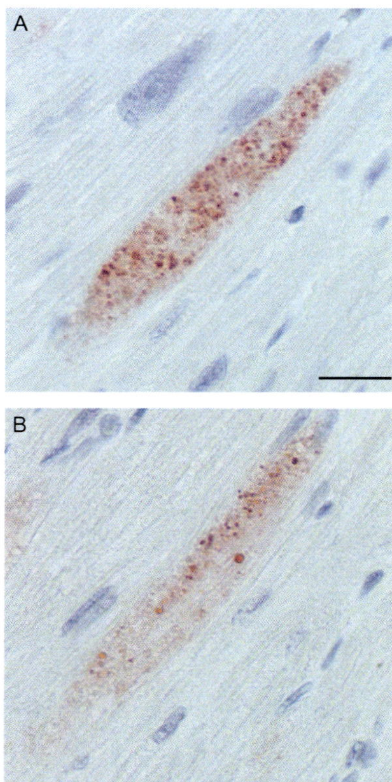

Wim Martinet *et al.*, Figure 5.12 *Immunohistochemical detection of ubiquitin (A) and SQSTM1/p62 (B) in cardiomyocytes of failing human heart*. Tissue samples from the heart were isolated from patients with heart failure. After fixation in neutral buffered formalin, tissues were paraffin-embedded and stained for ubiquitin and p62 using mouse monoclonal anti-ubiquitin (clone 6C1; Sigma, 1:300,000) and rabbit polyclonal anti-p62 (Sigma, 1:500), respectively, in combination with Vectastain ABC. Granular ubiquitin inclusions accumulate in the cytosol of degenerating cardiomyocytes during heart failure and colocalize with p62 staining. This finding suggests a malfunction in the autophagic pathway, rather than enhanced autophagic activity mediating cardiomyocyte death. Scale bar $= 20\ \mu m$.

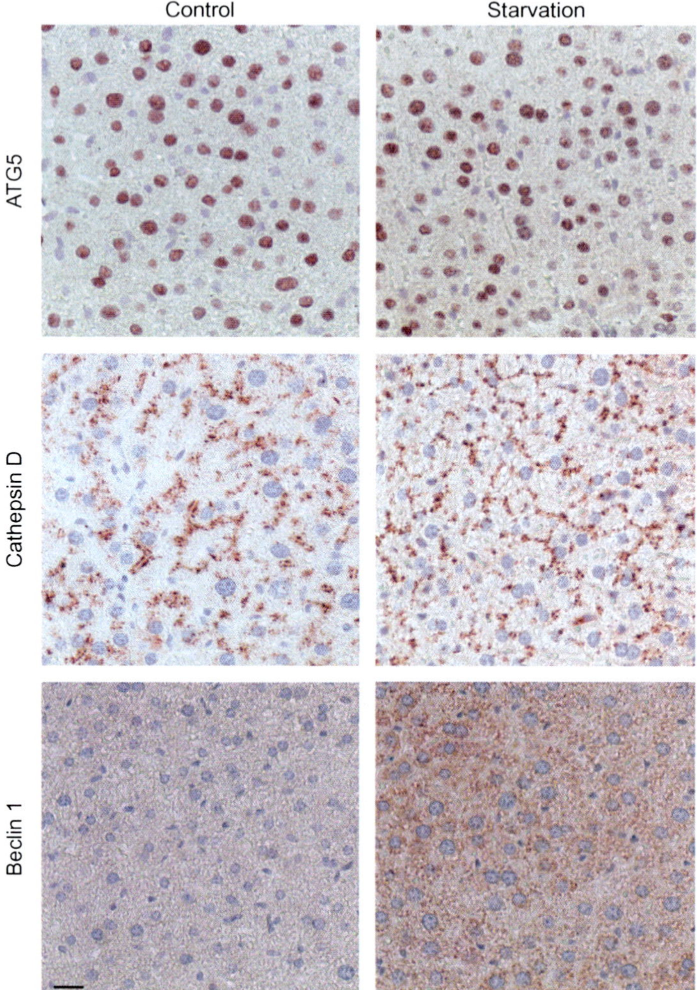

Wim Martinet *et al.*, Figure 5.13 *ATG5, CTSD/cathepsin D, and BECN1/Beclin 1 are not ideal targets for the immunohistochemical detection of autophagy.* Liver samples were isolated from control mice and mice that underwent starvation for 48 h to induce autophagy. After fixation in neutral buffered formalin, tissues were paraffin-embedded and stained for ATG5, CTSD, or BECN1 using rabbit polyconal anti-ATG5 (Abcam, ab78073, 1:100), rabbit monoclonal anti-CTSD (clone EPR3057Y; Abcam, ab75852, 1:1000), or rabbit polyclonal anti-BECN1 (Lifespan, LS-B3202, 1:1000) in combination with Vectastain ABC. Scale bar = 20 μm.

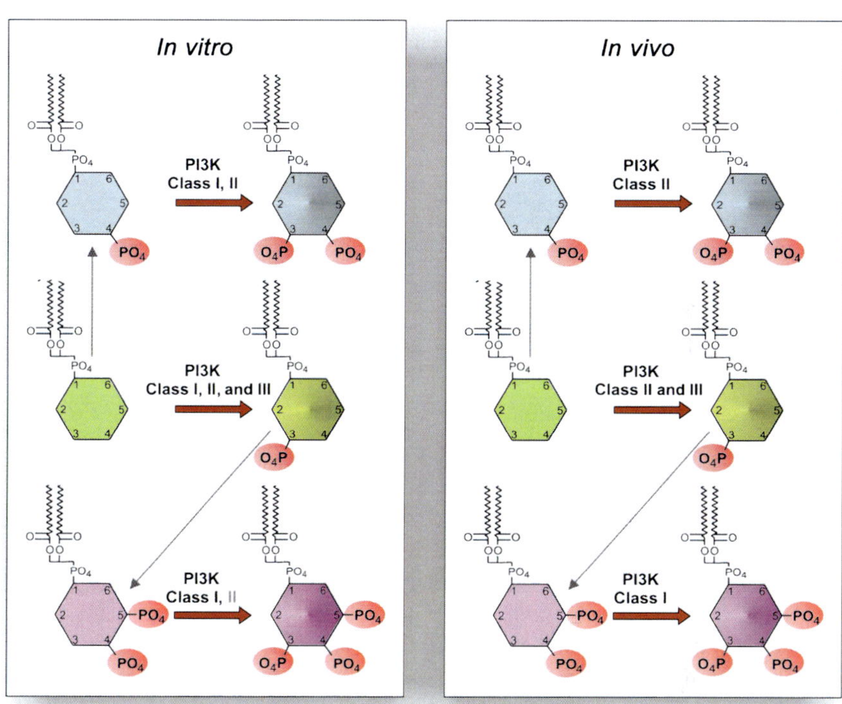

Elisa Ciraolo *et al.*, Figure 6.1 *PI3K-mediated production of phosphoinositides*. Position 3 on the inositol ring can be phosphorylated by PI3K family enzyme to produce PtdIns(3)P, PtdInd(3,4)P2, and PtdIns(3,4,5)P3. The three classes of PI3K can be distinguished by their specific product. *In vitro* class I PI3Ks are able to produce PtdIns(3), PtdIns(3,4)P2 and PtdIns(3,4,5)P3, while the PtdIns(3,4,5)P3 is their unique *in vivo* lipid product. Class II PI3Ks are able to produce, *in vitro* and *in vivo*, primarily PtdIns(3)P, PtdIns(3,4)P2 and PtdIns(3,4,5)P3 only *in vitro* in presence of specific substrates. The unique member of class III can produce uniquely PtdIns(3)P, both *in vitro* and *in vivo*.

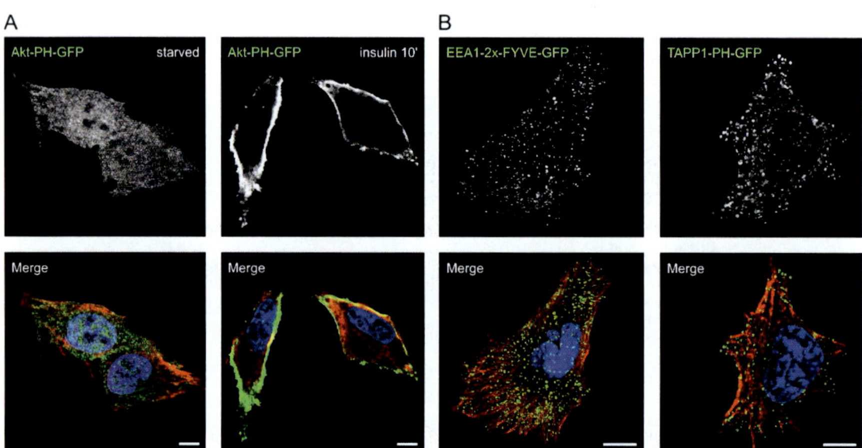

Elisa Ciraolo *et al.*, Figure 6.5 *Phosphoinositide-binding probes and their intercellular localization.* (A) Analysis of the intracellular localization of Akt-PH-GFP probe in HeLa cells during serum deprivation and after 10 min of insulin stimulation. After the addition of insulin, it is possible to appreciate the relocalization of the probe from cytosol (left panel) to plasma membrane (right panel) as a consequence of class I PI3K activation. (B) Analysis of the intracellular localization of EEA1-2x-FYVE-GFP (left panel) and TAPP1-PH-GFP (right panel) in HeLa cells cultured in Dulbecco's modified Eagle's medium (DMEM) supplemented with 10% fetal bovine serum.

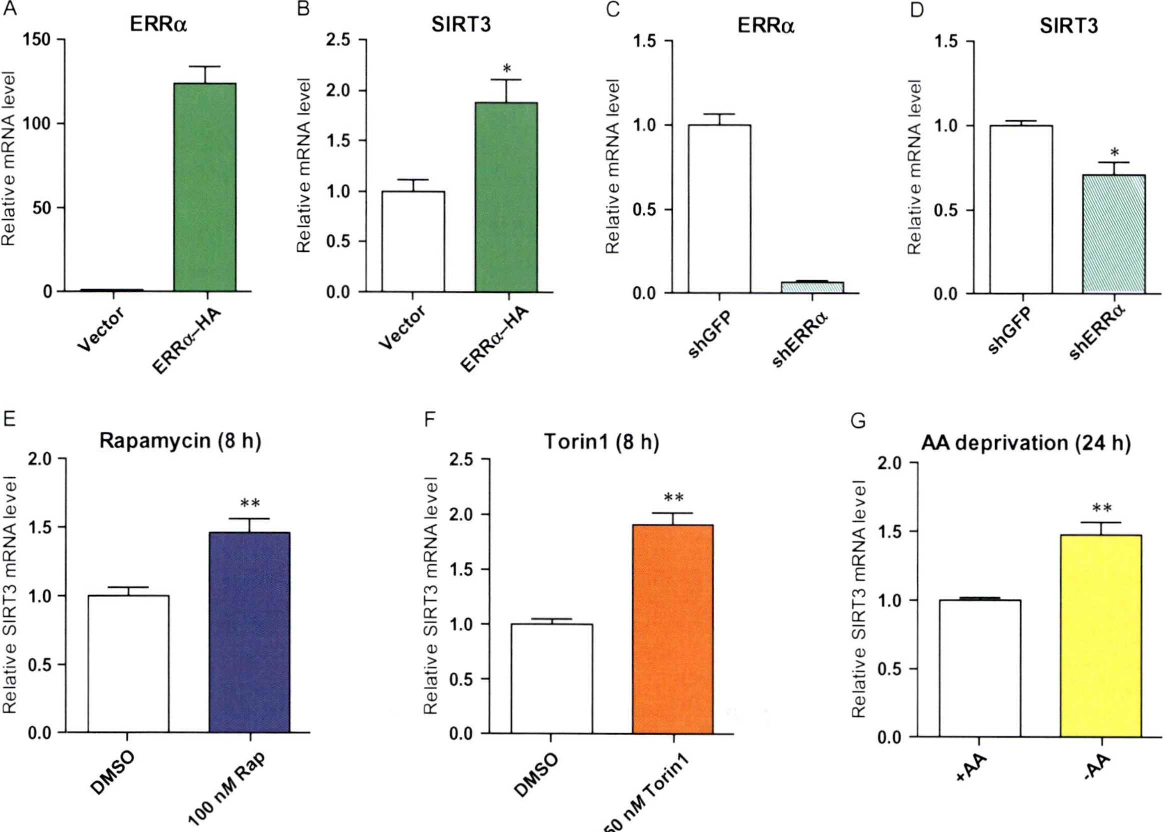

F. Kyle Satterstrom and Marcia C. Haigis, Figure 7.4 (A) Verification of ERRα overexpression in 293T cells and (B) effect on SIRT3 expression; (C) verification of ERRα knockdown in 293T cells, and (D) effect on SIRT3 expression; induction of SIRT3 expression by (E) 8-h 100 n*M* rapamycin treatment, (F) 8-h 250 n*M* Torin1 treatment, and (G) 24-h amino acid deprivation. *: $p < 0.05$, **: $p < 0.01$.

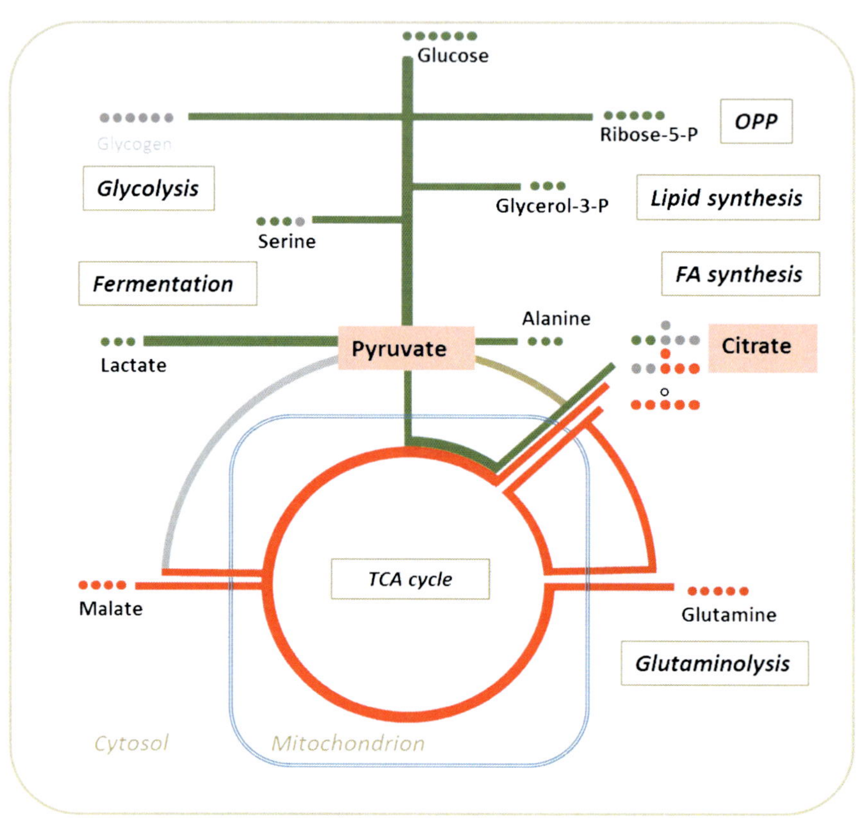

Matthias Pietzke and Stefan Kempa, Figure 9.5 *CCM including the carbon flow through glycolysis or glutaminolysis as observed in many cancer cells in vitro.* The TCA cycle is a central hub in the metabolism of the cell and can be replenished from different entry points. The scheme represents the metabolic map of a fast proliferating cancer cell with a high glutaminolytic activity.

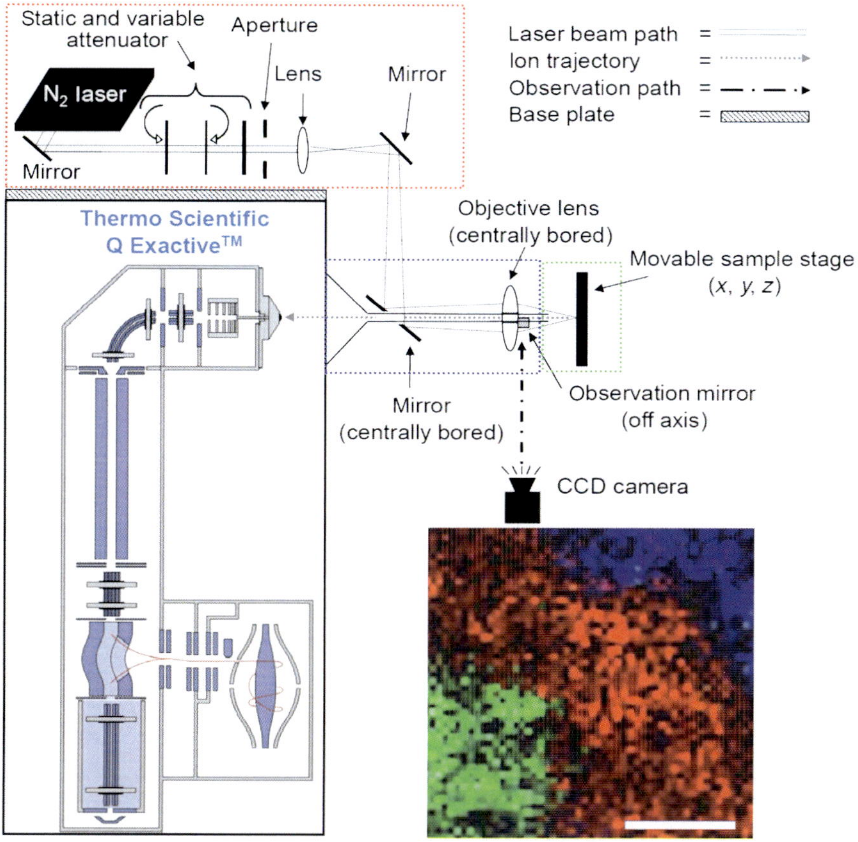

Aleš Svatoš and Alfredo J. Ibanez, Figure 10.1 *Image10 AP-MALDI source (TransMIT, University Giessen, Germany) mounted on hybrid Q-Exactive mass spectrometer (Thermo Scientific, Bremen, Germany).* Inserts document lateral resolution of 10 μm, scale bar: 50 μm.

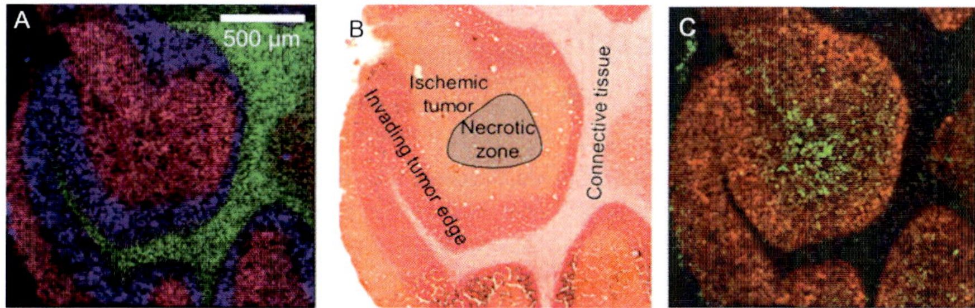

Aleš Svatoš and Alfredo J. Ibanez, Figure 10.2 *Distribution of lipids in human non-small cell lung carcinoma*. Carcinoma was induced into a severe combined immunodeficiency (SCID) mouse model. (A) MS image, 10 μm pixel size, with false-color-coded distribution of lipids: red sphingomyelin SM (36:1), green cerebroside Cer (42:2), blue phosphatidylcholine PC (36:4). (B) H&E staining after measurement. (C) MS image, 10 μm pixel size, with false-color-coded distribution of lipids: green lysophosphatidylcholine LPC (16:1), red phosphatidylcholine PC (38:6).

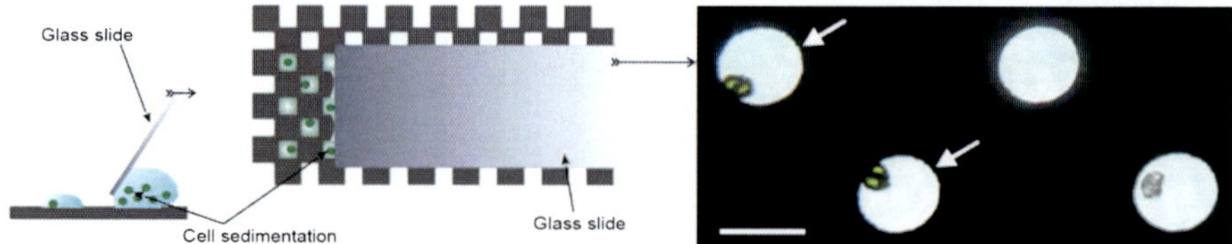

Aleš Svatoš and Alfredo J. Ibanez, Figure 10.3 *MAMS array*. MAMS substrate has the self-aliquoting properties. Cell suspension deposited on the MAMS array is dragged over the array surface (left) and cells are self-aliquoted to individual hydrophilic cells. Arrows indicate recipient sites with the cells of *Cosmarium turpinii*. Scale bar: 100 μm. *Modified from Urban et al. (2010).*

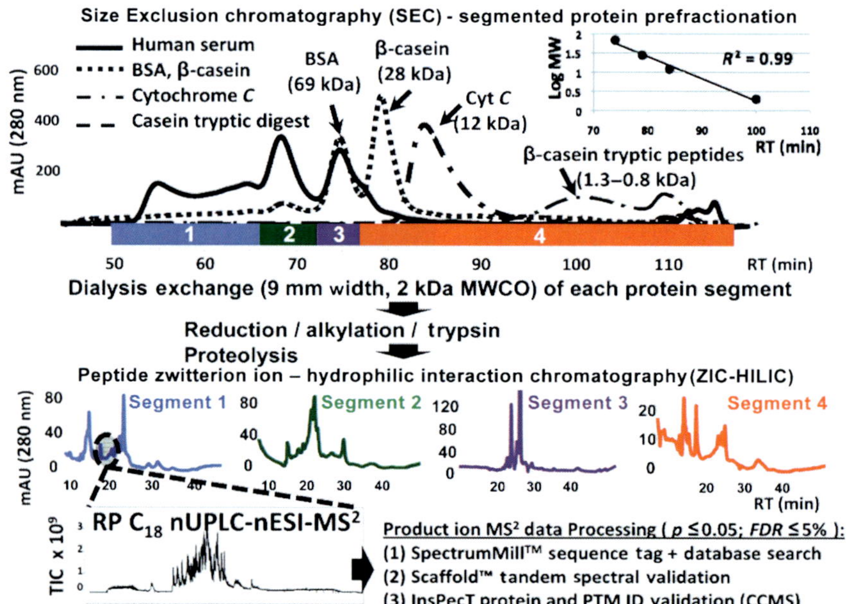

Claire Aukim-Hastie and Spiros D. Garbis, Figure 11.1 *iTRAQ 3D LC–MS method for the analysis of the serum samples from patients with benign prostate hyperplasia*. High performance size-exclusion chromatography was applied to 100 μL serum volumes from each specimen (6 ×) for the prefractionation of their protein content followed by their dialysis exchange purification, enrichment, reduction, and proteolysis. The resulting tryptic peptides are amenable to iTRAQ labeling and analysis to LC–MS analysis with reverse-phase nano-UPLC (0.75 mm × 25 cm × 1.9 μm particle) with nanoelectrospray ionization-tandem ultra-high-resolution mass spectrometry. *Reprinted with permission from Garbis et al. (2011), Copyright 2013 American Chemical Society.*

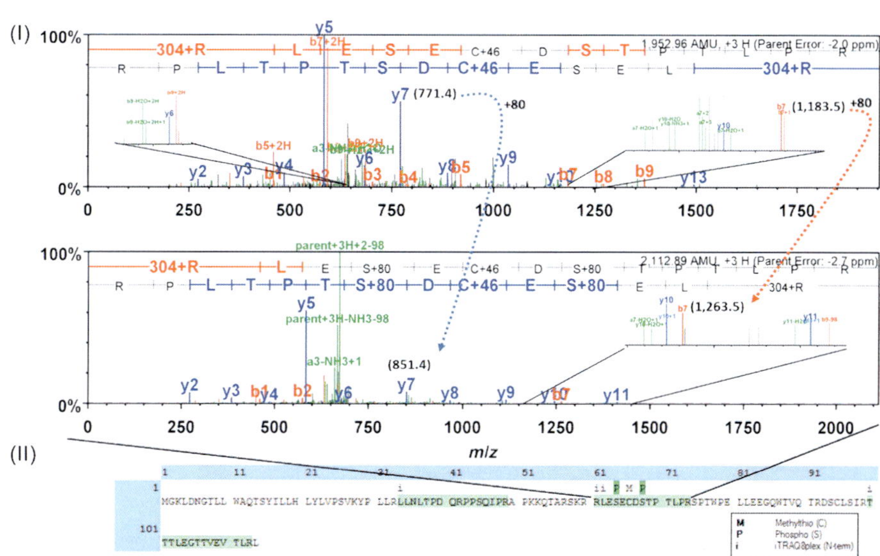

Claire Aukim-Hastie and Spiros D. Garbis, Figure 11.2 (I) Annotated ITMS CID MS/MS spectrum of the phosphopeptide R(iTRAQ8plex)LES(phospho)EC(methylthio)DS(phospho)TPTLPR with *m/z* 705.30310 Da (−2.61 ppm), $z = +3$, traceable to Protein E4 of HPV43 found in experiment A along with the annotated ITMS CID MS/MS spectrum of the native peptide. (II) The peptide sequence coverage of Protein E4 of HPV43. *Reprinted with permission from Garbis et al. (2011) and Papachristou et al. (2013), Copyright 2013 American Chemical Society.*

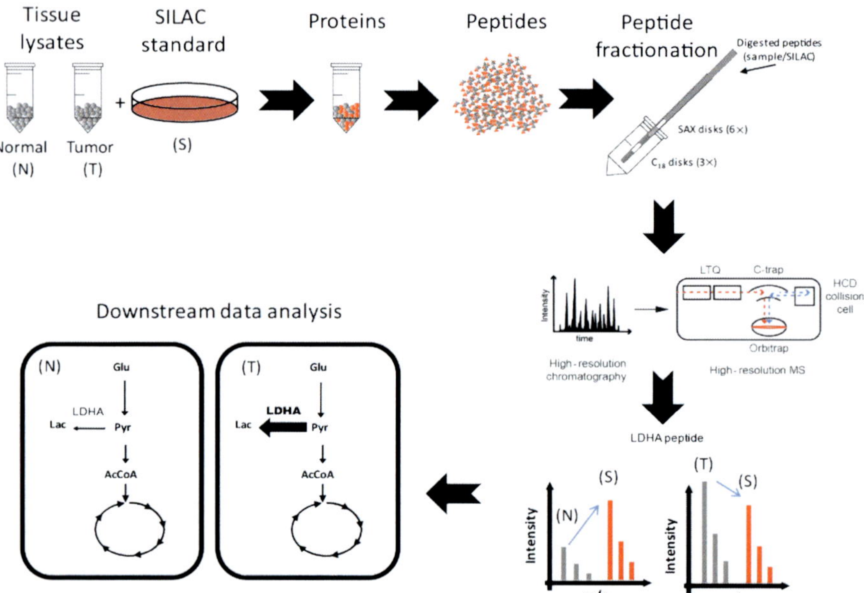

Steven Reid *et al.*, Figure 12.1 *MS shotgun proteomics workflow: from mouse tissue to metabolic pathway analysis.*

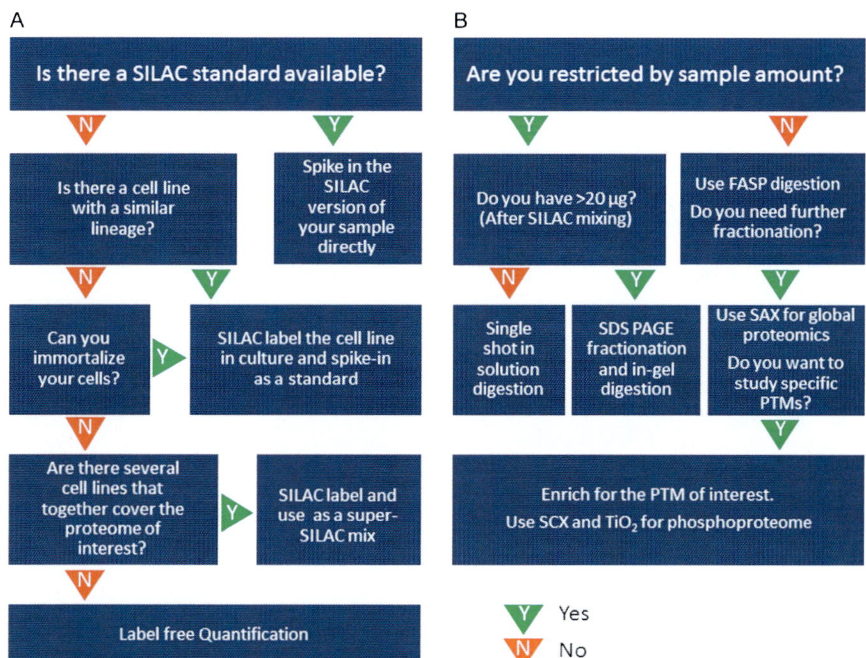

Steven Reid *et al.*, Figure 12.2 *Workflow for sample preparation*. Flow chart recommending which MS approaches to take depending on your experimental needs, answering the questions, follow (Y/N) for yes/no, respectively. (A) Helps readers choose between SILAC spike-in and label-free approach, (B) helps readers choose which sample preparation method to use according to sample quantity. Single shot, no fractionation at protein or peptide level before MS analysis. Super-SILAC mix, more than one SILAC-labeled cell lines mixed together to recapitulate the proteome to be analyzed at the MS.

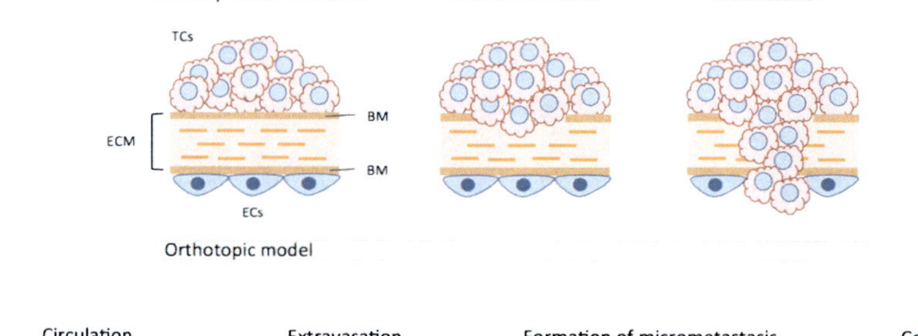

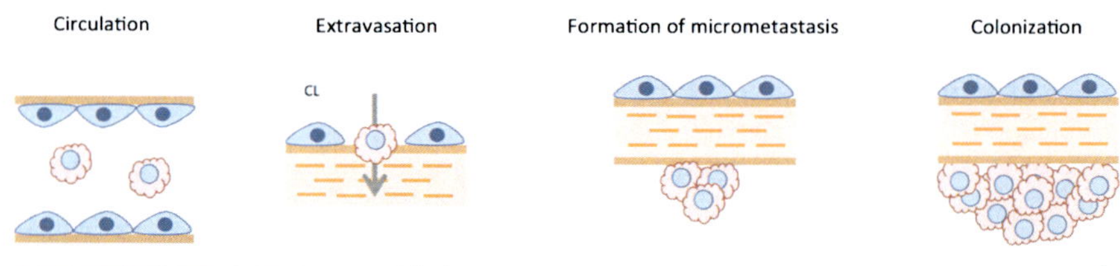

Experimental metastatic model

Hiromi I. Wettersten *et al.*, Figure 14.2 *Orthotopic models recapitulate the entire process of metastasis, while other experimental metastasis models cover only post intravasation.* Metastatic progression consists of primary tumor formation, localized invasion, intravasation, circulation, extravasation, formation of micrometastasis, and colonization. The tumor cells injected orthotopically go through all the steps above, while the tumor cells go through only post intravasation steps in experimental metastasis models. Tumor cell, TC; basement membrane, BM; extracellular matrix, ECM; EC, endothelial cell; CL, capillary lumen.

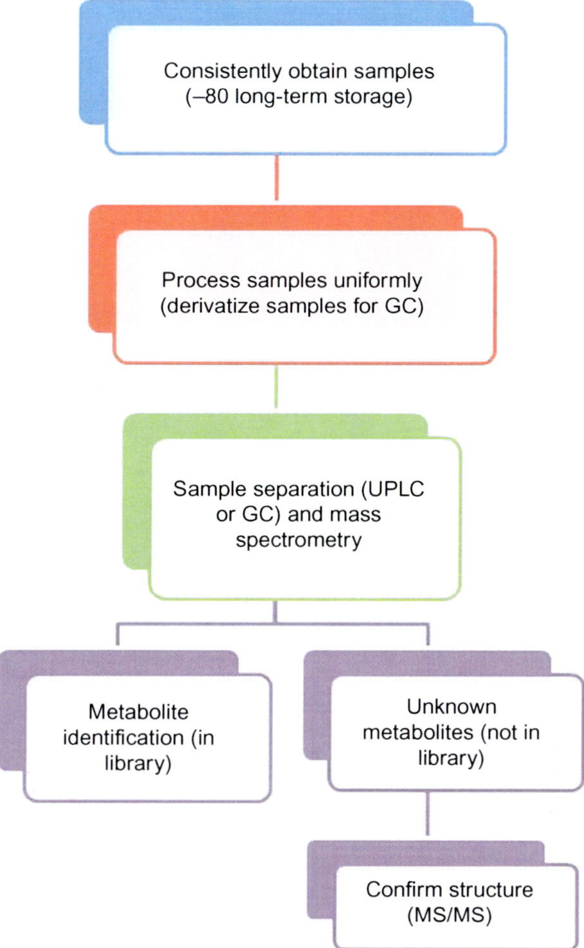

Hiromi I. Wettersten *et al.*, Figure 14.3 *Typical workflow for metabolomics analysis.* In a typical mouse metabolomics experiment, serum, tissue, and/or urine are obtained using consistent collection procedures, are uniformly processed, and then metabolites are separated either by LC or GC. Following mass spectrometry, metabolites are identified based on comparison to a known library. "Unknown" metabolites can be identified by further high-resolution MS/MS or NMR analysis.

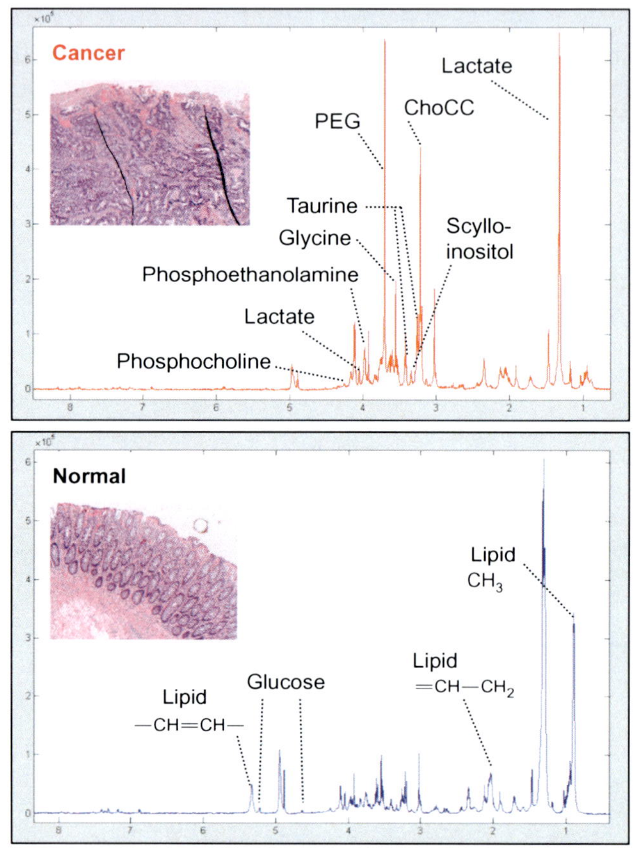

Hector Keun, Figure 15.1 *Typical ^{1}H MAS-NMR spectra of malignant and normal colon tissues.* ChoCC, choline-containing compounds. *Reproduced from Chan et al. (2009).*

CPI Antony Rowe
Eastbourne, UK
September 26, 2014